LABOR GEOGRAPHIES

A GUILFORD SERIES

Perspectives on Economic Change

Editors

MERIC S. GERTLER
University of Toronto

PETER DICKEN
University of Manchester

LABOR GEOGRAPHIES

*Workers and the Landscapes
of Capitalism*

ANDREW HEROD

THE GUILFORD PRESS
New York   London

For my parents,
who sacrificed so that I might learn.

©2001 The Guilford Press
A Division of Guilford Publications, Inc.
72 Spring Street, New York, NY 10012
www.guilford.com

Printed in the United States of America

This book is printed on acid-free paper.

Last digit is print number: 9 8 7 6 5 4 3 2 1

Library of Congress Cataloging-in-Publication Data
is available from the Publisher

ISBN 1-57230-685-8 (pbk.)

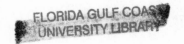

Someone once wrote "All the world's a stage."
But the truth comes written somewhat different on my page.
I know this script's fixed from the start,
because ordinary men and women only get supporting parts.

—CHRIS DEAN, *The Power is Yours*

"Remember," he was saying, "the facts about this church of Santa
Croce; how it was built by faith in the full fervour of medievalism . . ."
"No!" exclaimed Mr. Emerson, in much too loud a voice for church.
"Remember nothing of the sort! Built by faith indeed! That simply
means the workmen weren't paid properly."

—E. M. FORSTER, *A Room with a View*

Who built the seven gates of Thebes?
The books are filled with names of kings.
Was it kings who hauled the craggy blocks of stone?
And Babylon, so many times destroyed,
Who built the city up each time? In which of Lima's houses,
That city glittering with gold, lived those who built it?
In the evening when the Chinese wall was finished
Where did the masons go? Imperial Rome
Is full of arcs of triumph. Who reared them up?

—BERTOLT BRECHT, *A Worker Reads History*

Contents

Selective List of Acronyms

ACWA	Amalgamated Clothing Workers of America
AFL	American Federation of Labor
AFL-CIO	American Federation of Labor–Congress of Industrial Organizations
AIFLD	American Institute for Free Labor Development
ALJ	Administrative law judge
ASINCOOP	Alianza Sindical Cooperativa (Peru)
ATUC	Autonomous Trade Union Confederation (Hungary)
BGIWU	British Guiana Industrial Workers' Union
BIT	Building Investment Trust (AFL-CIO)
BKN	Consulting and Negotiating Bureaux (Poland)
BNS	Blocul National Sindical (Romania)
CIA	Central Intelligence Agency
CIO	Congress of Industrial Organizations
CISC	Confédération Internationale des Syndicats Chrétiens
CIT	Confederación Interamericana de Trabajadores
CITUB	Confederation of Independent Trade Unions in Bulgaria
ČMKOS	Českomoravská Komora Odborových Svazů (since 1998, the Českomoravská Konfederace Odborových Svazů) (Czech Republic)
CNSLR-Fratia	National Free Trade Union Confederation of Romania
CNV	Christelijk Nationaal Vakverbond (Netherlands)
CONASA	Council of North Atlantic Steamship Associations
CONATRAL	Confederación Nacional de Trabajadores Libres (Dominican Republic)
CROM	Confederación Regional Obrera Mexicana (Mexico)
CTAL	Confederación de Trabajadores de América Latina
CTC	Confederación de Trabajadores de Colombia
CTE	Confederación de Trabajadores del Ecuador
CTM	Confederación de Trabajadores de México
FDI	Foreign direct investment
FECESITLIH	Federación Central de Sindicatos Libres de Honduras

FESITRANH	Federación Sindical de Trabajadores Nacionales de Honduras
FGSRM	General Confederation of Trade Unions of Moldova
FIOST	Fédération Internationale des Organisations Syndicales du Personnel des Transports
FMC	Federal Maritime Commission
FMTI	Fédération Mondiale des Travailleurs de l'Industrie
FOECYT	Federación de Obreros y Empleados de Correos y Telégrafos (Argentina)
FSL	Full shipper load
FSZDL	Democratic League of Independent Trade Unions (Hungary)
FTUI	Free Trade Union Institute
FUTH	Federación Unitaria de Trabajadores de Honduras
GAI	Guaranteed annual income
GATT	General Agreement on Tariffs and Trade
HUS	Hrvatska Unija Sindikate (Croatia)
IBL	International Brotherhood of Longshoremen
ICEF	International Federation of Chemical, Energy and General Workers' Unions
ICEM	International Confederation of Chemical, Energy, Mine, and General Workers' Unions
ICFTU	International Federation of Free Trade Unions
IDB	Inter-American Development Bank
IFTC	International Federation of Textile and Clothing Workers
IFTU	International Federation of Trade Unions
IG Metall	German Metalworkers' Federation
ILA	International Longshoremen's Association
ILGWU	International Ladies Garment Workers' Union
ILO	International Labour Organisation
IMF	International Metalworkers' Federation
ITS	International trade secretariat
IUD	Industrial Union Department (AFL-CIO)
IUF	International Union of Food, Agricultural, Hotel, Restaurant, Catering, Tobacco and Allied Workers' Associations
J4J	Justice for Janitors
JIT	Just-in-time production
JSP	Job Security Program
KOK	Krestanska Odborova Koalice (Slovak Republic)
KOZ SR	Konfederácia Odborových Zväzov Slovenskej Republiky (Slovak Republic)
KNSH	Confederation of Independent Trade Unions of Croatia
LBAS	Free Trade Union Federation of Latvia
LCL	Less-than-container load
LDS	Lithuanian Workers' Union
LPSS	Lithuanian Trade Union Unification
LTL	Less-than-trailer load
MSA	Mobile Steamship Association

MSZOSZ	National Confederation of Hungarian Trade Unions
NLRA	National Labor Relations Act
NLRB	National Labor Relations Board
NMB	Nederlandsche Middenstandsbank (Netherlands)
NYSA	New York Shipping Association
OECD	Organisation for Economic Cooperation and Development
OIAA	Office of Inter-American Affairs
ORIT	Organización Regional Interamericana de Trabajadores
OS KOVO	Odborový Svaz KOVO (Czech Metalworkers' Federation)
OZ KOVO	Odborový Zväz KOVO (Slovak Metalworkers' Federation)
OSHA	Occupational Safety and Health Administration
PAFL	Pan-American Federation of Labor
PCH	Partido Comunisto de Honduras
PNC	People's National Congress (Guyana)
PPP	People's Progressive Party (Guyana)
PRI	Partido Revolucionario Institucional
RAC	Ravenswood Aluminum Corporation
SAENC	South Atlantic Employers' Negotiating Committee
SEIU	Service Employees' International Union
SITRAFA	Lard and Soap Workers' Union (Honduras)
SITRATERCO	Sindicato de Trabajadores de la Tela Railroad Company (Honduras)
SUTRASFCO	Sindicato de Trabajadores de la Standard Fruit Company (Honduras)
TGWU	Transport and General Workers' Union (Britain)
TIE	Transnationals Information Exchange
TOKG	Croatian Textile Workers' Federation
TTT	Transamerican Trailer Transport
TUC	Trades Union Congress (Britain); Trades Union Council (Guyana)
UATUC	Union of Autonomous Trade Unions of Croatia
UAW	United Auto Workers
UNITE	Union of Needletrades, Industrial and Textile Employees
USAID	United States Agency for International Development
USIA	United States Information Agency
USWA	United Steelworkers of America
USAS	United Students Against Sweatshops
UTAA	Unión de Trabajadores Azucareros de Artigas (Uruguay)
UTC	Unión de Trabajadores de Colombia
VOST	All-Ukrainian Union of Workers' Solidarity
WAC	World Auto Council
WCL	World Confederation of Labour
WFCW	World Federation of Clerical Workers
WFTU	World Federation of Trade Unions
WTO	World Trade Organization
ZSSS	Free Trade Union Association of Slovenia

Preface

Whatever else it may be, capitalism is a way of organizing social life according to various temporal and spatial requirements. Expressed another way, the geography of capitalism looks different than that of, say, feudal societies. The space–time organization of capitalism—or any other social system, for that matter—is both a reflection of, and a shaper of, how social life plays itself out. However, this space–time organization develops unevenly. For example, new transportation technologies may transform the spatial and temporal relationships between places, but they do so differentially. Thus, those places located on the railroad lines laid out during the nineteenth century were brought much closer together than were those that were bypassed. Indeed, as the transcontinental railroad was built across the western United States during the 1860s, many communities lobbied for it to go through their towns, so convinced were they that this would effectively bring them closer to East and West Coast markets and sources of investment (Ambrose, 2000). In the twentieth century, jet aircraft have similarly brought some places much closer together than others. Equally, contemporary economic restructuring has impacted how landscapes are organized in some parts of the globe to a much greater extent than it has in others. The transformation of huge swaths of the industrialized world that deindustrialization has wrought has significantly changed the space–time organization of global capitalism as capital has left behind—metaphorically and literally—many of the regions that once formed the base of the industrial revolution and has sought out new geographical realms.

Given, then, that its geographical organization seems to make a difference in the way capitalism works, the question that must be raised is this: how might different groups with different sets of interests seek to manipulate this geographical organization to their own benefit and to the disadvantage of those with whom they may be in conflict? This question forms the central theme of this book. More specifically, in this volume I seek to examine how workers and their organizations—principally labor unions—shape the ways in which the landscapes of capitalism are formed. My central assertion here is that workers and their organizations require certain configurations of the landscape—what I call, borrowing from David Harvey (1982), their "spatial fix"—in order for them to reproduce themselves (socially and biologically) from day to day and from generation to generation. In the process of trying to ensure that such configurations are, in fact, put in place, workers shape the very geography of capitalism. Of course, the particular configurations that one group of workers prefers may not be that preferred either by other workers or by capital. Thus, the process of establishing a particular spatial fix in certain places and at specific historical moments is always contested. However, the fact that it is indeed necessary for workers to develop such spatial fixes if they are to reproduce themselves on a daily and generational basis means that social conflict is also fundamentally spatial in nature. Consequently, the production of space and landscapes must be seen not just as secondary to the social relations of everyday life but as highly political acts that are central to it. This, then, is the broad ambit of what I seek to do in this volume.

*   *   *

I have been toying with such ideas for a decade or so now, and earlier versions of several of the chapters presented in this book first saw the light of day as stand-alone articles in various places. However, in drawing upon some of the material and ideas first explored elsewhere, I have tried to provide more than a collection of previously published articles where the total is simply the sum of the various individual parts. Instead, I have sought to provide a certain degree of "value-addedness." In particular, I have organized the chapters as an integrated whole around two central goals. The first of these is to understand how various groups of workers went about actually constructing spatial fixes in pursuit of their varied political and economic objectives. The second, and related, goal is to examine the significance of geographic scale in this process, particularly the ways in which social actors create not just spatial fixes but also "scalar fixes"—that is to say, how they organize their social and spatial activities at particular geographic resolutions at various times and how the ways in which social life is organized at one spatial scale may greatly affect how it operates at another. Thus, each of these chapters is not just (what I hope will be) an interesting empirical case, but each addresses different aspects of the issue of scale and landscape production, building on previous chapters by providing a different angle on the goal of the project as a whole, namely, understanding how workers and their organizations produce what I am here calling "labor geographies."

In developing such ideas, I seek to address two groups of scholars, these being, on the one hand, economic geographers (who know much about space and the economy but have often been negligent toward issues of working-class life) and, on the other, labor historians and those working in the field of industrial relations (who have often focused upon aspects of working-class life but have tended to have a rather undynamic view of space). Indeed, for the most part, these two groups, who by rights should have so much in common, have not had much to say to each other until very recently (for more on this situation, see Herod et al., in press, and Ellem & Shields, 1999). The engagement between economic geography, labor history, and industrial relations is an engagement very much waiting to happen, though I believe that all sides have much to learn from each other. It is my hope that this book may contribute in some small way to hastening this engagement.

Of course, while working through these ideas, I have collected many intellectual debts. Any book, even by a single author, is necessarily a collaborative effort, and this one is no exception. Given that I have been thinking about the ideas presented here for over a decade now and, in that time, have had conversations about issues of labor and geography with literally dozens of people who have helped me hone my own thoughts, it would be impossible to thank everyone individually. Nevertheless, there are a number of people I would like to thank specifically for the interest they have shown in this project in various ways (some of which they themselves may not recognize), for their inspiration when it comes to issues of labor and geography, and for their belief both that workers and their organizations are worthy of study by geographers and that such study may help bring about a more progressive and egalitarian world. In this regard I would like to thank Neil Smith, Don Mitchell, Rebecca Johns, and Jane Wills. I would particularly like to thank Robert Hanham for first giving me the opportunity at West Virginia University to pursue graduate study and to indulge my interest in labor issues. I would also like to thank Eric Sheppard, Dick Walker, Erica Schoenberger, Kevin Cox, Jamie Peck, and John Holmes for the professional support they have shown me.

I would like to thank the numerous people who have commented on earlier drafts (in various forms) of chapters in this volume, including Helga Leitner, Byron Miller, Steven Silvern, Jamie Peck, Dick Walker, Kevin Cox, Mike Renning, Dick Peet, Gearóid Ó Tuathail, Susan Roberts, Paul Garver, Anne-Marie Mureau, Poul-Erik Olsen, Erica Schoenberger, John O'Loughlin, Ikubolajeh Logan, Melissa Wright, Paul Plummer, Scott Salmon, Derek Alderman, John Britton, Andrew Kirby, Jane Wills, Roger Lee, John Pickles, and Adrian Smith, together with a host of anonymous reviewers. Peter Dicken and Neil Smith were obliged to read the entire manuscript (Peter twice!). Thanks also to the Department of Geography, Pennsylvania State University, for inviting me to present material from Chapter 7 as part of their "Glenda Laws Memorial Lecture Series" in February 1998. Glenda was a friend of mine and I was thus particularly honored to be part of the lecture series program.

I would also like to thank a number of individuals who have particularly helped me in the research process, including: David Fowler, former Assistant General Secretary of the International Metalworkers' Federation (IMF) Geneva, Switzerland; Anne-Marie Mureau, former Coordinator for Central and Eastern Europe, International Metalworkers' Federation Geneva, Switzerland; Poul-Erik Olsen, former Director, Education and Working Environment, Eastern and Central Europe, International Metalworkers' Federation Geneva, Switzerland; Jean Lapointe, former Regional Representative, Regional Office for Central and Eastern Europe, International Metalworkers' Federation, Budapest, Hungary; Paul Garver, International Union of Food, Agricultural, Hotel, Catering, Tobacco and Allied Workers' Associations (IUF) Geneva, Switzerland; Anna Oulatar, Head, Coordinating Unit for Central and Eastern Europe, International Confederation of Free Trade Unions (ICFTU), Brussels, Beligum; Peter Senft, Head, Economic Office, German Metalworkers' Union (IG Metall), Berlin, Germany; Jan Uhlíř, President, Czech Metalworkers' Federation (Odborový Svaz KOVO), Prague, Czech Republic; Lucy Zábranska, International Secretary, Czech Metalworkers' Federation (Odborový Svaz KOVO), Prague, Czech Republic; Dana Sakarova, Czech Metalworkers' Federation (Odborový Svaz KOVO), Prague, Czech Republic; Richard Falbr, President, Czech-Moravian Confederation of Trade Unions Českomoravská Konfederace Odborových Svazů [ČMKOS]), Prague, Czech Republic; Vlastimil Beran, Head of International Department, ČMKOS Prague, Czech Republic; Emil Machyna, President, Slovak Metalworkers' Federation (Odborový Zväz KOVO), Bratislava, Slovak Republic; Alexander Zharikov, General Secretary, World Federation of Trade Unions (WFTU), Prague, Czech Republic; James Holway, former director of the Social Projects Division, American Institute for Free Labor Development (AIFLD), Washington, DC; Armando Galdámez, General Secretary of the Federación Sindical de Trabajadores Nacionales de Honduras (FESITRANH) Executive Committee, San Pedro Sula, Honduras; Joseph H. "Polly" Pollydore, General Secretary, Trades Union Council (TUC), Georgetown, Guyana; Gordon Todd, General Secretary, Guyanese Commercial and Clerical Workers' Union, Georgetown, Guyana; Richard Satkin, New York City Department of City Planning; Dan Stidham, former president of United Steelworkers of America (USWA) Local 5668, Ravenswood, West Virginia; Joseph Uehlein, Director of Special Projects, Industrial Union Department at the AFL-CIO, Washington, DC; Joe Chapman, staff representative, USWA District 23,

Charleston, West Virginia; and Jane Mellow and Jennifer Frum from the Washington, DC, office of former U.S. Representative Robert E. Wise Jr. (D, West Virginia), in whose former district Ravenswood, West Virginia, is located. I do want to point out, however, that none of these individuals should be held responsible for any misinterpretations of facts I may have made, nor should my interpretation of particular events be taken as necessarily representing the opinions of any of the above.

Research trips to Switzerland and Central and Eastern Europe were funded by two faculty development grants from the University of Georgia Research Foundation. Research trips to Honduras and Guyana were funded by a grant from the Americas Council of the University of Georgia System. Funds for other research trips were also made available by the Department of Geography, University of Georgia; by the Institute for Behavioral Research, University of Georgia; and by the Center for the Critical Analysis of Contemporary Culture, Rutgers University. Parts of chapters have been presented at various conferences, including: several annual meetings of the Association of American Geographers; the 1994 Regional Conference of the International Geographical Union on "Environment and Quality of Life in Central Europe: Problems of Transition," held in Prague, Czech Republic; the 1995 Regional Conference of the International Geographical Union on "Latin America in the World: Environment, Society and Development," held in Havana, Cuba; the 1997 Conference of the Americas Council of Georgia, held in Savannah, Georgia; and the 1998 Annual Conference of the Royal Geographical Society/Institute of British Geographers, held in Kingston-Upon-Thames, UK. I would like to thank those who made comments upon these papers, both as copanelists and from the floor.

At The Guilford Press I would like to thank Peter Wissoker for his faith that I would actually finish this book and for his goodnaturedness when I continued to miss deadlines, together with Paul Gordon, Oliver Sharpe, Kristal Hawkins, and Kim Miller, who helped bring it to fruition. I would like to thank the Frum family for the kindness they have shown me over the years and especially Jennifer for tolerating my bad moods when I was working to finish the never-ending manuscript.

CHAPTER 1

Introduction
Labor and Landscapes

Open the business section of any newspaper, and the global power of
capital appears to be writ large. The triumphalism with which neoliberal
changes in Eastern Europe or Latin America are heralded by many is taken
to suggest that capital is the singular force shaping the contemporary economic geography of the planet at scales from the very local to the truly
global. Stories of billions of dollars flowing here, of factories moving there,
of venture capitalists "opening up" new markets, and of workers having to
"take it or leave it" are the staple diet of those who would have us believe
that the new global geography of capitalism is being forged solely by the
managers of transnational corporations, by government ministers of trade,
and by the faceless bureaucrats who run the World Bank, the International
Monetary Fund, and the World Trade Organization. Those on the receiving
end of such transformations are invariably portrayed as hopelessly divided
by language, nationalism, and political ideology, and powerless to do much
of anything as the juggernaut of global capital marches on, recasting in its
wake the planet's economic landscapes. However, this is clearly not the
whole story, despite what the imagineers of a new neoliberal global order
might seek to portray. There is always opposition to power and domination,
a fact that is seen everyday in countless workplaces, fields, offices, and elsewhere, but which was perhaps most amply demonstrated to a global audi-
ence by events at the 1999 World Trade Organization (WTO) meeting in
Seattle, Washington, where tens of thousands of unionists, environmentalists, students, and others challenged the power of corporate managers and
trade ministers to create a new global regime of neoliberal trade rules (*New
York Times*, December 1, 1999). Indeed, the protests against the WTO literally ringed the globe, occurring in numerous national capitals, regional cen-

1

ters, out-of-the-way villages, street corners, and across the vastness of cyber-space. Protestors have continued to hound the WTO, World Bank, and International Monetary Fund at subsequent meetings from Prague to Davos, and from Nice to Washington, DC.

In arguing that capital's power to shape the planet's economic land-scapes does not go unchallenged, in this book I want to explore the conten-tion that working-class people both have a vested interest in trying to en-sure that the geography of capitalism is produced in certain ways and not in others, and that they play active parts in seeking to bring this about. Al-though it would seem a simple proposition to suggest that working-class people and their organizations affect the ways in which the landscapes of capitalism are made, until recently there has been surprisingly little work within economic geography that has taken it seriously. For whatever reasons—and we can perhaps readily think of a few—economic geogra-phers have historically focused the bulk of their intellectual pursuits on ana-lyzing various fractions of capital when seeking to explain how the eco-nomic landscapes of capitalism are produced and reproduced. It is to the decision-making processes of capitalists and their managers, theory in eco-nomic geography has so frequently told us, that we must turn if we are to understand the geography of capitalism. Equally, industrial relations special-ists and labor and social historians have sometimes not been as attentive as they otherwise might have been to issues of geography and "spatial praxis"—that is to say, how social actors struggle to shape the geographical relationships within which they live their lives. While there have been nota-ble exceptions (e.g., Cowie, 1999), such scholars have generally tended to view geography in terms of how place functions as a "context" for social ac-tion rather than in terms of how space and spatial relations may serve as sources of power and objects of struggle (a good critique of such approaches—by two industrial relations scholars, no less—is provided by Ellem & Shields, 1999).

This book, then, is aimed both at geographers (who have often ne-glected matters of working-class life) and at industrial relations scolars and labor and social historians (who have often failed to appreciate how mold-ing the geography of capitalism is a key aspect of class struggle). Whereas aspatial class analyses have often treated geography as merely a pesky com-plication to analyzing class structures and processes—and despite the new-found interest in geography by economists such as Paul Krugman—concep-tualizing how space can be a source of power which may sometimes enable, and may sometimes constrain, both workers' and capitalists' social actions forces us to recognize that whatever else they may be, processes of class formation and class politics are fundamentally geographic at heart.[1]

In pursuing such arguments, I draw upon a certain tradition within critical social science in general and Western Marxism in particular. Although during the past hundred years or so social theory has often neglected space for a preoccupation with the temporal aspects of social life (see Soja, 1989, for an extended critique), a number of theorists have expended considerable efforts to spatialize Marx and to conceptualize how space plays a central role in the functioning of societies. Indeed, French theoretician Henri Lefebvre has gone so far as to argue that the way in which economic landscapes have been made has been crucial to capitalism's very survival. Capitalism, he has suggested (Lefebvre, 1976 [1973]: 21), has been "able to attenuate (if not resolve) its internal contradictions for a century . . . by occupying a space, by producing a space." Put another way, it has been capitalists' success in organizing the economic and social landscape in such a way as to facilitate the production and realization of surplus value that has been key to their ability to reproduce capitalism on a daily and generational basis. Thus, capitalists (either individually or collectively) have to ensure, for example, that rail and road systems, supply networks, housing, shopping centers, and myriad other aspects of the landscape are constructed in such a fashion that workers and raw materials may be brought together in specific locations, that the fruits of the production process may be distributed and consumed, that profit may thus be realized, and that the capitalist system of accumulation continues.

Within Anglo-American Marxist geography such ideas began to be developed during the 1970s, principally by writers such as David Harvey, Neil Smith, Doreen Massey, Manuel Castells, Ed Soja, Richard Peet, and Richard Walker. All of these authors, in different ways, attempted to theorize the importance of space for the functioning of capitalist society and to explore the ways in which the making of landscapes was connected to the machinations of capitalism. Whereas the "spatial scientists" who had dominated the discipline of economic geography since the 1950s had tended to explain the geography of one phenomenon by attempting to correlate it spatially with another—for example, explaining the geography of poverty by examining the geography of where racial and ethnic minorities live, an explanation that ignores how racism in the job and housing markets may shape the spread of poverty in different places at different times—the Marxist geographers who came to prominence in the 1970s found this approach wholly unsatisfactory and argued, instead, for one that sought to relate the geography of poverty, urbanization, industrial employment, and so forth more directly to the inner workings of capitalism. However, perhaps still suffering the intellectual hangover of the environmental determinism that had permeated the discipline in the first part of the century (Peet, 1985), early Marxist geographers' writings paradoxically tended to treat space in a very passive manner. In this first iteration of conjoining space and the social relations of capitalism, explanatory power was usually deemed to reside in examinations of the social organization of capitalism that were then seen to shape the unfolding of the economic landscape in a one-way direction—spatial pattern was to be understood by examining social organization, but there was little sense that the ways in which landscapes were produced might themselves affect the social organization of capitalism. All in all, it seemed, Marxist geographers were willing to go from a situation in which the cause of geographical change was theorized as laying within geography itself to one in which space appeared to play no role in shaping social organization.

If this first take on linking space and society, then, was one in which the spatial arrangement of capitalism was seen to be simply a reflection of its social organization, by the early 1980s a new theoretical direction had begun to emerge, one which talked of the need to theorize what Soja (1980) called the "socio-spatial dialectic" (see also the debate between Peet, 1981, and Smith, 1981, on this point). Marxist work in geography increasingly came to see the relationship between space and society as a dialectical one, one in which the social organization of capitalism certainly shaped its geographical organization but in which, also, capitalism's spatial structures shaped its social organization. Hence, Marxist geographers increasingly argued that spatial patterns were "not just an outcome . . . [but] part of the explanation" (Massey, 1995 [1984]: 4).[2] Infused by such theoretical insights, a relatively small group of Marxist geographers set about the none-too-small task of analyzing how geography made a difference in the ways in which capitalism (and other social systems, for that matter) functioned, a research project that led to a vigorous reconstruction of much previously aspatial theorizing about the world in what Soja (1989) has termed the "reassertion of space in critical social theory."

However, although this second iteration incorporated space as a central part of the explanation of how capitalism as a system operates, it did tend to focus primarily upon the activities of capital to explain the geography of capitalism, treating workers as largely passive in the process of making landscapes. This is not to say that it completely ignored labor, for the geographical structuring of capital was often explained in terms of how investments followed geographies of low wages, for example. Rather, in considering labor it tended to do so in terms of what I will call here a "geography of labor," that is, in terms of how *capital* makes use of the geographical differentiation of labor—such as by playing workers in different localities against one another through its whipsawing activities. What it largely did not do, though, was to seek to understand much of how workers' lives are structured and embedded spatially or how they may themselves attempt to shape the landscapes of capitalism to their own advantage, in either revolutionary or nonrevolutionary ways (i.e., in ways that may challenge extant class relations but also in ways which may reinforce them).

That such Marxist-inspired work tended to prioritize capital as actor is perhaps not surprising, given the heritage of Marx's own work. As Stanley Aronowitz (1990) points out, Marx himself was concerned with examining the processes of capitalist accumulation and developing a critique of bourgeois political economy as constructed from the viewpoint of capital. Hence, "the three volumes of *Capital*," notes Aronowitz (1990: 171), "are written from the point of view of capitalist accumulation." Nevertheless, this does little to mitigate the fact that such a marginalization of workers as (pro)active, sentient, geographical actors was limiting theoretically, for it presented a world in which workers' social and spatial practices remained unconnected to the process of the uneven development of capitalism, a world in which workers were not theorized as being present at the making of the economic geography of capitalism but, instead, were seen to struggle and live within the contours of an economic and social geography created by, and for, capital. While capital was viewed as capable of fashioning the geography of capitalism to suit its own needs, there was little sense that workers could also do the same.

The present volume fits within what might be called a "third iteration" examining the relationship between space and society. Beginning in the 1990s, a new group of Marxist and critical theory-inspired geographers began to examine how not just capital but also labor (broadly defined) shaped, through its actions, the geography of capitalism.[3] Principally, this work grew out of both a theoretical frustration with the extant work on understanding how the geography of capitalism is made—for many, a focus on capital as actor seemed to be inviting too narrow an explanation—but also a political frustration in that all the spatial action was seen to originate with capital, whereas workers were viewed simply as the passive bearers of the geographical transformations wrought by capital. Certainly, the bulk of work in economic geography still does not take labor very seriously, but the growing interest in the spatiality of labor—that is, in how workers live their lives spatially and how this affects the geography of capitalism—is beginning to provide something of a counterweight to the myriad studies of how capital structures the geography of capitalism. Although within this growing area of research there are definite differences in theoretical and empirical approach, all, I think it is probably fair to say, share a belief that understanding how workers' lives are structured spatially adds a great deal to studies of working-class life—which, in turn, might just make some difference in such people's lives.

In the chapters that follow, then, I argue both that geography plays a role in structuring workers' lives, and that workers and their organizations may play important roles in shaping landscapes as part of their social self-reproduction. The chapters, which are designed to be illustrative rather than comprehensive or representative, investigate how workers have been involved in two processes that are important in fashioning the uneven development of capitalism, these being the production of space and the production of geographic scale (by which I mean the very real spatial resolutions at which social, economic, and political processes take place). Although the "production" of geographic scale may at first seem a strange concept, upon reflection it becomes apparent not only that scale is a central spatial calculus that allows us to make sense of the landscape—in a very basic way it serves both to delineate and differentiate the landscape into areas that are similar to one another and areas that are different (as in the case of, say, economic regions), and also to indicate at what geographical resolution (local or national, for instance) certain processes, institutions, and social actors operate—but that it is not a fixed, pregiven metric. The scales at which social, political, and economic processes operate, and at which organizations may constitute themselves, together with the interconnections among such scales, are constantly made and remade through human action. How this is done, and whose interests are served by ensuring that certain processes play out at particular geographic scales, is a central focus of the cases that I examine below.

Such a focus on the production of space and geographic scale as part of what I am here terming a "labor geography" provides, I think, two useful theoretical points. First, while accepting the argument that the production of space in certain ways is not simply secondary to, but is an integral part of, how capitalist societies reproduce themselves, focusing on labor's role in this process provides a more complete picture of how capitalism works. Certainly I am not suggesting that theorizing how capital produces landscapes should be abandoned, but I am suggesting that explaining the geography of capitalism cannot be done through a focus on the activities and structures of capital alone. Second, such a focus on space and geographic scale allows a much less essentialist view of labor to be written. Rather than seeing "labor" as a spatially homogenous category, examining how different groups of workers face different challenges—and thus may seek to construct different types of geographies within which they wish to live—provides a means of recognizing that workers' interests may vary considerably depending upon the geographic context within which they find themselves. In turn, this may lead some groups of working-class people to attempt to construct landscapes that advantage themselves over other groups of working-class people, a situation that demands a more sophisticated understanding of how processes of class formation and reproduction are structured geographically than that provided by aspatial theorization.

This, then, is the broad scope of what I am trying to do in this book. Before proceeding further, however, there are a number of things I want to say about terminology and the book's organization. First, in using the terms "geography of labor" and "labor geography," I am not seeking to set up some kind of taxonomic structure through which work by various authors can be categorized (although, at the same time, it is clear that some work more readily falls into one category than the other). Instead, what I am trying to suggest is that there are distinct differences in perspective and focus about how labor is incorporated into explanations of why the geography of capitalism looks the way it does and how economic geography is written. Much as feminist writers have suggested that there is a difference between a "history of women" and a "women's history" (with the former simply focusing on how women have fared historically, whereas the latter attempts to write history specifically from the perspective of women), here I want to draw a similar distinction between approaches that incorporate labor as rather peripheral to explanations of the geography of capitalism and those that focus principally on the activities of working-class people and attempt to understand and write about the geography of capitalism from their perspective, as far as is possible.

Second, given that this book comes out of a debate within the Marxist literature about the role of space in the survival of capitalism, I draw upon Marxist theory to define what I mean by "labor." Principally, I use the term "labor" to refer to those individuals who do not own the means of production, from whom surplus labor is appropriated as part of the capitalist production process, and who do not occupy what Wolff and Resnick (1987) describe as a "subsumed" class position—that is to say, who are not part of either the "managerial" class involved in extracting surplus value or that part of the capitalist class involved in transferring or working with the surplus value that already exists (e.g., bankers). Thus, I adopt a definition that focuses on workers' class position in terms of "objective" classes-in-themselves rather than "subjective" classes-for-themselves, although I do recognize that how workers think of themselves "subjectively" will shape how they act (see Chapter 10 in Sheppard & Barnes, 1990, for a good discussion of different approaches to class within the Marxist tradition). Also, in what follows I tend to use the terms "labor," "workers," and "working class" interchangeably. While some may find problems with this, I do so largely to provide linguistic variety in the narrative. Hence, for example, when using "workers," I do not mean to confine this term to those who are currently in work but also use it to refer to those out of work yet who nevertheless conform to my definition laid out above.

Equally, although all of the case studies in this book examine the spatial activities of workers who belong to labor unions, I do not mean to suggest that a focus on "labor" as a spatial actor should just be a focus upon organized labor. Rather, the reason I have chosen to illustrate the theoretical arguments made in this book by an examination of the activities of several unions and workers' organizations is threefold: (1) I have had a longstanding interest in unions as social and political organizations, and so—not unreasonably, I hope—my empirical examples reflect my research interests; (2) I do believe that unions represent crucial, though not the only, entities through which workers may express their social, economic, and political aspirations (though, of course, this is not to say that unions have always pursued progressive policies on all matters!); and (3) as with capitalists, the power of workers to effect social change is usually greater when they act collectively than when they act as individuals, that is to say the geographical impacts of workers' spatial praxis is likely to be most evident (and thus most illustrative of the arguments I am trying to make in this volume) when we look at their collective, rather than their individual, actions.

Third, some might claim that the examples in the following chapters are exceptional empirically. This is a claim that I would vigorously contest, since there are myriad other examples of workers doing the kinds of things I recount here—struggling successfully to impose national contracts, building regional, national, or international links of solidarity, helping various national capitals to create new markets abroad and thereby to shape the unevenly developed geography of capitalism, engaging in debates over local economic development, and the like. Yet, even if the empirical examples *were* indeed unusual, the theoretical lessons that can be drawn from them are, I would argue, nevertheless important because they force us to consider seriously how workers and their organizations struggle to impose particular sets of spatial organization on the economic landscape and how these struggles, in turn, shape the geography of capitalism. Although the empirical research presented here is "intensive" rather than "extensive" in nature—that is to say, through detailed case studies it seeks to examine social and spatial processes and causal mechanisms rather than to present a large population of cases that may be subject to statistical generalization (see Sayer, 1984, and Massey & Meegan, 1985, on the differences between "intensive" and "extensive" approaches)—the theoretical principles that are elaborated upon are, I would argue, useful for examining other empirical situations and retheorizing how the geography of capitalism is forged.

*　　*　　*

The structure of this volume is fairly straightforward and follows a certain "scalar logic." Specifically, the initial chapters focus primarily on workers' and unions' activities at local and regional scales, while later chapters present analyses at more global scales. My reason for organizing the book in such a fashion was quite simple. Principally, I have chosen such a format because, typically, workers have been seen to exert their greatest power on the spatiality of capitalism at more local scales of activity, with their capacity for shaping landscapes frequently being assumed to vary inversely with the geographic scale of analysis. Thus, much of the literature on contemporary global economic restructuring tends to assume a priori that capital can act globally but that workers and their organizations must necessarily operate at subglobal scales (national, regional, local). In choosing the chapter order, then, I wanted to begin with examples of workers shaping landscapes at scales with which readers would perhaps be more familiar in their own work—that is, workers shaping local and regional geographies—before proceeding to examine larger-scale activities. At the same time, however, I seek to emphasize that activities at such different scales should not be seen as mutually exclusive but, instead, as intimately connected—what happens locally shapes what happens globally, and vice versa.

In light of this organizational logic, then, in Chapter 2 I lay out a theoretical argument concerning the role of space in workers' social activities and how work in economic geography has often tended to ignore labor's agency. At the same time, I argue that space does not simply provide a context within which social action takes place but that manipulating the geography of capitalism in certain ways can serve as a source of power, a reality that industrial relations practitioners and other students of working-class life might wish to consider when theorizing about social praxis. The subsequent chapters illustrate how workers have sought to shape spatial relationships and scales of organization in pursuit of their social, economic, and political goals. Thus, Chapter 3 shows how workers and their organizations and allies in New York City engaged in a very local zoning battle to control how the built environment was restructured in response to globally driven real estate pressures in traditional manufacturing areas of Manhattan. Specifically, they were successful in having a special preservation zone established in which conversion of loft manufacturing space to office space would be severely restricted. My purpose in examining this particular conflict is to suggest that, contrary to much of the rhetoric of neoliberalism, organizing locally can, in fact, be used effectively to shape the global processes of contemporary capitalism.

Chapter 4 examines how dockers were able to control the location of work in the East Coast longshoring and cargo handling industry through a series of work rules they managed to negotiate with their employers. In response to the threat of the relocation of employment inland that freight containerization augured, dockers forced the industry to agree to retain certain types of cargo handling work at the waterfront, thereby ensuring jobs for themselves. Again, my goal is to show how very local struggles enabled workers to negotiate national and global scale processes of economic and technological change in their industry. Following from this, in Chapter 5 I recount how, beginning in the 1950s, these same dockers were successful in imposing a new national level of contract bargaining against their employers' wills. Whereas previously contract negotiations had been conducted on a port-by-port basis, the union saw that in the age of containerization such a structure threatened to undercut the wages and work protection clauses they had managed to negotiate in New York, the East Coast port where dockers were most powerful. In securing multi-port bargaining, I argue, not only did the union and its membership limit the ability of employers to play dockers in different ports against each other but they also fundamentally shaped the geographic evolution of the industry in response to the global growth of new modes of cargo handling, particularly containerization.

Chapter 6 serves as something of an entrée to the remainder of the book, which focuses principally on unions' international activities and their shaping of the global political economy of capitalism. Through a recounting of the U.S. labor movement's official foreign policy and the activities of several international labor organizations during the twentieth century, the chapter challenges some of the assumptions that seem to have been accepted as part of what some have termed the "globalization debate." In particular, I argue not only that workers are not necessarily impotent in the face of global capital but also that they have frequently been actively involved in the very processes that are bringing about an (apparently) globalizing world economic system. Although this has been an uneven process, both historically and geographically, recognizing that workers have played active roles in processes of economic and political globalization forces us to reconsider many of the terms of this debate. In this light, Chapter 7 examines an international program initiated during the 1960s to encourage the building of worker cooperative housing units in Latin America and the Caribbean as a means of encouraging development and so of limiting the appeal of Communism in the region. Through the physical reconstruction of the built environment in numerous neighborhoods throughout the hemisphere, those involved in the projects hoped to shape ideological sympathies during the Cold War by transforming the spatial contexts within which many people lived their lives. Not only did the global ideological rivalry between Washington, DC, and Moscow shape international labor politics and, in turn, local labor politics, but, I argue, the outcomes of local struggles between Communist and non-Communist elements in various labor movements throughout the region also recursively shaped the global geography of the Cold War. Even in the context of a superpower rivalry that played itself out in the international politics of half a century, activities at the very local scale fundamentally impacted the landscapes of global ideological struggle.

Another aspect of globalization is the growing concern with international solidarity that appears to be pervading many labor movements. In light of this, Chapter 8 provides an account of an international solidarity campaign waged by U.S. unionists in Ravenswood, West Virginia, between 1990 and 1992 against the international financier and global commodities trader Marc Rich—the same Marc Rich whose pardon, granted just before president Bill Clinton left office, has been the source of much controversy. This campaign was one of the most important international solidarity campaigns waged by U.S. workers during the 1990s, both because it represented a significant win for these workers against a multibillion-dollar transnational corporation, but also because the almost two-year-long global campaign that secured this win made use of a number of innovative tactics that have subsequently become standard tools in similar campaigns. In the chapter I illustrate not only that the workers' campaign was successful because it managed to link together activities at a number of geographic scales, but also that through their actions the unionists shaped the foreign investment patterns of this multibillion-dollar corporation by preventing it from expanding its operations in Central and Eastern Europe and Latin America and the Caribbean.

The context for the final substantive chapter (Chapter 9) is, arguably, the most significant political and economic event of the second half of the twentieth century, namely, that of the collapse of Communism in Central and Eastern Europe and the end of the Cold War. Specifically, the collapse of the Soviet Union resulted in the implosion of many of the state-backed official union movements of the region and the rise of myriad "independent" unions in their stead. In the chapter I show how a number of Western-based international labor organizations are developing strategies to deal with these transformations currently occurring in this region. Through their international actions such organizations are helping to remake the geopolitical and geoeconomic situation in Central and Eastern Europe and beyond. The final chapter (Chapter 10) serves as a conclusion by highlighting some of the main conceptual issues drawn out of the case studies examined in the book.

Although the chapters making up this volume, then, have a quite varied empirical focus—conflicts over zoning regulations and restrictive work rules, international labor solidarity campaigns and patterns of foreign direct investment, geopolitics in the post-Cold War period, worker housing schemes, the geographic scale at which contract bargaining is to take place—in all of them I have tried to show how geographical context has shaped the actions of workers and their organizations and how, through these actions, workers and their organizations have helped shape the unfolding geographies of capitalism at scales from the local to the global. However, as will perhaps now have become evident, in addition to examining how workers shape the geography of capitalism as part of their social praxis and self-reproduction, another goal in this volume is to use the situations analyzed here as a means to confront the rhetoric of neoliberalism and capital-centrism that pervades our daily lives, particularly as it relates to the global economy. Specifically, there are three sets of assertions concerning political praxis in a global economy that run through the narrative as a whole that, drawing from my arguments about workers molding the geography of capitalism, challenge the view—paradoxically often put forward (though for different reasons) by both neoliberal and Marxist writers—that the global scale in particular is the unchallenged purview of capital.

    First, I seek to show how organizing locally can, in fact, be an effective
strategy for use against social actors (e.g., corporations) who are organized
at the global and other extralocal scales. Of course, how successful such local campaigns are will often be determined by the contingencies of history
and geography, and I certainly do not want to elevate the local to a privileged scale of activity in the way that some postmodernists have done. But,
I would argue, in certain instances the local can definitely serve as a powerful scale of activity in the face of larger-scale forces and processes (Chapters
3, 4, 5, and 7 emphasize this point). The second position that I seek to support through this work is the assertion that labor has been actively involved
in the very process of globalization itself (Chapters 6, 7, 8, and 9 examine
aspects of this). By focusing on workers' activities at the global scale, I want
to reclaim this part of labor's historical and geographical experience. Examining how workers and their organizations have been involved in the globalization of economic and political relationships suggests that they may
therefore be able to shape the development of the global economy in some
directions and not others. This clearly has both political and conceptual implications, for it takes us beyond the simple binary of globalization–not
globalization (i.e., should workers encourage globalization or should they
fight against it?) and towards a more textured and penetrating question of
"what type of globalization (proletarian internationalism? neoliberalism?)
could/should workers fight for?" Third, and relatedly, I seek to show how
labor can be effective against capital at the global scale by itself organizing
across national boundaries (Chapter 8, in particular, examines this aspect of
workers' spatial praxis, though Chapters 6, 7, and 9 also touch upon this issue). Despite the triumphalism with which capital appears to be claiming
the global as inherently its own reserved scale of organization, "international
solidarity" and "globalism" appear to be taking on added importance in
many labor circles these days, and workers are increasingly developing
links with their compatriots across the globe. In turn, this is having impacts
on the ways in which the new geography of global capitalism is being inscribed.

CHAPTER 2

Toward a Labor Geography

Making landscapes in certain ways is central to the ability of any social system to reproduce itself. Through his notion of the "spatial fix" David Harvey (1982) has shown how, under capitalism, capitalists must arrange the landscape so as to be able to ensure that profits are made. It is through their ability to make the geography of capitalism in some ways and not others that capitalists as individuals and as a class, together with capitalism as a social system, are able to ensure their survival (Lefebvre, 1976 [1973]). For example, capitalists need to make sure that workers can get to work on time, a concern that may necessitate ensuring transportation systems are built in certain ways and in certain neighborhoods. They also need to ensure that raw materials can reach factories, that finished commodities can reach consumers, and that information and capital crucial to profit making can flow to where they are needed. All of this requires locating factories in certain places or laying down particular networks of fiber-optic cables and other technologies so that information and capital may be disseminated—that is to say, it requires a certain spatial arrangement of investments in plant, infrastructure, and the built environment more generally. New ways of organizing production and consumption—such as the much-hyped shift from Fordist to post-Fordist methods of manufacturing (see Amin, 1994)—often require that landscapes also be restructured, to a greater or lesser degree.[1] Thus, the way in which the geography of capitalism is made appears to be of no mere peripheral concern to capital. Rather, it is central to the ability of capitalism to function and reproduce itself over time, and to the ways in which people live under capitalism, even if the particular ways in which space is made are historically and geographically contingent.[2] To borrow a phrase from Doreen Massey (1995 [1984]: x): "The geography of a society makes a difference to the way it works."

It is my contention in this volume that workers also need to ensure that

13

the geography of capitalism is made in certain ways and that, in the process of trying to do so, they and their organizations play important roles in forging the landscapes of capitalism. Although this statement would seem to be a truism that should be readily accepted—why would we expect workers not to play roles in making the economic landscapes of capitalism?— workers' roles in producing landscapes have, until fairly recently, largely been ignored within human geography. While geographers and others have long recognized that the geography of capitalism is contested and that individual capitalists do not always get their unfettered way when it comes to setting in place economic landscapes, they have frequently neglected labor's active role in shaping this geography. Whereas the notion of "spatial praxis" is now firmly on the intellectual agenda in human geography, and the making of the economic and social landscape in particular ways is recognized as being fundamental to the articulation of political power (see Harvey, 1982; Soja, 1989; Lefebvre, 1991 [1974]), many economic geographers and other theorizers of the geography of the capitalist space-economy either have tended to ignore the role of workers in making the economic landscapes of capitalism or have frequently conceived of them in a passive manner. Although during the past two decades economic geographers have generated a considerable literature that seeks to understand how *capital* attempts to make the geography of capitalism in particular ways to facilitate accumulation and the reproduction of capitalist social relations, there has been much less work that explicitly examines—either theoretically or empirically—how *workers* actively shape economic landscapes and uneven development. Labor's role in making the economic geography of capitalism has been rendered largely invisible by the analyses both of traditional mainstream neoclassical economic geographers but also, ironically, by many Marxists, for both approaches have primarily presented economic geographies devoid of workers as active geographical agents. In their explanations of the dynamics of the capitalist space-economy both neoclassical and Marxist approaches in geography have conceived of workers primarily from the viewpoint of how capital (in the form of transnational corporations, the firm, etc.) and, to a lesser degree, the state make investment decisions based on differences between workers located in particular places. Whereas for scholars trained in the tenets of neoclassical economics it is the relative importance of various factors of production and consumption that determines the location of economic activity or perhaps the structure of the firm that is significant, for Marxists it is capital that acts, capital that produces landscapes in its continual search for profit. In both views capitalists are theorized as capable of actively making economic geographies through their investment decisions, whereas workers are seen rather passively either as inert "factors" in the calculus of location or, per Harvey (1982: 380–381, emphasis in original), as little more than "*variable capital*, an aspect of capital itself."

It is not hard to understand why economic geography has developed this way. Geographers informed by neoclassical economics have largely been concerned with providing insights about the nature of the firm or enterprise, while Marxist work has sought to examine the expansion of capital during the accumulation process. However, this understanding does not mitigate the fact that both approaches as they have frequently been articulated have been unsatisfactory theoretically because neither has left much room to conceive of workers as active makers of the geography of capitalism. What I seek to do here, then, is to argue for a much more active conceptualization of workers as engaged in producing the unevenly developed geography of capitalism. In so doing, I am trying to do two things theoretically. First, I wish to question the primacy accorded the activities of the firm and capital in general in defining the field of economic geography (and particularly explanations of the location of economic activities) as it is usually practised. In this sense, I would align this work in the broader emergent literature that is beginning to question the very nature of what economic geography as a field of inquiry *is* (see Thrift & Olds, 1996). Traditionally, economic geographers have primarily focused on trying to understand the decision-making processes of managers and capitalists. Although workers' activities are sometimes seen as a "modifying" force that needs to be factored (and I use this terminology quite deliberately) into the locational equation, such an approach essentially places capital center stage both empirically and theoretically as the focus of research while banishing workers to the fringes of both the discipline and the process of landscape production. In essence, such approaches tell the story of the making of the geography of capitalism through the eyes of capital(ists). My goal here, in contrast, is to suggest that much insight into how the economic geography of capitalism is produced can be gained by greater analysis of the social and spatial practices of workers. Thus, while not rejecting the understanding into the production of the geography of capitalism that can be gained by examining the actions of capital, I would argue that at the same time it is important to recognize that workers, too, are *active* geographical agents whose activities can shape economic landscapes in ways that differ significantly from those of capital. Examining how workers actively shape economic space, then, significantly broadens the ambit of economic geography and efforts to understand how the geography of capitalism evolves.

Second, and following from the foregoing, I wish to return agency to workers in the literature on the production of economic geographies. This means conceptualizing labor not merely in terms of "factors" of location or the exchange value of "abstract labor" but to treat working-class people as sentient social beings who both intentionally and unintentionally produce economic geographies through their actions—all the while recognizing that they are constrained (as is capital) in these actions. Certainly, in suggesting that "capital is *not* all" I do not wish to argue that "labor *is* all." That is not my purpose, for such an argument would be equally problematic. Instead, I wish to assert that workers have a vested interest in making space in certain ways and that their ability to do so is a potent form of social power. Recognizing this fact raises important questions about the theoretical status of spatial relations in workers' everyday lives and the issue of workers' "spatial praxis." While capital's efforts to create landscapes in particular ways have been theorized as an integral part of its self-reproduction and survival (e.g., Harvey, 1982; Lefebvre, 1976 [1973], 1991 [1974]), in this book I seek to show how workers, too, make space in particular ways for their own ends and to ensure their own self-reproduction and, ultimately, survival—even if this is self-reproduction and survival *as workers in a capitalist society*. The economic geography of capitalism, I hope to show, does not simply evolve around workers who themselves are disconnected from the process. Rather, they are active participants in its very creation.

These assertions are important for both theoretical and political reasons. From a theoretical perspective a focus on workers' spatial praxis provides us, I would argue, with a more insightful look at the forces and struggles that shape the geography of capitalism. From a political perspective it provides us with a way to think of workers and their organizations as sentient and (pro)active social agents who have options—sometimes many, sometimes few—with regard to the spatial relations within which they live their lives, rather than conceiving of them simply as either dupes or instruments of capital. This is important both in terms of the implications it has for the possibilities of creating more emancipatory geographies—if we are to take seriously Castells's (1983: 311, emphasis in original) assertion that "space is not a 'reflection of society,' it *is* society," then any emancipatory political project must be geographical in nature—but also with regard to the politics of writing and the discursive representation of capital and labor. As Gibson-Graham (1996: 4) has argued, there has been a tendency by those on the political right but also, perhaps somewhat paradoxically, by those on the political left to present capital discursively as "large, powerful, persistent, active, expansive, progressive, dynamic, transformative; embracing, penetrating, disciplining, colonizing, constraining; systemic, self-reproducing, rational, lawful, self-rectifying; organized and organizing, centered and centering; originating, creative, protean; victorious and ascendant; self-identical, self-expressive, full, definite, real, positive, and capable of conferring identity and meaning." In the process, workers have habitually been represented as always being outmaneuvered by a seemingly omnipotent, omniscient, infinitely flexible capital, as unquestionably dupes when they side with capital on certain issues, as invariably divided in the face of an always more internally coherent capital, and as inevitably less important than capital in explanations of the making of the geography of global capitalism. A consequence of this, Gibson-Graham (1996: 4) suggests, is that "it is the way capitalism has been 'thought' that has made it so difficult for people to imagine its supersession." Arguably this is most obvious in the burgeoning literature on globalization, much of which portrays globalization as an inevitability and, therefore, unchallengeable, a process driven purely by the internal dynamics of the demands of capital accumulation (e.g., see Bryan & Farrell, 1996; Gates with Myhrvold & Rinearson, 1995; Kanter, 1995; and Ohmae, 1990, 1995). Certainly I do not want to succumb to the fallacy that simply by writing about capitalism in a different way we negate it. I do, however, want to suggest that by writing a more active role for workers concerning how capitalism functions, workers, activists, and progressive scholars may begin to identify geographical possibilities and strategies through which workers may challenge, outmaneuver, and perhaps even beat capital.

Implicitly, by focusing upon how different groups of workers and their organizations make landscapes and how, in the process, their activities are sometimes constrained by, and sometimes enabled by, particular geographies under capitalism, the work presented here is a critique of essentialist notions that have often dominated in analyses of the activities of capital and labor. Capital is neither omnipotent and omniscient, nor does it always present a common front in the face of challenges by labor. Likewise, labor is neither necessarily impotent in the face of, and inevitably more divided than, capital, nor would it necessarily be unified in some "heroic proletarian" fashion if workers would only recognize and abandon the "false consciousness" that they are often accused of possessing when, for instance, they ally themselves with local capitals (such as banks and utility companies involved in urban growth coalitions) to defend "their" particular spaces in the global economy at the expense of making common cause with workers located elsewhere. By examining how capital and, particularly, labor attempt to shape the spatiality of everyday life and the landscapes of capitalism, in what follows I hope to show both that geography may serve at times to bring workers together in the face of a more divided capital and to divide them in the face of a more unified capital, but also that in very real ways workers can get the upper hand over capital and force it to do things socially and spatially that it may not want to do. Put another way, I hope to show that workers and their organizations struggle spatially as well as historically and that in doing so they can have very real influence on how the geography of capitalism is made.

In serving as a theoretical framework for the empirical chapters that follow, the present chapter is divided into two main parts. The first of these is an examination of how labor has traditionally been conceptualized in both neoclassical and Marxist-inspired economic geography. I argue that the way labor has usually been thought of in these approaches is very much in terms of what I am calling here a "geography of labor," by which I mean that anal-

yses have largely examined the spatial distribution of workers across the
landscape to show how this affects the decision-making process of capitalists and, hence, the economic geography of capitalism. Such approaches,
while frequently providing important insights into the geography of capitalism, have by their very nature tended to portray labor in rather descriptive
and passive terms. (Readers who are already somewhat familiar with elements of this critique may choose to skip this segment and move directly to
the section titled ". . . To a Labor Geography" on page 33.) The second section attempts to outline a way in which a more active sense of workers as
being involved in the creation, manipulation, and use of space can be incorporated into economic geography. As a means of distinction I have termed
this alternate approach a "labor geography," by which I mean it is an effort
to see the making of the economic geography of capitalism through the
eyes of labor by understanding how workers seek to make the landscape in
their own image, as it were. More particularly, I argue that one axis of spatial struggle that is crucial to examine in any analysis of workers' geographical praxis is that surrounding what Neil Smith (1990 [1984]: 169) has called
"the production of [geographic] scale," that is to say, how workers may create different scales of organization and social structure and how they may
engage in social and spatial praxis at a number of different geographic resolutions as they attempt to make economic landscapes at scales that suit their
needs at particular times in particular places.

FROM A GEOGRAPHY OF LABOR . . .

"The history of capitalism," remarks Stanley Aronowitz (1990: 171), "has,
typically, been written as a series of narratives unified by the themes of accumulation." These narratives, he suggests, have been marked by a focus on

> mercantile and imperialist interests seeking fresh sources of investment;
> the scientific and technological revolutions that have driven growth; inter-
> national rivalries over territory and labor supplies and the multitude of
> conflicts among fractions of capital that take political forms, such as the
> struggles for power among capital's personifications or wars. . . . In these
> accounts, workers enter the theater of history as abstract labor, factors of
> production, dependent variables in the grand narratives of crisis and re-
> newal.

Much the same could be said of the writing of the geography of capitalism.
    Such treatment of labor by geographers has its roots in the very estab-
lishment of the modern discipline. Nineteenth-century commercial geography, notoriously linked as it was to imperial adventurism, was one of the

first branches of geography to give any sort of systematic accounting of labor.[3] Principally the practitioners of this subject sought to record the spoils of empire: raw materials for industry, climatic conditions affecting production, what times of the year riverways were navigable, the location of mountain pathways—in short, anything that would be of commercial use to an imperial power seeking to augment its economic and political might. In this type of geography labor was inert, a factor of production to be catalogued no differently than the various soil types or climatic conditions found in different regions of the globe. Effectively, workers were marginalized as sentient human beings. This is not to say, however, that commercial geographers considered labor unimportant. Far from it. Indeed, one of their greatest concerns was the relative productivity rates, skills, and "upkeep costs" of different colonized peoples. For an imperial power such as Britain, knowing which of its subjects could pick tea or produce textiles most efficiently was of vital concern.[4]

Similar theoretical tendencies were also evident in early-twentieth-century urban economic geography. In an approach that drew heavily upon environmental determinism (Peet, 1985), urban geographers such as Robert Park (1936) used a naturalistic analogy to explain the built environment's form in terms of a living ecosystem in which "biological economics" and the processes of ecological competition, dominance, and succession determined which population groups lived where in the city.[5] Although many urban geographers of the early twentieth century were undoubtedly concerned about the ills that affected working-class neighborhoods, most primarily conceived of workers' actions and living choices as being driven principally by some manner of sociobiological evolutionary process. In such an approach change in the physical form of the built environment resulted not from workers' active efforts to shape the landscape in particular ways but from the rhythmic pulse of ecological expansion and contraction that drove particular groups of working-class people ever outward from the city center, in the process changing the character of the various neighborhoods in which they lived.

Although both commercial geography and the urban ecology of the 1920s and 1930s incorporated quite passive views of labor into their theorizing, it is, however, neoclassical industrial location theory that has had perhaps the most significant impact on the way in which labor has been theorized in economic geography throughout much of the twentieth century.

Industrial Location Theory and the Neoclassical Approach

Arguably the dominant theoretical impulse in economic geography until the 1970s, neoclassical economics has served as the basis for four major strands of theorizing about how the economic landscape is made (Massey, 1973).

First, there is a large body of work based on Weber's (1929 [1909]) treatise that examined how transportation costs, labor costs, and what Weber called "agglomeration forces" affect the locational decisions of independent single-plant firms. Second, geographers have developed location theory stemming from Hotelling's (1929) examination of the spatial behavior of businesses whose location is dependent on the location of their competitors. Third, during the 1960s and early 1970s there was great interest in the "behavioral" approach derived from attempts made to understand how the internal operation of the firm influences its decision-making processes and thus its economic geography (e.g., Cyert & March, 1963). Fourth, economic geographers have followed the approach developed by Lösch (1954) that, while beginning with an examination of individual firm behavior, sought to broaden the scope of analysis to the development of entire economic landscapes. Although contemporary neoclassical location theory has been refined since these initial propositions were first presented, many of its underlying assumptions about how economic landscapes are made can be boiled down to the basic theoretical principles articulated in these four approaches. Certainly this brief summary does not exhaust the literature. Nevertheless, the four approaches are, I would argue, broadly emblematic of the major traditions in mainstream neoclassical economic geography and of how labor is conceived within these traditions.[6]

These approaches have been fairly thoroughly critiqued on a number of grounds: for too often being subservient to the economic and political ambitions of capital and the state; for operating under largely aspatial assumptions; for their positivist methodology and philosophical underpinnings (particularly attempts at "value-freedom"); and for assuming that economic exchange takes place between social actors who come to the market as equals. These criticisms have been well articulated elsewhere, and I do not wish to rehash them here.[7] However, I do want to reiterate one particular point concerning neoclassical economic geography's marginalization of labor, namely, that neoclassical economic geography is fundamentally about how *firms* make locational decisions—*it is firms' behavior and investment decisions that are both the activities to be explained but also the activities that define economic geographies.* Certainly workers are not always totally ignored. Bid-rent models of land use, for example, focus upon how consumers—who may be assumed to be members of the working class, though this is rarely made explicit—affect the structuring of economic and urban geographies through the choices they make as consumers of commodities and services. However, it is only in this very narrow sense of their role as consumers of land in the marketplace that working-class people are theorized as capable of shaping the economic landscape. Although such "choice-driven" models appear to offer a role for working-class agency in shaping the geography of capitalism, such agency is incorporated only in this very circumscribed way. Ultimately, in such approaches there is little room for conceiving of working-class people as sentient geographical actors whose broader political and cultural praxis directly shapes the geography of capitalism and for whom the making of the economic landscape in ways that further their own economic and social agendas is integral to their ability to reproduce themselves as workers on a daily and generational basis.

Arguably such a view of workers stems from the guiding principle of neoclassical economics, which sees labor as just another factor of production and labor organizations as simply impediments to the smooth operation of labor markets. Even the field of "labor economics" is dominated by many of the neoclassical assumptions that see labor as a commodity no different from any other that may be bought and sold in the market.[8] Conceptualizing labor as merely a factor of production that is subject to the same market dictates as any other commodity has a number of implications for considering how economic geographies are made and how workers fit into this.

First, within the neoclassical framework a particular industry's or region's economic geography becomes defined by the spatial distribution of raw materials, transportation nodes and networks, labor, the market, and so forth, and how these various factors interact spatially through the processes of production and consumption, all of which are determined by consumer choice. There is no sense that workers may play active roles in shaping economic landscapes other than in their capacity to act as consumers. Second, by reducing everything, ultimately, to market forces of supply and demand, neoclassical approaches tend to lose historical and geographical specificity—Weber (1929: [1909]: 10), for instance, talks about the need to develop "pure laws of industrial location . . . independent of any particular kind of economic system." The historical waxing and waning of workers' collective power and their ability thus to shape economic landscapes in different ways in response to pressures that they may face is written out of the story of how capitalism's economic geography is made. Third, because neoclassical economic geography draws explanatory power from its claimed ability to evaluate the importance of various locational factors in the siting decision of a particular economic activity, any account of labor necessarily reduces it to the mere categorization of skills, wages, levels of unionization, proclivity to striking, and how these are spatially differentiated. Thus, neoclassical economic geography can never hope to incorporate the rich diversity and complexity of working people's experiences. Rather, labor is stripped of many of its most analytically telling qualities and presented merely in the stark terms of its cost to any particular purchaser.

Such problems are true even for human capital theory, which focuses most particularly upon labor force characteristics. Although human capital theory argues that by investing in things like education and skills training workers may improve their chances of finding employment, may become more productive, and so may indirectly affect the evolution of the economic landscape (Becker, 1975), it actually theorizes workers as capable of playing an active role in the making of the geography of capitalism only in this narrow sense of how they may make themselves more attractive to capital, which ultimately makes the decisions regarding how the economic landscape is to be fashioned. Certainly it is the case that under human capital theory workers are not theorized as capable of engaging in spatial praxis through direct interventions into making the geography of capitalism in any broader, more holistic, proactive sense. Furthermore, the use of such terminology in economic theory takes on grave political and theoretical implications. Referring to workers' various attributes and skills in terms of "human capital" simply reinforces labor's theoretical invisibility, for it treats workers not as social agents capable of making landscapes in their own right but, rather, as simply an aspect of capital. Such language effectively collapses workers' geographic agency into that of capital. As a result, there is little conceptual impetus for seriously considering how workers themselves make economic spaces. In such a reductionist world everything becomes an aspect of capital, and so only capital's activities need be theorized. Workers' agency is ignored.

Although the inability to account for "noneconomic" issues in explaining economic geographies led to the rise of "behavioralism" in economic geography during the 1960s, such a focus on behavior did little to increase the possibility for a more active incorporation of workers' political and cultural practices into location theory in economic geography.[9] Given behavioralism's origins in the management and administrative sciences, this is, perhaps, hardly surprising (Carr, 1983). Indeed, as Hayter and Watts (1983: 164) have argued, the "main emphasis in [behavioralist] studies of locational choice has been in providing practical guidelines for managers concerned with finding new locations." Again, the assumptions underlying this new approach were defined by the needs of capital. Hence, for instance, explanations of the emergence of a new international division of labor based on the product cycle model (e.g., Vernon, 1966, 1971, 1979) were largely predicated on the argument that, as production becomes more standardized, corporations will seek out locations with low labor costs and/or high labor productivities, whereas the literature on branch plant location in peripheral regions frequently emphasized the importance of lower labor costs relative to those found in core regions.[10] While both these sets of literature considered labor an important influence on enterprises' spatial structures, the point is that they still did so from the explicit perspective of how *capital* can best exploit different labor markets.

In sum, then, the standard neoclassical approach as interpreted in economic geography has at least two important consequences when considering the conceptual marginalization of the geographic power of working-class people in the literature. First, clearly, the point of view presented is that of capital. Second, it presents an economic geography devoid of workers, both as individuals and as members of social groups. Because it is the firm that acts, neoclassical economic geography does not need to theorize workers as active makers of economic geographies. Thus, there are no people in neoclassical explanations of the production of economic landscapes, merely crude abstractions in which labor is reduced to the categories of wages, skill levels, location, gender, union membership, and so forth, the relative importance of which is weighed by firms in their locational decision making. Working-class people are stripped of many of their most analytically telling qualities and presented in the stark terms of how the cost and quality of their labor power to any particular purchaser affects that purchaser's investment decision-making process. In perhaps the clearest expression of such workerless landscapes, Weber (1929 [1909]: 95) even goes so far as to suggest that "labor costs can only become factors in location by varying from place to place." Taking this argument to its logical conclusion, it quickly becomes evident that, in Weber's mind at least, it is theoretically possible (in cases where costs do not vary spatially) for labor to play absolutely no role in explanations of the economic geography of a capitalism!

Marxist-Inspired Approaches to Understanding the Geography of Capitalism

The influx of Marxist thought in the 1970s transformed the way in which economic geography had traditionally been done and led to the emergence of what has become known as the "new economic geography" (see Walker, 1998).[11] In particular, it sparked a welter of theorizing about how capitalism as a system worked geographically and what connections there were between the uneven development of space and the broad forces of capitalist accumulation. Much of this work focused upon the activities of multinational corporations and the role of urbanization under capitalism.[12] Although there was much early debate about whether or not focusing on patterns of geographically uneven development was simply fetishizing space and diverting attention away from the *social* relations of capitalism, by the 1980s the notion of the "social production of space" had become common parlance in the Marxist literature in economic geography.[13]

Whereas neoclassical approaches tended to see space as an ontologically prior stage upon which economic transactions took place—so that space was theorized as a "container" of social life in which "human activity does not restructure space; it simply rearranges objects in space" (Smith, 1990 [1984]: xi)—Marxist approaches sought to portray the actual *production* of space in particular ways as *integral, rather than contingent,* to the continuation of the accumulation process. Marxists argued that not only is

space socially produced but also it does not have an ontological status separate from, and prior to, the economic and political relationships that produce it. In this view space was not simply a reflection of society but was
more fundamentally imbricated in its very constitution and functioning.
Thus, Harvey (1982: 416–417, emphasis added) pointed out that the geography of capitalism took on particular spatial forms at particular historical
junctures and that it was important to recognize that "the territorial and regional coherence that . . . is at least partially discernible within capitalism is
actively produced rather than passively received as a concession to 'nature'
or 'history.' " Placing spatial relations centrally in his analysis of the dynamics of capitalism, Harvey asked rhetorically whether there was indeed a
"spatial fix" to capital's crises, a particular way of organizing the landscape
to overcome—if only temporarily—some of "the contradictions within the
value form." Clearly, for Harvey and others, the making of space in particular ways had come to be seen as a key part of the circulation of capital,
while processes of capitalist accumulation were argued to take on specific
spatial forms.

The idea that capital must actively produce space in particular ways
owes much to the work of French theoretician Henri Lefebvre, whose ideas
have provoked a great deal of recent theorizing in critical human geography
(Soja, 1989; Smith, 1990 [1984]; Merrifield, 1993).[14] Central to Lefebvre's
work was his (1991 [1974]: 53) argument that "every society produces a
space, its own space." Hence, medieval European feudal society produced a
panorama in which "manors, monasteries, [and] cathedrals . . . were the
strong points anchoring the network of lanes and main roads to a landscape
transformed by peasant communities," whereas in capitalist society the
space of everyday living "is founded on the vast network of banks, business
centres and major productive entities, as also on motorways, airports and information lattices" (Lefebvre, 1991 [1974]: 53). Essentially, Lefebvre's thesis
argued that under capitalism capital needs to create particular landscapes for
accumulation to occur and that the production of space in certain ways is
central to the reproduction of capital and capitalist social relations (Lefebvre,
1970, 1976 [1973], 1991 [1974]). Thus, he suggested (1976 [1973]: 21, emphasis in original), in a now somewhat overused turn of phrase, that

> capitalism has found itself able to ˙attenuate (if not resolve) its internal
> contradictions for a century, and consequently, in the hundred years since
> the writing of *Capital*, it has succeeded in achieving "growth." We cannot
> calculate at what price, but we do know the means: *by occupying space,*
> *by producing a space.*

For Lefebvre, the secret to capital's success lies in its ability to construct
the appropriate material geographies that it can use to facilitate the extrac-

tion and realization of surplus value during the accumulation process, that is to say, the material landscapes that capital must build to ensure the tacit renewal of capitalism. However, whereas landscapes produced by capital in the pursuit of profit must have use value for capital—that is, such landscapes must in some way facilitate the production and realization of surplus value—they may have no use value for labor. Herein lies a source of contradiction and (potential) struggle, for workers may have very different geographical visions with regard to how the economic landscape should look and function than do capitalists, and may need very different types of landscapes in order to facilitate their own social and biological reproduction on a daily, generational, or any other basis.[15] Such a conclusion would appear to suggest that in Lefebvre's schema there is great theoretical room for incorporating workers as (pro)active shapers of the geography of capitalism in the first instance. However, this is not the case. Despite his admonitions that struggle is important for shaping the production of space, Lefebvre actually has trouble incorporating labor theoretically as an active geographical agent whose activities can directly shape the geography of capitalism. This problem stems from his initial failure to locate struggle at the center of his analysis. Let me explain this contention in more detail.

The dynamic nature of capitalist accumulation, Lefebvre argues, means that space is always in a state of flux, constantly being made and remade. This renewal is contested, for, in the reproduction of capitalist social relations, "there is no purely repetitive process" (Lefebvre, 1976 [1973]: 11). With this statement we clearly see that Lefebvre means to argue that struggle shapes the production of space. He further states (1991 [1974]: 55, emphasis added):

> As for the class struggle, its role in the production of space is a cardinal one in that this production is performed solely by classes, fractions of classes and groups representative of classes. Today more than ever, the class struggle is inscribed in space. Indeed, it is that *struggle alone which prevents abstract space from taking over the whole planet* and papering over all differences. Only the class struggle has the capacity to differentiate, to generate differences which are not intrinsic to economic growth *qua* strategy, "logic" or "system"—that is to say, differences which are neither induced by nor acceptable to that growth.

This is an illuminating statement for several reasons. In particular, although he appears to emphasize the production of space as emanating from the dynamics of class struggle, Lefebvre in fact conceives of workers' struggles as being somewhat secondary to the actual process of producing space under capitalism. Instead, for him, it is capital as maker of abstract space (which he conceives as the "space of the bourgeoisie" [1991 (1974): 57]) that is assigned primacy. Class struggle on the part of labor merely serves to modify these spaces, to prevent abstract space from taking over the whole planet. In effect, class struggle is seen to play its part only *after* abstract space has already come into existence. Furthermore, Lefebvre adopts (1991 [1974]: 392) a dubious dualism in suggesting that "business and the state, institutions, the family, the 'establishment,' the established order, corporate and constituted bodies of all kinds" seek to "dominate and control space," whereas "various forms of self-management or workers' control of territorial and industrial entities, communities and communes, elite groups striving to change life and to transcend political institutions and parties" merely "seek to appropriate space." Implying that capital and the state may dominate and control space but that workers may only appropriate space suggests that the former may produce the landscapes that they desire at particular times and places, whereas the latter may only ever hope to occupy such landscapes for certain periods of time, never actually capable themselves of making their own landscapes.

Such a theoretical privileging of capital as primary maker of the geography of capitalism leads Lefebvre to argue that the downfall of abstract space results, ultimately, from its own internal contradictions. In producing abstract space "contradictions . . . come into being which are liable eventually to precipitate the downfall of abstract space" (Lefebvre, 1991 [1974]: 52). Not only are these contradictions (what Lefebvre calls "differential space") internal to abstract space, but they simply "come into being." There is only the most abstract sense that they are actively created through struggle. Despite his earlier assertion that struggle prevents abstract space from taking over the whole planet, Lefebvre (1991 [1974]: 393) indicates that abstract space's "falsification is self-generated." In the final analysis such a view provides very little theoretical room to conceive of workers as active participants in the production of the geography of capitalism, or as capable of challenging capital's spatial practices and the landscapes that they produce.

Within Anglophonic geography David Harvey has, without question, been the leading theoretician attempting to conceptualize the role that the production of space plays in the workings of capitalism. In early work Harvey (1972, 1973, 1976, 1978) sought to understand the geography of capitalist urbanization and how this geography was reflective of the dynamics of accumulation. In particular he argued (1978) that three circuits of capital could be identified, namely a "primary circuit" (in which surplus value is extracted during the production process), a "secondary circuit" (representing the built environment), and a "tertiary circuit" (representing investments in science, technology, education, and other social expenditures such as a police force). The form of the built environment (the "secondary circuit"), Harvey suggested, was shaped by the flows of capital into the landscape to produce those structures necessary either for production to take place (factories, roads, etc.) or for consumption (housing, sidewalks, shops, and the like). The result, Harvey (1978: 124, emphasis added) argued, was a situation in which "*capital* builds a physical landscape appropriate to its own condition at a particular moment in time, only to have to destroy it, usually in the course of a crisis, at a subsequent point in time." In his book *The Limits to Capital*, Harvey (1982) took this argument one step further to provide in more detail a theoretically integrated account of the production of space and how this is linked to the accumulation process. In particular he suggested that capital must create particular "spatial fixes" in the landscape to allow accumulation to proceed. These spatial fixes are not only integral to the circulation of capital but, Harvey (1982: 426) argued, they constitute the very basis for the uneven development of the geography of capitalism, which is never static but is constantly remade during "the continuous restructuring of spatial configurations through revolutions in value."

Following in this tradition, Neil Smith (1990 [1984]), in his book *Uneven Development*, laid out a powerful analysis of the dynamic of the geography of capitalism and the production of uneven development. For Smith, uneven development is not merely a historical "accident," nor simply the result of the impossibility of "even development," but is, instead, integral to the accumulation process and is the very "hallmark of the geography of capitalism" (p. xiii). Smith's work makes clear the links between the accumulation process and the unevenly developed geography of capitalism in which, he argued, capital produces space "in its own image" (p. xv). Thus, he maintained, the geography of uneven development "derives specifically from the opposed tendencies, *inherent in capital*, towards the differentiation but simultaneous equalization of the levels and conditions of production" (p. xv, emphasis added). Furthermore, Smith averred that understanding how the uneven geography of capitalism is made requires a more sophisticated analysis of how the actual scales of capitalist development are also socially constituted. Whereas social scientists and others have often conceived of scales as little more than a convenient metric for ordering the world (as in when they have decided to conduct research at the "regional scale" or "national scale" of analysis), Smith argued that scales were, in fact, socially constructed as part of the process of capitalist accumulation and that, therefore, there was a material economic basis to their construction (we shall return to this issue in the following section).

In Britain perhaps the most influential theorist of the 1970s and 1980s concerning the uneven spatial development of capitalism was Doreen Massey, who, along with her colleague Richard Meegan, sought to place the locational dynamics of firms within the broader social forces of capitalist political economy and what she called the spatial division of labor (Massey, 1995 [1984]; see also Massey & Meegan, 1982, 1985). In this work Massey sought to make theoretical connections between production, social structures, and the geography of accumulation as a means of understanding the uneven development of the British space-economy. In her provocative book *Spatial Divisions of Labour*, she argued that the geography of capitalism could be thought of in terms of a geologic-type continuous depositing across the landscape of layers of investment—sometimes deep, sometimes shallow—that interacted with preexisting spatial arrangements of production and consumption to produce new geographies of uneven development. By conceiving of economic geographies as the product of such historical layerings of differential capital investment, Massey illustrated how the changing spatial organization of capital and the social relations of capitalist accumulation systematically structures the geography of manufacturing.

Whereas writers such as Harvey, Smith, and Massey largely concentrated on understanding the broad parameters of uneven development and the implementation of spatial divisions of labor under capitalism, others were more interested in theorizing specifically the role of urbanism in the functioning of capitalism. In his writings on U.S. urban development Gordon (1978), for instance, illustrated the links between urban spatial form and the level of class struggle in the paid workplace at particular historical junctures. His analysis of the establishment during the first two decades of the twentieth century of suburban manufacturing towns by large industrial concerns as a means of escaping unionized workforces in older urban centers clearly shows how capitalists deliberately produced new urban spaces in response to working-class agitation (see also Page, 1998). Likewise, Walker's (1981) examination of what he called the "suburban solution" showed how federal stimulation of suburbanization was used to prime the economic pump in the United States after World War II by encouraging working-class homeownership on a massive scale (see also Davis, 1986; Moody, 1988). Equally, Castells's (1977, 1978) earlier work examined issues of class and power, particularly as they revolve around urban collective consumption, while his later (1983) study of urban social movements focused, in part, on examining how worker-led rent strikes have shaped urban politics in several advanced capitalist and Latin American countries.

These works were truly pathbreaking in the way in which they have encouraged social scientists to think about the dynamics of uneven development under capitalism. However, despite providing important insights into the making of the geography of capitalism, such works were also problematic for the way in which they conceived of and/or marginalized conceptually the roles of workers in actively shaping the economic geography of capitalism. Certainly all argued for the importance of class struggle in capitalist society. Some, such as Castells's study of rent strikes, even focused to a degree upon worker urban politics. Yet, in terms of their theoretical analyses, these approaches paid relatively little attention to conceptualizing how workers' activities can directly and significantly shape the economic geography of capitalism, or to the role played by space in workers' self-reproduction as social and geographical actors. Instead, the Marxist work of the 1970s and 1980s tended to focus primarily on the geographical structure of capital and how capital structures landscapes through its activities (such as the pursuit of profit). While this was considered by many to be an improvement on the neoclassical approaches that had dominated the discipline until the 1970s and that had assumed that landscapes operate in some kind of equilibrium, like the neoclassical approach Marxist work left little room for linking the social practices of workers to the production of the geography of capitalism.[16] Capital as actor was prioritized, with workers entering analyses either as abstract labor, aspects of capital in the form of variable capital, or else in terms of rather vague references to the importance of class struggles.

For example, in his magisterial *The Limits to Capital* Harvey (1982: 380) tended to conceive of workers' roles as shapers of the economic landscape only in rather limited terms of how the migration of labor affects the accumulation process (see also Katz, 1986). Elsewhere, in viewing class struggle in terms of "the *resistance* which the working class offers to the violence which the capitalist form of accumulation inevitably inflicts upon it," Harvey's (1978: 124, emphasis added) epistemological priority allowed labor to resist capital but apparently not to take the initiative in class struggle (see also Parson, 1982). Likewise, his comments that "capital represents itself in the form of a physical landscape *created in its own image*" and that "*capital* builds a physical landscape appropriate to its own condition at a particular moment in time" (Harvey, 1978: 124, emphasis added) highlight the extent to which he saw the geography of capitalism as largely the product of capital itself. In this schema labor may "*us[e]* the built environment as a means of consumption and as a means for its own reproduction" (Harvey, 1976: 265, emphasis added), but workers are not theorized as (pro)active agents actually capable of *shaping* the built environment themselves as part of the process of their own self-reproduction. Instead, workers are conceptualized as reliant for their daily and generational reproduction on the utility presented them by the landscapes produced by capital in the process of its own reproduction. Although Harvey (1982: 412) suggested that his analysis allowed us to think about "how capital and labour can use space as a weapon in class struggle," in practice his efforts to theorize how the geography of capitalism is made principally focused on the former.

Likewise, Smith's theoretical work on uneven development, while forging important conceptual links between the active uneven development of space and the reproduction of capitalism, also presented an exegesis that focused almost exclusively on the activities and structure of capital. Although Smith was sensitive to issues of class struggle, ultimately his explanation for the process of uneven development rested on an analysis of the internal dynamic of capital. Thus, for Smith (1990 [1984]: xiii, emphasis added), "un-

even development is the systematic geographical expression of the contradictions inherent in the very constitution and structure of *capital.*" In such a
conceptualization there was, then, little theoretical examination of how
workers may also be actively involved in the uneven geographical development of capitalism as part of their own social and spatial praxis designed to
facilitate their own self-reproduction (e.g., by engaging in boosterism or
protectionist policies to ensure that particular places and groups of workers
enjoy privileged positions in the local, regional, national, or global economies).

Equally, while Massey was certainly correct to argue (1995 [1984]: 7)
that "the world is not simply the product of capital's requirements," the foremost concern of her book was to illuminate the connections between the
activities of capital and the spatial structures of production. The theoretical
framework upon which she drew to understand the geography of uneven
development relied principally on an analysis of the social and spatial organization of *firms* for its explanation, particularly concerning how they restructure their production through intensification and rationalization of the
production process and investment in technology. While arguing that landscapes are contested objects, Massey's early work, perhaps understandibly
given her interests at the time, tended to pay less attention to the geographical practice of labor than it did to that of capital.

Castells, too, tended to ignore the roles that workers and their organizations may play in making the geography of capitalism. Although his book
The City and the Grassroots is one of the few pieces of neo-Marxist research
to document empirically in much detail specific examples of working-class
people actively shaping the built environment, in framing his work he re-
verted to a theoretical position in which he argued that "spatial forms . . .
express and perform the interests of the dominant class according to a given
mode of production and to a specific mode of development [and it is the]
resistance from exploited classes, oppressed subjects, and abused women"
that modifies them (Castells, 1983: 311–312, emphasis added). In such an
ontology, workers and others may resist change, but it is the dominant social groups (i.e., capitalists) who orchestrate such change in the built environment. Despite the voluminous empirical evidence to the contrary presented in *The City and the Grassroots*, Castells's ontological prioritization of
capital as actor, and workers as reactors, essentially denied working-class
people their positions as subjects of their own history and geography.

Certainly this is not to say that labor was ignored completely in Marxist
economic geography during the 1970s and 1980s. Richard Walker and
Michael Storper (Walker & Storper, 1981; Storper & Walker, 1983; 1984;
1989), for instance, argued for a greater recognition of the role of labor in
locational analysis. In a series of publications they criticized both neoclassical economic geography and some of its Marxist variants for a number of

flaws, not least of which is that labor was invariably conceptualized in the same terms used to describe inanimate commodities such as cars, shirts, grapes, and the like, a conceptualization that had the effect of denying the human element in the production process. Principally Walker and Storper argued that, although labor power does indeed take a commodity form, it is quite different from a true commodity because workers are not the same as objects of work or machines used in production—and worker behavior can drastically affect the labor process. In contrast to neoclassical economics, which they maintained had considered labor as a factor of location only in labor-intensive industries, Walker and Storper averred that labor is an important locational factor even in highly capital-intensive industries and "that a strong argument can still be made for the primacy of labour over all other market factors influencing industrial location" (Walker & Storper, 1981: 497).

Nevertheless, despite arguing that a greater emphasis should be placed on labor as a factor of location in locational analysis, Walker and Storper's approach did not link explicitly workers' own economic and social practices to the production of their own spatial fixes and thus their shaping directly of economic landscapes. While they may have refocused attention on labor, the point is that they did so from a perspective that still examined the evolution of the economic landscape in terms of how *capital* makes locational decisions. Although they touched briefly on the issue of worker migration and how this affects the geography of industrial development (e.g., see Storper & Walker, 1989: 157), they still were primarily concerned with how the supply and demand of particular types of labor are important to capitalists in making their investment decisions and how, through these decisions, capitalists shape the economic landscape. In the final analysis, there was still little consideration of how workers themselves actively make space and shape the economic geography of capitalism in ways not dictated by capital.

The 1980s also saw much work within human geography on the connections between the spatial aspects of labor regulation and the making of economic geographies. Arguably Gordon Clark was the leading scholar in this regard. His work focused particularly on the role of labor law and state institutions (such as the National Labor Relations Board) in shaping processes of industrial restructuring (Clark, 1985; 1988; 1989a; 1989b; Herod, 1998a, especially pp. 6–9). This represented a vital contribution to the literature on the location of economic activity, for previously both neoclassical and Marxist brands of economic geography had virtually ignored the spatial implications of the law and how varying legal interpretations can have crucial implications for the geography of economic development. However, while providing tantalizing insights into the geography of labor law, Clark stopped short of examining how working-class people themselves construct landscapes and shape uneven development as part of their spatial praxis, focusing instead on how the activities of the state structure the economic landscape. Although a strand of this type of work on unions was followed up on in the early 1990s in a debate about the changing landscape of organized labor in Britain, this focused largely on methodological issues related to studying the spatiality of labor unions rather than on workers' spatial praxis as such.[17]

Richard Peet's (1983) analysis of the geography of class struggle in the United States also sought to highlight the role of labor in the location of economic activities. By analyzing state-level data concerning numbers of strikes, percentage of the workforce represented by labor unions, wage levels, and an inverse measure of a state's business climate, Peet made connections between the geography of class struggle and the location of manufacturing industry. By classifying those states with high numbers of strikes, high wages, a large percentage of workers represented by unions, and a poor business climate (as seen through the eyes of employers) as areas of "high" class struggle, he showed how manufacturing capital has relocated in the post-World War II period to states with "low" levels of class struggle (defined as states with the opposite characteristics). "Overall," he concluded, "the data reveal a definite relation between the level of class struggle and the changing geography of employment" (Peet, 1983: 130). Again, this represented a significant contribution to understanding the thoroughly political and class nature of economic restructuring and the location of economic activity. Ultimately, however, Peet's approach also remained largely descriptive of labor. Although in one very restricted sense workers' activities were seen as shaping the geography of United States capitalism (in that capital stayed away from areas of "high class struggle"), the story presented was primarily one of how *capital* has used the uneven spatial distribution of these four factors to restructure the post-World War II economic geography of the United States as part of a geographic strategy for its self-reorganization.[18]

Of course, such a brief examination can never hope to do justice to the depth and breadth of insight that these and other analyses have provided concerning how the geography of capitalism is made. However, by focusing on these extremely influential works, which were key in setting the agenda for the "new economic geography" of the 1970s, 1980s, and early 1990s, it is possible to get a sense of the major thrust of such work. Specifically, whereas neoclassical approaches conceived of space as a stage upon which economic actors simply interact, Marxist work focused on the processes whereby geographies are actively produced as social constructions integral to the reproduction of capitalist social relations and the continuance of the accumulation process. And yet, despite its claims that workers enjoy a privileged position within analysis because value is derived from labor, ironically Marxist work during this period tended to overlook labor's role in actively making the geography of capitalism in favor of the analysis of the dynamics of capital. In such work it was capital (together, sometimes, with the state) that was conceptualized as the active agent structuring economic landscapes and that received the greatest theoretical and empirical attention. Although there was some interest in issues related to labor law and patterns of trade union density, workers' roles as active makers of the geography of capitalism were decidedly undertheorized.

By way of contrast, in what follows I sketch out a framework for thinking about how and why workers may seek to shape the landscape for their own purposes and how, in so doing, they participate in forging the uneven development of capitalism. The central thrust of my argument is that, in addition to theorizing how capital produces space as part of its spatial fix, it is important to theorize how workers shape space as a *fundamental* part of their own social praxis. In making such an argument I certainly build on the highly influential Marxist work conducted during the 1970s and 1980s concerning the relationship between the production of space and the functioning of capitalism, but I also hope to advance this theory to expand our understanding of capitalism's geography. Whereas examining how capital makes use of the underlying geography of labor in the landscape provides a crucial half of the story of the geography of capitalism, it is, in fact, just that—half the story.

. . . TO A LABOR GEOGRAPHY

So far I have suggested that both neoclassical and Marxist-influenced theorizing about the geography of capitalism have tended to be rather capital-centric in nature. Recently, however, such capital-centric approaches have come under criticism as a small group of writers have begun to argue for a more nuanced understanding of how workers' lives are spatially embedded and how their actions impact the form of the economic landscape (see Herod, 1995; Wills, 1996; Mitchell, 1996; Holmes & Rusonik, 1991; Painter, 1991; Southall, 1988). In line with such work, in the remainder of this chapter I outline my argument for an approach (what I am here calling a "labor geography") to the study of working-class life that recognizes that the production of space in particular ways is not only important for capital's ability to survive by enabling accumulation and the reproduction of capital itself, but it is also crucial for workers' abilities to survive and reproduce themselves. Just as capital does not exist in an aspatial world, neither does labor. The process of labor's self-reproduction (both biological and social), I argue, must take place in particular geographical locations. Given this fact, it becomes clear that workers are likely to want to shape the economic landscape in ways that facilitate this self-reproduction. Recognizing that workers may see their own self-reproduction as integrally tied to ensuring that the economic landscape is made in certain ways and not in others (as a landscape of employment rather than of unemployment, for instance) allows them to be incorporated into analyses of the location of economic activities and the production of space in a theoretically much more active manner than heretofore has been the case.

For example, workers who are relatively geographically immobile owing to kinship ties or perhaps the inability to sell their homes in a depressed regional economy may feel the need to engage in local boosterism to encourage investment in their communities in much the same way as do local capitalists whose sunk costs preclude them from relocating from one place to another when a particular locale begins to experience economic hard times (see Cox & Mair, 1988). However, whereas workers' involvement in local boosterist campaigns is often portrayed simply as a bad dose of false consciousness in which they help to sustain local capitals, in fact most workers would probably see retaining and/ or attracting investment to their particular communities as integral to their ability to sustain their own livelihoods. Thus, not only may workers take on very active roles in helping to ensure that investment flows into, rather than out of, their communities, but they do so as active economic and geographic agents rather than as class dupes.[19] In the process they may play significant roles in shaping the uneven geographic development of capitalism, such that uneven development can be seen to result not only from the activities of capital but also from those of labor. Likewise, just as capital may find itself constrained by the structure of landscapes created during previous periods of accumulation (see Harvey, 1982), so also may workers find that the landscapes that facilitated their social and biological reproduction in earlier times are no longer appropriate for doing so, and thus they may struggle to transform them in ways not necessarily controlled by capital itself.

Suggesting that workers have a vested interest in making the geography of capitalism in some ways and not in others allows us to make five interrelated theoretical points. First, it suggests that, even if workers' struggles are less than revolutionary and even if they are still bound within the confines of a capitalist economic system, the production of the geography of capitalism is not the sole prerogative of capital. Understanding only how capital is structured and operates is *not* sufficient to understand the making of the geography of capitalism. For sure, this does not mean that labor is free to construct landscapes as it pleases, for its agency is restricted just as is capital's—by history, by geography, by structures that it cannot control, and by the actions of its opponents. But, it *does* mean that a more active conception of workers' geographical agency must be incorporated into explanations of how economic landscapes come to look and function the way they do. As we shall see in the case studies to follow, capital is not the only actor actively shaping the geography of capitalism *or even, in some places and times, the most significant one,* and labor is not simply a "factor" of location in the sense in which it is so often conceived.

Second, such a conceptualization allows us to begin thinking about how the social actions of workers relate to their desire to implement in the physical landscape their own spatial visions of a geography of capitalism that is enabling of their own self-reproduction and social survival. Following Harvey's (1982) argument about how capital seeks to make a "spatial fix" appropriate to its own condition and needs at particular times in particular locations, it is also necessary to see workers' activities in terms of their desire to create particular spatial fixes appropriate to their own condition and needs at particular times in particular locations. In addition to theorizing how capital attempts to create a spatial fix in its own image, we should also think of how workers attempt to create in their own image what we might call "labor's spatial fix." Such a conceptualization places the production of space at the center of workers' social praxis, so that the making of the spatial relations of capitalism becomes not secondary to their social activities but at their very heart, even if it is not conceptualized as such by those workers involved.[20] Recognizing that workers may try to make their own spatial fixes, then, allows us to link forcefully their spatial practices with the evolving social and spatial relations of capitalism.

Third, following from the foregoing, understanding that different groups of workers may, depending upon the contexts within which they find themselves, prefer quite different types of spatial fixes—both of their own and of capital's—allows us to view "labor" not as an undifferentiated mass in some "heroic proletarian" sense but as a social group that may have significant cleavages within itself, depending upon workers' relationship to the means of production, their industrial sector, the region of a particular country within which they live, whether they live in the global north or global south, whether they are male or female, young or old, and so forth. Adopting such a "nonessentialist" view of how workers seek to impose their spatial visions on the landscape enables us to recognize that different and competing groups of workers may, in fact, have vested interests in generating quite different spatial fixes. Whereas one group, for instance, may seek to keep employment in their home community or country, another may encourage capital flight to theirs by agreeing to work for less or with fewer restrictive work practices. Thus, in debates over globalization and the establishment of the new regulatory bodies of the global economy such as the World Trade Organization, what workers in the global north see as efforts to save their livelihoods from the effects of the unleashing of neoliberalism many workers in the global south have seen as little more than efforts by these workers to protect their privileged places in the new international division of labor by preventing capital flight to the less developed countries.

Adopting such a nonessentialist view of labor, then, is important not only theoretically—for it allows a more rigorous and realistic understanding of workers' varied sociospatial existence—but is important also in terms of how workers are represented discursively (see Gibson-Graham, 1996) because it allows for greater agency to be written into accounts of their activities rather than simply assuming that all workers will react in the same way because they share the same relationship to the means of production (i.e., they do not own them). Recognizing that different groups of workers may have, in very real ways, different options as a result of differences in where they quite literally stand in the world—global north or global south, for instance—provides something of a corrective to accounts that present workers either as inherently powerless and condemned only to follow the dictates of (global) capital or as simply dupes of capital.

Fourth, conceiving of how labor might seek particular spatial fixes at particular historical junctures—fixes that are sometimes coincident with, but frequently different from, those favored by capital—allows a much less mechanistic and more deeply political theorization of the contested nature of the production of space under capitalism for, ultimately, *it is the conflicts over whose spatial fix (capitalists' or workers') is actually set in the landscape that are at the heart of the dynamism of the geography of capitalism.* This means that understanding processes of class formation and inter- and intraclass relations is fundamentally a geographical project. Workers often succeed in constructing landscapes in certain ways that augment their own social power and undercut that of capital—as, for example, when they manage to unionize across particular geographical regions to limit capital's ability to whipsaw different plants into making wage concessions. Despite neoclassical and many Marxist portrayals of capital as the maker of economic landscapes into which workers are merely deposited as if dropped from the air, workers are, in fact, intimately connected to the production of space through their efforts to develop geographical solutions to the problems of ensuring their own self-reproduction. Even when they are defeated in this goal, the very fact of their social and geographical existence and struggle means they shape the process of producing space in ways not fully controlled by capital.

Fifth, in addition to recognizing the active roles played by working-class people in making the geography of capitalism, there is also the need to recognize their roles in producing the very real geographic *scales* that, as Smith (1990 [1984]: xv) has argued, "give uneven development its coherence." To talk of the "production of scale" may at first seem strange, though perhaps no more so than to talk of the "production of space." Yet, contrary to the assumptions of those neoclassical models that assume that people live in isotropic planes, space is not in fact produced as a uniform surface but is pockmarked at a number of geographic resolutions by the scars of uneven development. We cannot, therefore, consider the production of space without also considering geographic scale, for it is geographic scale that gives space much of its structure. For instance, the notion of an "industrial region" as a coherently produced space is meaningless unless we have some way of identifying the region and what holds it together, of distinguishing how it is geographically bounded, and of discerning where it ends and different regional space-economies begin.[21] Likewise, it makes little sense to talk about "local" or "global" processes without a fuller understanding of what, precisely, is meant by the "local" or the "global"—if processes occur in geographically circumscribed areas but have impacts that reach the other side of the globe, are these local or global processes, or both, or neither? Concomitantly, as the experiences of the new international division of labor and the implementation of global assembly lines have shown, the scale at which space is socially produced is thoroughly implicated in the historical geography of capitalism. Thus, the geographies produced under nineteenth-century conditions of nationally organized capitalism were quite different from the geographies produced by late-twentieth-century globally organized capitalism. Equally, the scale at which power is to be exercised may be vigorously fought over, for example when workers with different agendas argue over whether it is the provisions of local or national labor agreements that should take precedence in any particular situation.

Given that geographic scale plays such a central role in the ways in which the geography of capitalism is made, it makes sense to question how such scales come about in the first place and what the role of working-class people is in this process. In other words, if workers play important roles in shaping patterns of uneven development, then the question must also be asked as to what role they play in making the very scales that define such uneven development.

The Production of Geographic Scale

Scale is intimately connected to the spatiality of the human experience, and geographers and others have long used scale as the conceptual mechanism to frame research questions spatially. Myriad studies in the social and behavioral sciences have employed a multiscale approach to investigating particular social processes or institutions with the aim, ultimately, of fitting together insights drawn from these different scales of analysis to reveal more holistic conclusions. Whether it concerns examining the impacts of industrial restructuring at a number of different geographical resolutions (why is it that some manufacturing communities may be booming at the very moment the larger region is deindustrializing?) or the knotty issue of theorizing the relationship between different levels of the state (just how is the local state articulated within the larger national entity?), such studies invariably seek to show that the scale at which researchers examine an issue can fundamentally shape the conclusions they draw.

In such analyses, however, scale itself has usually been given ontological priority over the processes that are being examined, with scale being conceptualized either as little more than simply a pregiven "natural" metric by which to order the world for the purposes of analysis or—and this particularly concerns the regional scale—as a purely mental construct for dividing the world, with one researcher's determination of the spatial extent of a particular region seen to be as good as the next (e.g., see Hart, 1982).[22] In other words, it is not the genesis of scale that researchers generally have sought to explain but, instead, the manner in which scales define and circumscribe the particular processes being studied. Traditional multiscale studies do not question how certain scales come about but focus, instead, upon how changing the scale of analysis can reveal different insights into particular social processes. Such a recognition is not to deny that the act of revealing is itself a highly political one that can result in the jarring juxtaposition of contradictory understandings drawn from different geographic scales, but it is to argue that, paradoxically, by initially conceiving of those very scales either as taken-for-granted devices for ordering the world or as purely mental constructs, simply revealing the workings of social relationships at different geographic resolutions actually closes off some avenues for political action because it accepts those very scales as pregiven and inviolable.

In contrast to these more traditional understandings of scale, here I want to argue that recognizing that scale is itself socially constructed as a material entity through social praxis *opens* possibilities for political action, for acknowledging that geographic scales are materially constituted by social actors means that there is a politics to this constitution. Clearly, industrial regions, metropolitan areas, national boundaries, and the global scale of contemporary capitalist production are historically and geographically created human constructions, subject to reformulation as economic and political conditions change. There is, in other words, nothing inherent about scale, and there are no "natural" scales by which to order and organize human geographies. Rather, scales are historically and geographically negotiated. For example, the scale of the "home" is dependent upon notions of the family (nuclear or extended) and the lines of demarcation between public and private spaces—is the home a private space and refuge from the public sphere (and, if so, for whom?), or may these enclaves of the "private" be transcended and made public by the intervention of the state, which may seek to regulate to a greater or lesser degree certain activities conducted within the "privacy" of one's own home?[23] Likewise, the "national" scale certainly became more important with the transition from feudalism to industrial capitalism and the growth of imperial rivalries, though in the contemporary period it may well be declining in some aspects under the dual assault

of devolution and supranationalism. Given such realities, the questions geographers (and others) should therefore ask, perhaps, are not how scale or-
ders social processes but, rather, how social actors create geographic scales
through their activities.

During the past decade or so, the notion of scale as a social construction whose form can be manipulated for political purposes has become central to a number of debates in human geography. Two of the first attempts
to develop a more critical understanding of geographic scale were those by
Peter Taylor and Neil Smith, although their approaches in this regard were
quite different. Adopting an analysis that drew on world-systems theory
(Wallerstein, 1979), Taylor (1981, 1982, 1987) suggested that the urban, national, and global scales have occupied critical and particular roles in the
historical geography of industrial capitalism. Thus, the global scale, he argued, should be considered the "scale of reality," for it is the scale at which
capitalism is organized; the urban scale should be considered the "scale of
experience," for it defines the arena within which everyday life is conducted; and the national scale should be considered the "scale of ideology,"
because it is the scale at which the capitalist class promulgates such ideologies as nationalism. Furthermore, he indicated (1987) that there was a fundamental contradiction under capitalism around the issue of the scale at
which workers have historically organized politically, with classes "for themselves" (*für sich*) tending to organize nationally (at least for most of the
twentieth century) whereas classes "in themselves" (*an sich*) have been defined globally. While offering important insights, this approach nevertheless
was problematic for two reasons. First, by allocating particular roles to particular scales, it was rather functionalist in its approach. Second, it failed really to examine how scales are themselves actively created. Although Taylor
indicated that scale is crucial for the maintenance of capitalist social relations, his approach did not allow for oppositional politics to shape, and
even sometimes dictate, the production of geographic scale. Ultimately this
schema denied the richness of political struggle and the power of working
people to create geographic scale.

Neil Smith's (1990 [1984], 1988; Smith & Dennis, 1987) work represented the second attempt to develop a political economy of scale. In his
book *Uneven Development*, Smith argued that two tendencies within capital
could be identified, tendencies that emerge out of a fundamental contradiction between capital's need for both immobility and mobility in the landscape at different historical junctures. Capital, he suggested, must simultaneously be able to fix itself in the landscape so that commodities may be
produced (i.e., production has to occur somewhere in physical space), yet it
must retain sufficient mobility to be able to relocate should conditions
somewhere else prove more appealing. Thus, on the one hand the (spatial)
nature of competition leads capital to seek to level economic space by

equalizing the rate of exploitation and profit across the surface of the planet (what we might think of as a geographical expression of the tendency for competition to equalize the rate of profit), while on the other hand the nature of production means capital is simultaneously forced to immobilize itself in the landscape in particular "spatial fixes" (cf. Harvey, 1982) in order for accumulation to occur. For Smith, then, scale is produced through the geographical negotiation of these opposing forces within capital. The tension between mobility and immobility, equalization and differentiation, cooperation and competition is literally inscribed in space by the material construction of scale. Put another way, the production of various scales of organization is the manner through which capital negotiates these opposing forces within itself.

Using such a formulation, Smith (1990 [1984]) identified four principal spatial resolutions at which this negotiation is played out. The urban scale, he suggested, is defined by the spatial coherence of labor markets and daily commuting patterns, for it represents the geographical limit to the day-to-day migrations of labor between home and places of paid work—within urban space there is a certain coherence to labor markets, whereas beyond the outer limits of daily "Travel-To-Work-Areas" (TTWAs) and the urban scale this coherence collapses and the labor market fragments.[24] The regional scale, he argued, is defined by the spatial concentration of capital to form particular territorial divisions of labor based on differentially developed sectors of the economy. For its part, the national scale is derived from the need of different capitals competing in a world market to retain political and economic control of markets—thus, within a particular nation-state capitals might agree collectively to cooperate on certain issues such as a common currency, basic laws of property ownership, and defending against working-class agitation while simultaneously both competing with one another domestically and with other nationally organized capitals internationally. The global scale, Smith maintains, results from capital's quest to universalize the wage–labor relation. This means that, although the physical limits to the planet are geologically given, the development of an integrated economy organized at the global scale is the historical and geographical product of capital's expansionist nature, with "capitalism inherit[ing] the global scale in the form of the world market . . . based on exchange [but transforming it] into a world economy based on production and the universality of wage labour" (Smith, 1990 [1984]: 139).

Such a theoretical interrogation of the production of scale was truly original. Nevertheless, this early take on the production of scale was also problematic. First, it tended to subordinate other social relationships upon which capital relies to produce space and scale (Herod, 1991a). For instance, daily commuting patterns are crucially shaped by the asymmetrical gender division of domestic responsibilities and the differential wages resulting, in part, from occupational segregation, such that women who engage in paid employment frequently work closer to home and in more spatially circumscribed labor markets than do men (Madden, 1981; Hanson & Johnston, 1985). Second, in this formulation it is the power of capital solely that creates geographical scale. Scale is seen to be produced simply by the internal requirements of capital and the logic of accumulation. Indeed, Smith concluded (1990 [1984]: xv, emphasis added), "not only does *capital* produce space in general, it produces the real spatial scales that give uneven development its coherence." For him, "the vital point [was] not simply to take these spatial scales as given . . . but to understand the origins, determination and inner coherence and differentiation of each scale *as already contained in the structure of capital*" (Smith, 1990 [1984]: 136, emphasis added). This resulted in a position that seemed to suggest that it is possible to "read off" the production of scale simply by understanding the logic of capital dynamics.[25]

Whereas, then, such a conceptualization of scale provided a very useful framework for theorizing geographic scale as an actively produced social construction rather than as a static category for organizing space, it did tend to conceive of scale's production somewhat mechanistically as coming out of the internal dynamics of capital. In response to some of these criticisms, Smith later moved away from such a narrow vision of the production of scale to adopt a richer view of scale as fashioned out of political struggle.[26] Focusing particularly on antigentrification struggles in New York's Lower East Side and the politics of organizing by the homeless, he showed how activists attempted to win their conflict with the city government and real estate speculators concerning the right of homeless people to sleep in public parks by trying to expand their struggle from one of controlling Tompkins Square Park (a public park in which homeless people had built a tent city and from which they had been evicted) to controlling the whole Lower East Side within which the park is located. For Smith, such an ability to "jump scales" to a higher spatial resolution of organization was a central element of activists' and homeless people's spatial and political praxis aimed at establishing links with supporters located in other places, and added another element to understanding how scale may be produced as an integral part of social and political struggle.

One of the few nongeographers to write at this time (mid-1980s) about the social significance of scale, although not in such explicit terms, was Anthony Giddens. In his theory of structuration, Giddens posited an important role for geographic differentiation of the landscape and thus for scale, which is the mechanism by which space is differentiated. For Giddens (1984: 119), the differentiation of space—what he called its "regionalization"—is to be understood "as referring to the zoning of time-space in relation to routinized social practices" that define such regionalization. With such a view, Giddens saw the differentiation of space as very much connected to the social practices of everyday life. However, although such a statement clearly implies that he viewed the production of regionalization (that is to say, "scale") as resulting from everyday social practices and thus open to constant renegotiation, there were several problems with his approach. To begin with, Giddens did not really theorize the *process* of scale *production* and, particularly, did not relate it to the processes that seek to differentiate and/or equalize space. There was little sense that the production of scale itself is actively fought over. Instead, scale remained a largely descriptive device for differentiating among regionalizations. There was also not the same powerful link to the making of material geographies provided by Smith's approach. Although Giddens viewed social practices as being choreographed in time and space, there was little explicit connection in this work to the actual production of economic landscapes. Space was seen merely as a container for such practices. Furthermore, Giddens's conception of scale adopted a reductionist dualism— viewing regionalization simply in terms of "core" and "periphery"—while also being somewhat imprecise, and therefore of questionable theoretical value, with regard to the geographical extent of such regionalizations (which were seen to range in physical extent from rooms in houses [p. 119] to global patterns of uneven development [p. 130]). Nevertheless, the notion that scale might be produced and reproduced out of people's everyday activities (their "routinized social practices," as Giddens puts it), even if not explicitly explored by Giddens himself, does provide important insights into how workers might construct scale during the course of their daily lives.

Following such early work, the 1990s saw a veritable flurry of writings regarding the theorization of geographic scale.[27] Much of this focused upon defining what Smith had earlier called the "rules of translation," by which he meant how the connections between different scales were understood. One useful attempt to further the theoretical debate about this issue has been Kevin Cox's (1998a) distinction between what he calls "spaces of dependence" ("those more-or-less localized social relations upon which we depend for the realization of essential interests[,] for which there are no substitutes elsewhere[, and which] define place-specific conditions for our material well being and our sense of significance" [p. 2]) and "spaces of engagement" (the spaces in and through which social actors construct associations with other actors located elsewhere). Principally Cox argued that, while they have acknowledged that such scales are socially constructed rather than naturally given, most theorists had tended to characterize scale in areal terms, with scale itself seen to represent a boundary around a set of closed spaces defined at different spatial resolutions (the local, the regional, the national, the global, etc.). This was especially so, he maintained, when it came to the notion of "jumping scales"—scale jumping was viewed simply as a process of moving up the spatial hierarchy from one areal unit (e.g., the local) to another (e.g., the national) (see also Herod, 1997b: 163 on this point). Instead of viewing scale in terms of boundaries around certain territorial units, Cox suggested that a more useful metaphor to use is one in which the spatiality of scale is seen in terms of a network of interactions that link different actors' spaces of dependence (which may be constituted at various spatial resolutions) as part of a political strategy of engagement that is contingent upon particular historical and geographical constellations of social relationships.

Although there seems to be some concern as to what Cox means exactly by terms such as the "local" and the "global" (see Jones, 1998; Judd, 1998; Smith, 1998; also Cox, 1998b), the result of such a theorization, he suggests, is an understanding in which the production of scale is seen as emerging contingently out of the ways in which actors build spaces of engagement that link various spaces of dependence. Hence, " 'jumping scales' is not a movement from one discrete arena to another" but a process of developing networks of associations that allow actors to shift between spaces of engagement (1998a: 20; see also Herod, 1991a, 1997b, for similar arguments). Furthermore, whereas the notion of social actors "jumping scales" has usually assumed a jump "upwards" to a larger scale of organization as a means to outmaneuver or "trump" more local scales of organization, Cox argues that this need not necessarily be the case and that the contingent nature of such spaces of engagement means that frequently actors may attempt to "go local" to outmaneuver more globally organized opponents. Indeed, as we shall see in Chapters 5 and 8, in the United States some groups of workers have sometimes seen the breakup of national collective bargaining agreements as an opportunity precisely to engage in local bargaining that they perceive to be more conducive to their goals than may be national bargaining.

Whereas much of the theorizing about scale, then, has attempted to show that scales are created socially and have a material existence, a number of authors have also sought to show how scale may be used ideologically as part of what we might call a politics of enabling or constraining. Thus, in a case involving a land-use conflict in England over the mining of sand and gravel deposits that threatened to damage a historic site, Murdoch and Marsden (1995) demonstrated how local activists successfully constructed a discourse of scale surrounding the issue that pitted the "local" concerns of the mining company against the "national" cultural heritage interests of the British public. Whereas in this case activists appealed to national-scale discourses to defeat the mining company, Bridge (1998) has shown how the obverse may also occur, as local interests are constructed discursively as more important than are national interests. Specifically, he shows how multinational mining companies in the southwestern United

States have forged discourses that purport to defend local mining communities throughout the region against the interferences of "nonlocal" federal and state government "bureaucrats" and "outside" environmentalists. Such a construction by mining companies of environmental concerns "as *non-local*, external, bureaucratic, and elitist," Bridge argues (1998: 234, emphasis added), serves to undermine opposition to mining in the region. Thus, Bridge contends, "by delegitimizing *local* environmental voices and representing environmental concerns as an *external* threat to livelihoods, labor and communities can be maintained in a state of compliance."

Similarly, Mitchell (1998) has examined how agribusiness interests in 1920s California constructed ideologically and valorized a "scale of legitimacy" for involvement in farmworker politics that was decidedly local. Seeking to limit the ability of migrant farmworker pickets (whom the growers labeled "outside agitators") to link up strikes and other harvest disturbances by moving throughout the state, the growers (who were themselves often part of nationally organized food companies) went about trying to construct a scale of legitimacy that allowed only "local" interests a say in "local" disputes.[28] By building a series of spatially grounded ideological constructions concerning the legitimate scale of local interests, growers attempted to construct migrant workers' identities as "legitimate locals" or "illegitimate non-locals" based solely upon whether or not they were supporters of the farmworkers' union: union supporters were viewed, by definition, as "non-local outside agitators" and therefore not entitled to play a role in local politics, whereas nonsupporters were considered "faithful local employees" regardless of their actual geographic origins. In a slightly different take on this issue, Alderman (1996, 1998) has analyzed how, in their efforts to memorialize Martin Luther King, Jr., through renaming streets after him in a number of southern U.S. towns, some African-American groups have had to confront the issue of whether he should be honored at the scale of the "black community" (i.e., by renaming streets that run within certain neighborhoods with high populations of African-Americans) or at a much broader spatial and social scale (i.e., by renaming more prominent streets, perhaps streets running through "white" sections of communities). In seeking to address such issues, those who would like to see King memorialized by the naming of a prominent street in the business or "white" section of town have frequently had to face opposition from whites and business owners who would prefer the renamed street to be located wholly within the confines of African-American neighborhoods. Such a situation raises the issue not simply of how geographic scales—in this case, scales of memorialization—are constituted through social action but, equally important, how such scales may be *prevented* from being constituted by those social and political forces that oppose their creation (see Judd, 1998).

As this brief discussion suggests, conflicts over the material and ideological construction of scale can be both quite intense and play dramatic
roles in the ways in which the geography of capitalism is made. Bearing this
in mind, in the empirical research that forms the remainder of this volume I
focus particularly upon conflicts over the social construction of scale. Specifically, I seek to draw upon, and make connections between, Smith's powerful theoretical statements about the production of scale in the material
landscape (particularly notions of how scale is created in processes of cooperation and competition among different social actors), Giddens's concern
for the way everyday social practices define "regionalizations" (i.e., the
scales of social life), and Cox's notion of "spaces of engagement" and
"spaces of dependence," in which scale is thought of as a series of networks
that are constituted at different spatial resolutions (see also Low, 1997). In
so doing, I examine how workers and their organizations make and contest
scales as part of their political praxis. This will help elucidate, I hope, an understanding of the production of scale that has clear and direct theoretical
links to the making of the economic geography of capitalism, yet that is also
sensitive to how scale is made and remade through ordinary, everyday social practices. In other words, rather than theorizing the production of scale
in terms of broad and abstract forces (e.g., the internal contradictions of
capital), I wish to examine how ordinary people can significantly shape its
production through the practices of their everyday lives. For instance, when
workers engage in activities such as contract bargaining, they create very
real structures that have material impacts on the ground (e.g., in terms of
wage rates, the geography of work, etc.) but that are subject to constant renegotiation on a daily basis as routinized social practices are transformed by
workers to fit new contexts.

In drawing upon both Smith's and Cox's work (which I do not see as
necessarily being mutually exclusive), I want to understand the creation of
scale *both* in terms of the production of relatively coherent bounded areal
units such as the home, the urban, the region, the nation, the global, etc.,
and in terms of the ability of social actors to build networks of interaction at
various geographical resolutions. By so doing, I want to retain the notion
(which is, I think, a useful one) that scale can serve to delimit areally different parts of the economic landscape as part of a technology of spatial control during the exercise of power (such as when social actors control certain
bounded spaces that they may imbue with particular meanings or to which
others' access is limited), yet also to think of the construction of scale as a
process of constructing networks of interaction that link places (and actors
within them) across space at different spatial resolutions. In addition, by
thinking of scale in terms of networks of interactions across space, I hope to
avoid problematic notions that see the construction of "spaces of engage-

ment" at different geographical resolutions as a process in which actors simply "jump" between what are conceived of as preexisting (even if socially constructed) scales, a metaphor that gives ontological priority to scales themselves rather than to the processes that create them. Finally, such an approach will allow me not only to examine conflicts that result in the production of scales of organization but also to examine how powerful social agents may attempt to prevent others from making the scales of organization that they might prefer.

RETHEORIZING WORKERS' ROLES IN MAKING THE GEOGRAPHY OF CAPITALISM

In summary, what I have sought to do in this chapter is both to question the theoretical and empirical primacy usually given capital when seeking to understand the making of the economic geography of capitalism and to provide a conceptual framework for the chapters that follow. Principally I am arguing that geographers (and others) need to move beyond writing studies that incorporate working-class people in explanations of the geography of capitalism simply in descriptive, capital-oriented taxonomic terms of their wage rates and the like (what I have termed a geography of labor) and to include a more serious consideration of how they *actively produce economic spaces and scales* in particular ways (both directly and indirectly, consciously and unconsciously) as they implement in the landscape their own spatial fixes in the process of ensuring their own self-reproduction (what I have termed a labor geography). Equally, I want to encourage social and labor historians, scholars of industrial relations, and other social scientists to think about landscape and the spatial relationships within which people live their lives in a much more sophisticated manner than has perhaps been the case previously. Geographies can be manipulated for political ends, and landscapes are not merely the stages upon which social action simply takes place. Thus, in working out their own spatial fixes, working-class people's spatial practices shape the location of economic activity and thus the economic geography of capitalism. Just as the production of particular spatial fixes is integral to capital's ability to reproduce itself, working-class people's abilities to produce particular spatial fixes at particular geographical scales at particular historical junctures is an integral part of their self-reproduction on a daily and generational basis. As such, it demands a greater consideration of how the implementation of "labor's spatial fix" shapes the unevenly developed geography of capitalism.

Certainly some excellent empirical work along these lines has been conducted in economic geography. Cooke (1985), for example, has shown how Welsh miners' and steelworkers' cultural and work practices have been key to defining the regional division of labor in the South Wales coalfields. Mitchell (1996) analyzed how farm workers in California during the 1920s and 1930s shaped the agricultural landscape through their physical labor. Holmes and Rusonik (1991) illustrated how the changing economic context of the auto industry led Canadian and U.S. workers to adopt quite different geographical strategies during the 1980s that ultimately led Canadian locals to break away from the United Auto Workers (UAW) and that have had wider impacts on the industry's geographical structure. Wills (1998a) has shown how the migration of trade union activists has shaped the geography of industrial activity and union activity in Britain, while Cope (1998) recounted how the implementation of union work rules in the Lawrence, Massachusetts textile industry in the 1930s led to a restructuring of local labor markets and helped break down the spatial barriers that had once divided workers of different ethnicities among different neighborhoods.[29] Yet, these are few and far between in economic geography. Furthermore, of that work that has focused more specifically on the activities of workers, very little has sought specifically to link workers' activities to broader philosophical debates about the ontological status of space in the survival and reproduction of capitalism.

In arguing for an approach in economic geography that recognizes working-class people's capacity for proactive geographical praxis, I do, however, have three caveats to make before concluding this chapter. First, the discussion so far has tended to concentrate on the realm of production and, in particular, industrial production by waged labor. This is not because I necessarily consider production more important than issues of, say, consumption. Rather, because my critique is aimed at the failure of traditional neoclassical economic geography and of much Marxist work (both of which have tended to focus disproportionately on the world of industrial production) to incorporate labor in any proactive manner, it is a reflection of criticizing what has gone before. I do not mean, therefore, to imply that any labor geography need be confined to examining workers' roles in shaping the geography merely of industrial production. It is also important to consider how workers' activities shape the geography of, for example, consumption (see Frank, 1994; Marston, 2000). Such an approach, then, should not be defined so much by its focus on particular sectors of the economy as it should by its concern in showing how the actions of working-class people and their institutions actively shape the geography of capitalism in toto.

Second, although trade unions are significant expressions of working-class organization, any labor geography should not be so narrowly defined that it is concerned simply with their activities. Unions are certainly important, and often powerful, workers' institutions, but they do not hold a monopoly as instruments of the expression of working-class people's interests. A labor geography should therefore recognize that working-class people—

waged and unwaged—organize along many crosscutting political, social, racial, gender, and cultural lines. And yet, the fact that they do so as working-
class people rather than as capitalists distinctively shapes their spatial praxis
and, hence, their production of geographical landscapes.

Third, I do not simply want to replace a geography of labor with a labor geography. Both types of approach are important for understanding
how the geography of capitalism is made. Rather, I see the notion of a labor
geography as adding a new dimension to understanding the geography of
working-class life, a dimension in which workers' agency is specifically recognized and in which the production of space is seen (as much as is possible) through their eyes. Thus, it is certainly important to understand how
capital may use variations in labor's spatial organization (e.g., differences in
wage rates across the economic landscape) to shape the geography of capitalism, but it is also important to examine how workers actively make landscapes and shape the geography of capitalism. In so doing, workers' agency
is restored, and it becomes possible to view "labor" as a social category in
nonessentialist terms (see Gibson-Graham, 1996) because the activities of
different groups of workers can be seen to be shaped by, among other
things, their varying spatial contexts and the different pressures that such
varying contexts place on them. Recognizing that different groups of workers may adopt different solutions and practices in different geographical
contexts as they seek to develop their own spatial fixes provides one means
to move beyond monolithic notions of a working class that is always and
necessarily unified in its opposition to the predations of capital. Focusing
upon workers' efforts to build their own spatial fixes that are appropriate to
their concerns at different times and in different places not only enables us
to see labor as capable of acting proactively but also brings with it an appreciation of how the spatial context within which workers find themselves can
lead them to develop vastly different geographical solutions to particular
problems that they face. Thus, I would argue, workers must seek to produce
space as a necessary part of their condition as social agents and their ability
to reproduce themselves socially, though the particular ways in which they
do so are always subject to the contingencies of the historical and geographical contexts that enable and constrain their lives.

In part, then, my argument here about the neglect of working-class
people as sentient spatial actors in economic geography is a theoretical one.
Examining how particular groups of workers create their own spatial fixes—
either in collaboration with or in opposition to both capital and other
groups of workers—can provide significant insights into how the unevenly
developed geography of capitalism comes to be produced. But it is also a
political argument (to the extent that the theoretical and the political can
ever be separated) about whose interests economic geographers and other
social scientists choose to understand and represent. The social production

of knowledge is a political process. Conceiving of workers as simply factors of production or as "variable capital" is to tell the story of the making of economic geographies through the eyes of capital. By proposing the notion of "labor's spatial fix" and arguing for a more active and central role for workers in theory building and empirical investigations of the creation of economic landscapes, it is also possible to tell the story of the making of economic geographies through the eyes of workers. Whereas geographers, especially those in the "locational analysis" tradition, have often used their skills to help corporations locate and relocate new facilities, understanding how workers use and create space as part of their daily and generational self-reproduction can, perhaps, provide a way for geographers also to use their skills to aid the dispossessed and the oppressed.

CHAPTER 3

Challenging the Global Locally

Labor in a Postindustrial Global City

According to the popular wisdom of neoliberalism, the increasingly globally organized nature of capitalism is undermining the power of locally organized groups of workers and others to secure their interests. With economic globalization, so this wisdom goes, has come a world in which local differences will increasingly be effaced and places will come to look more and more alike. As the growing speed with which capital, information, goods, and people can get from one place to another results in a seemingly inexorable "shrinking globe," the importance of local scales of political action, and even of geography itself, it is often argued, are steadfastly being diminished. Such processes of change are not only resulting in the famous annihilation of space by time identified by Marx 150 years ago when the new transportation and communication technologies of Victorian capitalism were transforming the geographical relationships between places (see also Kern, 1983), but, if we are to believe the hype of the neoliberal literature, the dynamic nature of global capital(ism) is also leading us toward a "borderless world"—the phrase is Ohmae's (1990)—in which places and populations must get on board the free market economic train or be left at the station.[1] In such a worldview, we are told, locales and peoples must prostrate themselves in the face of "global economic realities" if their economies are to remain viable and if capital is to serve as the liberator of populations and places still held back by outmoded rules, regulations, and institutions. According to the gurus of neoliberalism, capital must be allowed flexibility and mobility if economic prosperity is to be secured, even if this is at the social expense of allowing corporations to play different places against one another in a "race to the bottom." By definition, such a model of economic development—and such an understanding of global capitalism—is one in which global capital determines what happens locally and local actors are (and should be!) relatively powerless to challenge this prerogative.

Although sharing a different politics from neoliberalism, somewhat paradoxically much of the radical political economy literature has adopted a similar theoretical stance toward globalization, with the power of global capital to restructure places frequently being taken as a given. In turn, a sense of despair and defeatism has often pervaded much of the analyses of the position of workers in the global economy, and many commentators appear to have accepted that it is relatively pointless for local social actors to struggle against what are portrayed as the omnipotent and inevitable forces of global capitalism. Thus, local actions mounted to defend communities against the ravages of, for example, deindustrialization are often seen as hopelessly parochial and destined to be outmaneuvered and defeated by a hypermobile and flexible capital that can always choose to abandon them and relocate to virtually any place on the planet. The only logical response, so it is claimed, is that workers and community activists must develop global campaigns to match the global reach of transnational corporations and global capital more generally (see Herod, 2001). Such arguments, I want to suggest here, are, however, problematic.

Certainly, I would not want to argue against the proposition that in particular circumstances workers and community activists may indeed have to develop transnational links with their confederates in other parts of the world, or that sometimes these are the only viable ways of organizing against the machinations of global capital.[2] However, at the same time as recognizing that workers sometimes may need to develop transnational contacts and strategies if they are to achieve their goals, I think that it is conceptually and politically problematic to suggest that even in an increasingly global economy this is the *only* option open to workers—or even, sometimes, the most efficacious—because to do so privileges the global in such a way as to suggest that if workers cannot develop transnational connections, then they might as well give up on struggle at any other scale, particularly the local.

With this in mind, in this chapter I examine how local labor activists, garment workers, their union, and various supporters in New York City's garment district were able to organize successfully to confront the loss of jobs caused (at least in part) by the transformation of the area's physical built environment from a manufacturing center to a postindustrial officescape during the 1980s. Specifically, the chapter examines the role played by the International Ladies Garment Workers' Union (ILGWU) in a series of conflicts over land use in and around the garment district as pressure built during the late 1970s and early 1980s to allow the conversion of manufacturing lofts into office space so as to accommodate various financial services activities that were expanding in the wake of the globalization of New York

City's financial services markets.[3] What is of particular importance for the theoretical argument laid out in Chapter 2 is the way in which the union developed an explicitly geographical strategy to limit loft conversions in the face of encroachment by financial and business services. Through consciously manipulating the geography of the built environment, the union sought to protect the jobs of its members—at least in the immediate future. I show, then, how workers, organized around a very local set of concerns, were able to confront successfully the "global forces" of the contemporary international economy. If workers were able to achieve such success at the very heart of the global financial system, then perhaps there is hope for workers everywhere.

As with all of the empirical studies presented in this volume, I hope to do more than simply tell stories that are specific to certain groups of workers. Instead, I hope to use these studies to illustrate broader conceptual and theoretical issues. With regard to building the argument outlined in the preceding chapter about how workers may go about building spatial fixes that suit their needs at particular times and places, the case presented here does two things conceptually. First, it suggests that in some instances, at least, organizing around local concerns may indeed be a viable strategy for confronting global capital. Local activities can and do impact global processes. Consequently, the global should not be privileged a priori as *the* scale at which activities to confront global capital must be carried out. Whereas in many cases it may be necessary for workers to organize globally, in others it is workers' local activities that may give them purchase upon global economic and political processes. At the same time, though, this should not be taken as a privileging ipso facto of the local. Rather, it is a recognition that workers may be able to organize themselves at one geographic scale in order to confront processes that operate at different scales, and that there is no necessary "scalar mirroring" of capital and labor (i.e., the fact that particular fractions of capital operate on a certain geographical scale does not mean that labor must necessarily do so also). Such a recognition raises important questions about the interconnections among different scales of organization (the local and the global, for example) and how, and under what conditions, processes and events at one scale impact those at others—whether it be global processes shaping local landscapes or locally organized actors impacting global flows of capital.

Second, the narrative presented here challenges traditional thinking in much urban geography about how the built environment is structured. Typically the major force seen to drive the internal structure of cities has been the division of labor (e.g., Scott, 1988). It is through the spatial expression of the division of labor, so it is argued, that particular sections of the city become focal points for certain types of economic activity—sweatshop manufacturing, office headquarters, service provision, etc. In turn, this is seen to play a significant role in shaping residential patterns. Given that it is capital that almost invariably is conceived to be the singular determiner of the division of labor, in this model of urban development it is capital that is implicitly portrayed as the structurer of cities' built environments and economic landscapes. Although workers may be thought of as shaping the social and ethnic geography of the city through their residential choices (to the extent that they have such "choice"), there has usually been little sense that they play much of a hand in shaping its underlying economic and industrial structure, except in the rather passive way in which their place of residence can influence the locational decisions of capital—areas of low-waged or highly skilled workers may attract different types of investment to different parts of the built environment, for instance.[4] By examining how workers and their organizations have shaped the spatial evolution of one industry (garment manufacturing) in one city (New York) as the city's urban fabric has experienced the impacts of recent economic globalization, I want to begin challenging such a capital-centric understanding of the evolution of the built environment. Although what is presented here is a quite specific case study, it is my hope that it will nevertheless provide sufficient theoretical and empirical inspiration to encourage researchers to challenge dominant narratives in economic and urban geography, together with those in other disciplines, which assume a priori that capital and the state are the only significant actors structuring the fabric of the built environment.

The chapter itself is in four parts. The first briefly examines the garment industry's place in the historical development of New York City's urban landscape and how this was challenged by processes of economic globalization during the 1970s and 1980s. The second analyzes how the real estate pressures brought about by the globalization of New York's economy and the influx of massive amounts of international financial capital have helped transform the industrial character of this part of Manhattan. The third describes the creation of a "special preservation zone" in New York City's Garment Center designed to limit loft conversions and preserve space for garment manufacturing. The fourth section draws out some broader conceptual links between the restructuring of the built environment and the political practice of the ILGWU, which sought to protect workers' livelihoods through such a conscious defense of the historical geography of the industrial landscape as the city adopted an increasingly postindustrial form.

GLOBALIZATION, THE RAG TRADE, AND REAL ESTATE DEVELOPMENT DURING THE NEW YORK "RENNAISSANCE"

Garment manufacturing and the fashion industry have long been associated with New York City and have had a significant impact on the evolution of the city's industrial landscape. Prior to the 1880s, garment manufacture was still largely a domestic occupation. Massive immigration during the last two decades of the nineteenth century, however, brought both a dramatic increase in demand for ready-made clothing from a rapidly urbanizing America and the cheap labor to satisfy that demand (Waldinger, 1986). New York's teeming supply of immigrant labor and its status as the nation's leading port gave the city's garment producers a great competitive advantage. Within three decades a relatively insignificant industry had turned into a crucial component of New York's economy (Pope, 1905). By 1920 women's apparel production, the leading sector within the city's garment industry, employed over 125,000 workers, one-sixth of all manufacturing workers in New York City (Heiman, 1988). Even into the 1980s the garment industry was the city's largest manufacturing employer, with 166,000 predominantly female Asian, Hispanic, and black registered apparel manufacturing and wholesaling workers in 1986 (Mazur, 1987; New York City, 1986).

If manufacturing activity—particularly of garments—had been the foundation for much of New York's urban development during the first three-quarters of the twentieth century, by 1975 the city's industrial future looked grim, as its manufacturing base was being wracked by deindustrialization. Declines in the garment industry were particularly worrisome. Although the city's importance as a center for garment production had steadily declined since the 1930s, during the 1970s and 1980s the industry experienced a dramatic employment and geographic restructuring in response to mounting international competition from lower-cost producers in such countries as South Korea, Thailand, and the People's Republic of China. Between 1973 and 1983 imports rose from 21% of U.S. clothing sales to 55% (Lochhead, 1987). Concomitantly, between 1970 and 1984 citywide registered employment in garment manufacturing fell 42%, while the trend of decentralizing large-batch standardized production out of central Manhattan to the fringes of the Greater New York region intensified under these new international competitive pressures (Conway Co., 1986; Scott, 1988). Simultaneously the number of industry sweatshops employing nonunion workers (frequently undocumented immigrants) for low wages and long hours without the protection of health or safety regulations, and often in cramped, noxious conditions, grew from an estimated 200 to over 3,000, many in areas of the city that had not previously had any garment shops. It is estimated that by the mid-1980s such sweatshops were employing upward of 50,000 workers, while at least an additional 10,000 "homeworkers" were thought to be manufacturing garments in domestic residences (Sassen, 1988a), although, given the difficulty of locating and monitoring such homeworkers (Herod, 1991b), the figure could actually have been much higher.

Whereas in 1975 the city's economic future looked bleak, by 1985, with tremendous changes taking place in the global economy and high interest rates attracting millions of dollars in foreign capital, New York's future looked quite different as a veritable explosion of investment in financial and business services had turned it into, arguably, the primary world center for international financial and business services functions. This economic renaissance brought with it both new employment patterns and a real estate boom that led to the restructuring of large portions of the city's built environment. As New York's real estate and financial sectors rode a wave of economic globalization, a series of struggles over land use increasingly came to dominate city politics during the 1980s. Many working-class neighborhoods—particularly in Manhattan's Lower East Side, in Brooklyn, and even in Harlem (Schaffer & Smith, 1986; Smith, 1996)—came under attack from developers interested in gentrifying old brownstones and tenement buildings to satiate demand from the yuppie (young urban professional) elite working in the city's business district. At the same time, the internationalization of New York's economy unleashed struggles between, on the one hand, declining manufacturing interests who were trying to hold on to traditional manufacturing spaces and, on the other, increasingly ascendant financial and business services interests who were looking to expand office space away from Manhattan's expensive and crowded downtown office districts. Nowhere, arguably, was this conflict over the reshaping of the built environment more intense than in the Manhattan Garment Center district, where struggles took place over the issue of converting into offices large numbers of lofts traditionally used for light manufacturing.

The growth of the financial and business services sector of New York City's economy posed problems for many manufacturers and their workers. Although it was the rising competition from imports that was undoubtedly the leading cause of employment loss in the city's garment industry during the 1970s and 1980s, the office boom that accompanied New York's renaissance reinforced these competitive pressures. The booming real estate market during the early 1980s tempted many landlords to convert their lofts into office space, thus limiting the space available to manufacturers unable to match the higher rents that office use could command. Between 1980 and 1985, the area between 14th and 42nd Streets in midtown Manhattan saw 20,017,823 square feet of loft space converted into office space (Conway Co., 1986: 19).[5] At the same time, the bourgeois chic among many of the new urban professionals for loft living further annihilated space for light manufacturing activity in traditional industrial lofts (Zukin, 1982). As the globalization of New York's financial and business service sector during the 1980s proceeded, many different kinds of manufacturers began to find that the manufacturing space that they had traditionally used was increasingly threatened with conversion to office and/or residential use. Such pressures for loft conversion were particularly hard felt in the Garment Center, an area lying between Fifth and Tenth Avenues from 34th to 41st Streets in Manhattan's midtown office market and which in 1986 employed 61,000 registered manufacturing and wholesaling garment workers (see Figure 3.1). It is these conversions within the Garment Center to which I now turn.

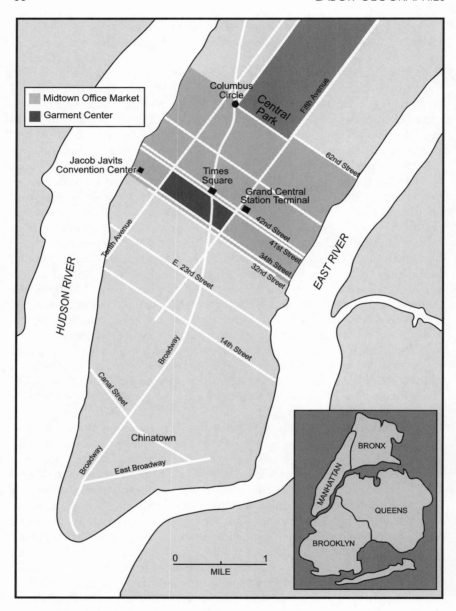

FIGURE 3.1. Map of Manhattan showing location of the Garment Center.

RECENT LAND USE CHANGES
IN MANHATTAN'S GARMENT CENTER

In the last decades of the nineteenth century the garment industry had located almost exclusively in the Lower East Side south of 14th Street and east of Fifth Avenue. However, it did not stay there for long, and its center soon began to migrate north. By 1922 nearly two-thirds of the industry was located between Fourth and Eighth Avenues from 14th to 38th Streets, giving manufacturers convenient access to out-of-town buyers staying in hotels close to the midtown passenger rail terminals (Heiman, 1988). Yet, in moving out of the Lower East Side, garment manufacturers soon came into conflict with corporate office and retailing interests. In particular, the city's patrician elite, fearing that the expanding garment industry threatened to encroach upon prime midtown real estate, sought to bring political pressure to bear to disperse the unsightly manufacturing activities, together with the immigrants they employed, to the outer boroughs of Brooklyn, Queens, and the Bronx—far from the fashionable Fifth Avenue residences of New York's leading society members (Toll, 1969). In drawing up a series of proposals culminating in the 1929 Regional Plan, New York's elite set out its agenda to "upgrade" Manhattan by using zoning changes and restrictive investment practices to remove manufacturing activity and its associated working-class housing (Zukin, 1982). Although these attempts failed to banish manufacturing entirely from Manhattan, they did create a special zone for the garment industry in the area that has since become known as the Garment Center (for more on the history of New York City's special zoning districts, see Barnett, 1982: 77–98).

The Garment Center contains approximately 29 million square feet of building space, 20 million of which was occupied in 1985 by apparel establishments. Apparel manufacturing and production-related activities such as cutting, sewing, design, storage, and shipping occupied 9.2 million square feet of space, 8.5 million of which was located in loft buildings. A Garment Center census conducted for the city shows that in 1985 35% of the apparel firms located there were manufacturers, 34% were showrooms, 13% were suppliers, 10% were service firms, and 6% were contractors, while other unclassified types of apparel firms made up the remainder (New York City, 1985). Manufacturing workers accounted for the highest proportion (41%) of workers in all apparel firms. Center producers tended to be engaged more in the high-fashion side of the garment industry, with larger-batch production generally subcontracted out to cheaper locations (Scott, 1988). Thus, whereas much routine sewing—particularly in women's and children's apparel—was being sent to Chinatown in the Lower East Side, high-fashion design and merchandising activities remained firmly fixed in the Garment Center. Given the structure of the industry, then, the Garment Center not only provided significant levels of employment in midtown Manhattan but, because it lay at the heart of a wide subcontracting network, it also served as the generator of work and employment for garment manufacturers throughout the city (Mankoff, 1998).

Within the Center there existed a core 10-block area between Sixth and Eighth Avenues from 35th to 40th Streets in which over 80% of all establishments, employments, and space-use were apparel-related. Although most garment firms tend to be fairly footloose, generally they have tried to remain within the Garment Center, locating wherever cheap interstices open up in the rental market. According to the census conducted by the city government, two-thirds of apparel manufacturing firms in the Garment Center were single-location establishments in which all activities from design and manufacture to sales and shipping were carried out in the same building. Throughout the Center as a whole, 53% of apparel manufacturing establishments had occupied their current locations for fewer than five years, while for suppliers the figure was 41% (New York City, 1985).

Three main factors explain the industry's continued centralization in the Garment Center despite the deindustrialization of large parts of Manhattan. First, the nature of the production process and commercial ties are powerful forces encouraging geographical clustering. The need for a "spot" market to respond quickly to rapidly changing fashion trends—particularly in the high-fashion end of the market—has encouraged the development of a highly integrated local network of wholesale and showroom activities, service firms (e.g., patternmaking and embroidery, sewing machine mechanics), suppliers (e.g., fabric and buttons), and contractors (e.g., sewing and cutting) servicing the needs of apparel manufacturers. Waldinger (1986) found, for instance, that 75% of manufacturers primarily used suppliers located within the Garment Center. Face-to-face contact is paramount. As industry and union representatives point out, it is impossible to describe precisely a color or pattern over a telephone.

Second, in a market (fashion) so prone to rapid change, firms have reaped several advantages by remaining relatively small in size and flexible. Indeed, large firm size has often been something of a liability that has produced few economies of scale and a tendency to react slowly to, and thus miss the boat on, changing consumer fads and tastes (Waldinger, 1986). Consequently, because they generally do not require large amounts of high-quality space, such small firms have historically been able to locate close to the city's central core. As Zukin (1982: 5) has pointed out, with loft rents fairly low and stable (at least until the mid-1970s) and with a plentiful and cheap supply of labor, "as long as these businesses could withstand the buffets of the national economy and the decentralization of industrial production, they stayed in Manhattan." Those garment shops that decentralized out of the Center have tended to exhibit different characteristics from those that remained, being instead higher-volume, standardized production units that are less dependent upon the close linkages essential to Garment Center operators (Abeles Schwartz Associates, 1986).

Third, the influx of new immigrants from Asia, Latin America, and the Caribbean since the 1960s has helped keep labor costs from rising too much and has made the city's industry more cost-competitive. The end result of all this was that, even with a loss of 3,400 jobs between 1977 and 1986, the Garment Center actually increased its share of the city's apparel manufacturing during this period (New York City, 1986).

During the 1970s, however, manufacturing lofts in the Garment Center came under increasing pressure of conversion to both residential and general office uses. In particular, the Center's eastern section, which provided ready access to work for professionals employed in East Midtown offices lying to the northeast, had steadily been invaded by those seeking to realize the growing fashion for loft living. The Real Estate Board of New York (a developers' trade association) had been heavily involved in creating this trend through the real estate section of the *New York Times*. In 1975 it successfully lobbied City Council to amend New York's Administrative Code and extend J-51 tax subsidies to developers and owners converting commercial and manufacturing lofts to residential use.[6] Although many illegal conversions had, in fact, previously been conducted, the granting of J-51 subsidies provided the impetus to rezone the Center's eastern section in 1981 to legalize residential lofts, a move which substantially increased conversion pressures. At the same time, the city prohibited all but manufacturing or office use in the remainder of the Garment Center (Zukin, 1982).

With average annual rents per square foot for office space in downtown Manhattan growing by nearly 170% between 1979 and 1984, the midtown office market (historically defined by the Real Estate Board as lying between 32nd and 62nd Streets from the Hudson River to the East River) proved to be an increasingly desirable location for corporate offices, and indeed most of Manhattan's new office construction during the 1970s and 1980s occurred there (*Business Week*, July 23, 1984). Already high rents in the traditionally favored East Midtown district encouraged the splitting off of some office functions toward other new or revitalizing office submarkets located nearby (Conway Co., 1986). This trend was reinforced by the city's adoption of zoning amendments in 1985 that were designed to encourage new office development in the West Midtown area just north of the Garment Center. In addition to new construction, between 1980 and 1985 over 38 million square feet of office space was created in these other submarkets through loft conversions and renovations of older buildings, allowing tenants relocating from the East Midtown area to save in some instances from $15 to $25 per square foot in rents (Conway Co., 1986; New York City, 1986).

The Garment Center's stock of loft buildings made it an attractive area for corporate offices, and by the early 1980s it had already come under pressure from firms beginning to relocate there. Relocating firms typically included advertising, publishing, consulting, and communications companies whose operating margins frequently prevent paying the level of rents necessary to support new office construction. Financial and business services firms were also among those locating in the Garment Center, causing financial, insurance, real estate (FIRE), and associated service-sector employment in the Center to rise 150% (21,000 new jobs) between 1977 and 1984. These pressures were further exacerbated by the spillover effects of two major development projects along the Garment Center's periphery—the Jacob Javits Convention Center to the west and the Times Square–42nd Street Redevelopment Project to the north—and rezoning amendments adopted in 1985 to encourage new office development in West Midtown (*City Almanac*, 1985; New York City, 1986).

Such encroachments were reflected in both rising land costs and rising rent levels in the Garment Center. Sales of loft buildings multiplied, with 1.8 million square feet of space changing hands in 1985 alone. Average sales prices for Garment Center lofts increased 387% per square foot between 1980 and 1985, indicative of growing market interest in anticipation of higher rents and returns subsequent to conversion. Although the highest rents were found in office-class buildings used for apparel showrooms at certain prestigious Broadway addresses, the largest rent increases during this period were recorded for general office space located in lofts. Rents for lofts marketed for general office occupancy jumped by 70% to an average for the Center as a whole of about $17 per square foot. Concomitantly, average asking rents for manufacturing space in lofts rose only about 20% during the early 1980s but then experienced a sharp 29% rise between 1984 and 1985 to about $7.50 per square foot (Conway Co., 1986).

Hoping to cash in on the boom for office space, many Garment Center landlords began converting their lofts to higher-rent office use. By 1986, 5.4 million square feet of loft space converted from both apparel and nonapparel activities in the Center was already being marketed for general office use. Loft owners upgrading their lofts for office use were then typically able to charge double the rents that manufacturing lofts could command. Demand for office space was reflected in the relatively low (5%) vacancy rate for office buildings, with slightly higher rates (7%–14%) for converted lofts, depending on whether the conversion had entailed major renovation (higher rates) or had been conducted without substantial renovation (lower rates). By comparison, the vacancy rate for manufacturing lofts ranged from 8%–20% (with the higher rates applying to areas west of Eighth Avenue), although it was not clear how much of this was attributable to lofts being held off the market in anticipation of conversion (Conway Co., 1986).

As a result of these developments, in the early 1980s the ILGWU increasingly began to concern itself that, as one union official put it (Mankoff, 1998), the added effects of the proposed Times Square–42nd Street Redevelopment Project (see Fainstein, 1985, for more details) were going to "put pressure on buildings to convert to commercial space, particularly those that were heavily occupied by garment manufacturers and were used by them for office and manufacturing purposes, [and that] the destruction of the Center would have had a major impact throughout the city." Such developments, union officials worried, were occurring "at the same time we were being hit by imports." The result was that "those in the industry could pay the high rents for showrooms but not for manufacturing, so we were concerned manufacturers would be forced out. You could see the writing on the wall. Times Square would be the last straw that would lead to the type of development we were worried about" (Mankoff, 1998). For the union, clearly, the forcing out of garment production shops would mean a loss of manufacturing jobs in the industry (at this time some 25,000 of the approximately 61,000 garment jobs in the Center were in direct manufacturing [McCain, 1987]). Fearing such consequences, and concerned to maintain space in the Garment Center for manufacturers, the ILGWU determined to develop a plan to restrict conversion in this part of Manhattan.

SETTING UP THE SPECIAL GARMENT CENTER DISTRICT

In response to these growing real estate pressures, in the early 1980s the ILGWU approached the office of Mayor Edward Koch with the goal of establishing a "garment manufacturing preservation zone" in the Garment Center. Having had some experience with the problem of loft conversion in other parts of the city—the union had recently been involved in a plan to protect loft space from conversion in the area south of 32nd Street and had also pressured the city to establish some protective zoning in Chinatown, where much sewing was then being done (Mankoff, 1998)—union officials indicated to the Koch administration that implementing a preservation zone to aid garment manufacturers in the Garment Center would be the ILGWU's quid pro quo for not opposing the Times Square–42nd Street Redevelopment Project that city planners, politicians, and developers were pushing as part of a plan to clean up the area around Times Square, long a center of prostitution and other activities associated with the sex industry (McCain, 1987). By limiting industrial loft conversion activity within this zone, the ILGWU hoped to preserve garment industry jobs. If the Mayor's Office did not acquiesce in this regard, the union would bring its resources to bear to stall the Project. Under such pressure from the ILGWU, in November 1984 Koch directed the Department of City Planning, the Office of Economic Development, and the Public Development Corporation to study the potential impacts of the Redevelopment Project on loft conversions in the Garment Center.

As part of its effort to come to some political agreement with the union, the city commissioned independent consultants to undertake three separate studies to detail whether the Center's character was changing and, if so, to determine possible causes. The first of these (New York City, 1985) was a census of apparel and nonapparel firms in the Center that sought to identify the types of activities they carried out, together with their respective space needs. Second, a more detailed survey of a representative sample of apparel firms was conducted that was designed to illuminate the problems firms faced, their linkages to other garment-related firms in the Center, and their future plans (Louis Harris Associates, Inc., 1986). The third study analyzed real estate activities in the Center and their potential impacts on garment activities (Conway Co., 1986). These studies revealed that 85% of establishments found that the need to be close to suppliers, contractors, and buyers made it essential to locate in the Center (New York City, 1985). Furthermore, 33% mentioned their top-ranked problem as the cost of space, closely followed by international competition (30%). Space costs were particularly worrisome for apparel manufacturers because 80% of establishments' leases would expire by 1990, and many feared their leases would either not be renewed or else they would not be able to pay the higher rents charged under new leases. Such fears were well grounded in many establishments' prior experience. In the early 1980s, rent pressure had already been reflected in shifting geographical patterns of activity within the Center, as some manufacturers attempted to cut costs by moving production out of more expensive spaces on the main thoroughfares (the "Avenues") and into cheaper mid-block locations (Satkin, 1987).

The studies also showed that the majority of manufacturing space in the Center was located in lofts suitable for office conversion. Indeed, the Conway real estate report noted that landlords were increasingly "actively showing [loft] space for office use and offering leases to office type tenants" (Conway Co., 1986: 52). Furthermore, city officials had evidence of manufacturers being denied leases by loft owners who wanted to warehouse their buildings in readiness for office conversions (Satkin, 1988). Officials determined that some 5.5 million of the 8.5 million square feet of apparel loft production space in the Center was suitable for conversion (Satkin, 1987).[7] Of this, 500,000 square feet had already been vacated. It was projected that future declines in apparel manufacturing due to foreign competition would cause a further 1.5 million square feet to be vacated by 1995. This left a total of 3.5 million square feet potentially at risk, although whether conversion would in fact occur would be determined by the future tightness of the Manhattan office market (New York City, 1986).

Fearing that without zoning changes much-needed manufacturing space could be converted, in 1986 the Department of City Planning (DCP) proposed to rezone part of the Center to create the Special Garment Center District. The Special District would include 8 million square feet of space and cover the area from 35th to 40th Streets between Seventh and Ninth Avenues and from 35th to 37th Streets between Seventh Avenue and Broadway (Figure 3.2). This area was chosen because it then contained 40% of Garment Center apparel production space and because two-thirds of all business activity within it related to apparel production. However, the Special District excluded the frontages on all the avenues (including Broadway) because apparel showrooms located there could pay higher rents and so were able, it was felt, to compete with general office uses. City planners anticipated that, as showroom activity expanded along the avenues, single-location manufacturers (who constituted two-thirds of all Garment Center firms) would increasingly be pushed into the midblock areas. Within the Special District loft conversion to office space would be circumscribed. Conversion would be allowed only upon certification that an equal amount of comparable floor space, either in the same building or in a similar building within the District, would be preserved for manufacturing, a formula that

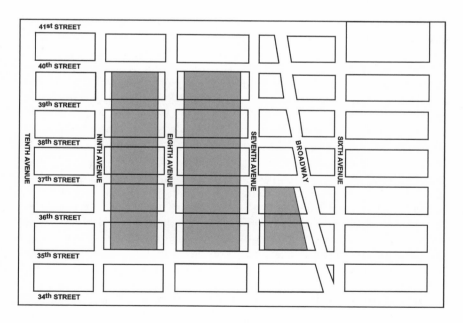

FIGURE 3.2. Map of the Garment District, showing location of Special Garment Center District preservation areas.

followed the union's earlier efforts to curtail loft conversion in Chinatown in Lower Manhattan (Mankoff, 1998). The DCP did not restrict this preservation to apparel production, but it specified a number of manufacturing activities that could fulfill the requirement. Despite its restrictions, the rezoning would still allow for extensive office conversion and the growth of showroom space. The preservation requirement would be triggered only by conversion of floor area to the city's Use Group 6B office category, which includes all business, professional, and government offices. Of the 8 million square feet making up the Special District, in 1986 5 million square feet were in use for apparel production. City planners estimated that, even if half that space were preserved under the rezoning, this would still fall 1 million square feet short of the 3.5 million square feet of production space city officials estimated the garment industry would require in 1995.

In the summer of 1986 the DCP applied to the City Planning Commission (CPC) to set up the Special District. The city's Department of Environmental Protection determined the rezoning would not require an environmental impact statement (EIS) because the district's creation involved only a zoning change rather than any physical construction (Satkin, 1989). Following a series of public meetings, the rezoning was referred to the Board of Estimate and approved in March 1987.[8] However, whereas ILGWU President Jay Mazur saw the rezoning as an "important first step to ensure that New York City's economy provides opportunities to all of our people, blue-collar as well as white-collar, skilled and unskilled [with the alternative being] the loss of thousands of irreplaceable blue-collar jobs from New York's working people" (Mazur, 1987), it was immediately challenged in court by the Real Estate Board of New York, the Midtown Property Owners' Association, and two property owners. Opponents leveled three complaints against the city, arguing that, (1) contrary to the city's original determination, the Department of Environmental Protection should indeed have conducted an EIS and was thus potentially in violation of state and city environmental regulations; (2) the city had failed to consider alternatives to rezoning; and (3) the rezoning represented an unconstitutional taking of property. In August 1988 the court ruled against the city. City officials, though, appealed the decision, during which time the Special District remained operational. On May 31, 1990, the Manhattan Appellate Division overturned the lower court's ruling and approved the rezoning. The Special District's provisions have remained in place since that time (Satkin, 1998).

THE POWER OF THE LOCAL IN THE FACE OF THE GLOBAL

Clearly, the internationalization of New York's economy has fundamentally impacted both the physical structure of the city's urban geography and its concomitant employment structure. The influx of massive amounts of speculative capital, the loss of manufacturing jobs, the juxtaposition of abandoned factories and sweatshops largely employing the new immigrants from the Caribbean and Asia (Sassen, 1988b), the growth of female "pink-collar" employment servicing the corporate sector, and the physical destruction of many working-class neighborhoods through the process of gentrification have all done much to remake New York's labor landscape. The crucial task for the garment industry and its workers has been to redefine both their place and their space within this global city.

It is significant for our purposes here that the strategy adopted by the ILGWU (together with those garment producers who also supported the implementation of the preservation zone) to counter the "industrial gentrification" of manufacturing loft space was explicitly geographical in nature, focusing upon the control of space through zoning regulations. The union had already used similar tactics to protect manufacturing space in Chinatown by pressuring the city to rezone a mixed-use manufacturing area so as to prohibit conversions of lofts from manufacturing to residential use (Abeles et al., 1983; Waldinger, 1991; Mankoff, 1998). Examining the ILGWU's political strategy involves understanding how the production and manipulation of geographic space and scale can act as a form of social power. In particular, as the actions of the ILGWU show, the production of landscapes clearly is not solely the dictate of capital, for even within the totality of capitalist social relations political struggles open up possibilities for multiple and different geographies to be constructed. Much as Gibson-Graham (1996) has argued that capitalism should not be seen as an all-or-nothing system and that different types of economic and social relations may coexist under it, so, too, we should not think that all landscapes created under capitalism are necessarily those favored by a hegemonic capital and have been developed in its own image. To think in such terms would portray labor as always outmaneuvered by capital, no matter what workers manage to achieve "on the ground" through their struggles. At the same time, it would portray capital in the infinitely flexible and adaptive terms that Gibson-Graham rightly criticizes.

The case study outlined here highlights three specific issues that are of broader theoretical importance for the argument I am pursuing in this volume. First, it shows how one group of workers and their union, organizing around very local issues (land use conversions in a small part of Manhattan), were able to address and overcome some of the challenges posed by the globalizing forces of the contemporary economy. This raises questions about the efficacy of local strategies to counter global (or any other extralocal) processes and about the "politics of scale" more generally—such as what is the relationship between the local and the global (e.g., how might such local activities to limit office conversions drive up the price of office space elsewhere in Manhattan, thereby perhaps limiting its attractiveness to foreign investors if the cost of doing business in the Manhattan office markets is perceived to be too great?), and how does this vary in time and space?[9] Particularly, it highlights what Jonas (1998) calls the "local–global paradox," a term used to describe the phenomenon whereby economic relationships have become ever more global in scope, yet political responses to this have often been articulated at the very local level. Thus, the union made little effort to seek support from outside the local area (e.g., by getting help from other ILGWU locals as part of a solidarity campaign or seeking intervention by the federal government) but focused, instead, upon engaging with the regulatory powers of the local government and bringing pressure to bear upon local elected representatives in city government.

Such a "localist" strategy has significant implications, for it shows that, given the right context and mix of actors, local struggles can indeed be successful against the supposedly omnipotent forces of global capital. This is crucial to bear in mind when listening to the rhetoric of neoliberals who suggest that all that labor can do in the "new global economy" is accept its fate at the hands of globally organized capital. Rather than accepting at face value neoliberal arguments (or even those of some Marxists!) about local labor's structural or "natural" weakness in the face of global capital, I want to argue that any such weaknesses are, in fact, a contingent matter, dependent upon a range of issues—such as whether workers are in a locale receiving capital investment or suffering from capital flight, how they are organized, what relationship they have to the local and/or national state, and so on. Indeed, if we accept, following Harvey (1989), that globalization allows capital increasingly to make investment decisions on the basis of often minutely differing local conditions that may exist in different parts of the globe—in other words, that in a hypermobile global economy in which capital can choose to locate investment in virtually any part of the planet local geographies and differences between places become more, not less, significant for capital—then we must also accept that, perhaps paradoxically, in some instances local struggles may have greater impacts upon the way in which global processes play out in a highly interconnected global economy than they would in a less interconnected national or regional one.

Second, the case study highlights the important role that geography and attachments to place may play in shaping political praxis. In particular, it illustrates how class politics cannot be divorced from the politics of space. The fact that several garment manufacturers joined forces with the ILGWU to support the establishment of the special preservation zone in the hope of retaining space for garment manufacturers and their workers is suggestive of the role played by spatial relations in shaping the political praxis of both employers and workers. As Cox and Mair (1988) have illustrated, employers' and workers' spatial entrapment in particular places means that they both may often see it in their own best interests to collaborate with each other across class cleavages to defend particular places against the decimation of their local economy by "external forces," however defined (see Hudson & Sadler, 1983, 1986). Whereas aspatial class analyses often tend toward the conclusion that workers who "collaborate" in such geographical coalitions with their employers must be deluded by "false consciousness," an analysis rooted in the geography of the situation shows that particular groups of workers may, in fact, gain more in real terms (e.g., continued employment opportunities) by organizing around spatial concerns than around class ones. The result is that they may frequently adopt strategies that pit "their" space against that of other workers (in this case, perhaps those janitors, office staff, and others who might have been employed by the business services firms that would have relocated into the Garment Center had the rezoning not gone ahead). Many workers, then, see such cross-class coalitions as very real opportunities to shape local economic geographies—even if they do not necessarily conceive of their actions in such academic terms— and engage in local boosterist activities of this sort out of a genuine conviction that this best serves their own interests. In the process, of course, they play an active role in shaping the landscapes of capitalism.

Third, through its activities the union clearly played an important role in shaping the evolving spatial division of labor in this part of the city, a fact that challenges explanations of urban form that are rooted in the internal dynamics of capital and one that should be borne in mind when theorizing how work under capitalism is organized spatially. Whereas much writing in urban analysis frequently explains the internal spatial structure of cities by examining, predominantly or solely, the investment decisions of various segments of capital and how these affect the spatial division of labor, the implementation of the Special Garment Center District suggests that such a capital-centric approach is too narrow. Indeed, in this case it was the actions of the ILGWU that played a substantial part in shaping the future spatial division of labor in this part of the city rather than the investment decisions of various local or transnational firms.[10] Clearly, an analysis that ignores the actions of the ILGWU would not be able to account fully for the emergent industrial geography in this part of Manhattan. By pushing the city to undertake the rezoning, the ILGWU was able to preserve space for garment producers in the heart of Manhattan and so to help maintain the social and spatial integrity of the industry. Had rezoning failed and loft conversions went ahead unencumbered, then it is likely that many garment producers would have gone out of business. In turn, this would undoubtedly have had major implications for other producers, given the highly networked and interconnected nature of their business relationships, and would probably have encouraged even greater migration out of the Garment District of those involved in the industry. Indeed, the ILGWU's activities were central to retaining the garment industry in this part of Manhattan, at least for the immediate future—a fact which theorizers of industrial location would do well to bear in mind.

In summary, then, while even the staunchest advocates of the preservation zone hardly expected that such geographical manipulation would prevent the long-term decline of garment manufacturing in the Center, their strategy was geared more toward slowing the rate at which employment declined, particularly given the dramatic changes that had taken place in the built environment during the early 1980s. By means of such an explicitly geographical praxis to defend the traditional spaces of garment manufacturing, proponents of the Special District sought to extend the time horizon of the industry's decline, thereby allowing a longer period of adjustment to the new historical and geographical imperatives of the global economy. This strategy could mean the difference between absorbing a loss of many thousands of garment jobs over, say, a 20-year period instead of a much shorter period of time.[11] With the evaporation of many of the city's traditional manufacturing jobs, such an employment cushion was thought to prove particularly important for minority workers who make up the majority of the garment industry's labor force and for whom alternative employment options are rapidly disappearing. Through their manipulation of space, the union and other proponents of the Special District were effectively able to buy time for those garment workers threatened with unemployment.

As it turns out, during the late 1980s much of the pressure to convert lofts into offices in the Garment District was relieved by the cooling of the real estate market, and much future demand may be accommodated by the redevelopment of the Times Square area (Satkin, 1998). Nevertheless, the preservation zone is still in effect, and union representatives have indicated that, as New York's financial sector rode the wave of a seemingly ever expanding stock market in the late 1990s, some conversion pressures began to revive (Mankoff, 1998). What is important to bear in mind for the theoretical arguments laid out in this book, though, is that the ILGWU's political strategy to defend the historical geography of the urban landscape as a means of protecting garment jobs illustrates how unions' spatial practices can shape the local geography of capitalism. By their actions the proponents of the preservation zone were able to present a barrier to the restructuring of the built environment brought about by the internationalization of New York's economy. Not only did the union carve out a *space* of resistance to the encroachments of speculative financial capital, but it also constructed a *scale* of resistance, as defined by the geographical perimeters of the Special Garment Center District. Indeed, creating both a material scale of identity (the Special Garment Center District as codified by New York City zoning laws) and a discursive scale of identity (the Garment District as a coherent "industrial," rather than "postindustrial" or "service-sector," entity) was crucial to the union's goals. The ILGWU's attempts to safeguard garment jobs did not merely entail manipulating space but also were intimately bound up with the issue of controlling the geographical scale at which political struggle is articulated and landscapes are thus produced—a theme we will pursue further in subsequent chapters. The union's ability to construct such a scale of local opposition to the impacts of globalization was an integral part of its strategy to differentiate the urban landscape into areas where speculation and conversion could proceed unfettered and other areas where it would be regulated. As the effects of the Special Garment Center District's preservation requirement are felt (particularly when the real estate market is hot), this scale of opposition will increasingly be made visible in the material landscape through the patterns of conversion and nonconversion of manufacturing loft space.

CHAPTER 4

Spatial Sabotage

Containerization, Union Work Rules, and the Geography of Waterfront Work

The introduction of new technologies has often dramatically transformed traditional ways of organizing work and led to job loss for those workers who are subjected to them. As we might expect, such job loss is frequently met with opposition by those affected. From the early days of the industrial revolution—when handloom operators threw their wooden clogs (*sabots*) into the gears of the mechanized textile looms that were being introduced across Europe—the history of capitalism has been replete with examples of workers seeking either to resist the introduction of new technologies that might put them out of work or to minimize the negative consequences of such technologies. Typically the struggles between workers and capitalists over the introduction of new technologies and ways of organizing work have been examined in temporal terms, that is to say, analyses have tended to focus upon how these struggles have played out over time, with one side or the other gaining the upper hand at different historical moments. In contrast, considerations of struggles over the introduction of new technologies have tended not to focus much upon their geographical aspects, except perhaps in rather general terms of references to the locations in which they are actually taking place.

In this chapter I want to explore the issue of how, in some circumstances, workers may manipulate the geography of capitalism either to resist the introduction of job-destroying new technologies and ways of organizing production or to negotiate the implications that such transformations in the nature of work may have for them—that is to say, how they engage in what I would like to think of here as forms of "spatial sabotage." Rather than thinking about struggles over new technologies in terms simply of where these struggles are taking place, below I want to consider how workers may seek to restructure the spatial relations within which they live their lives as part of a strategy of resisting change and job loss. In some ways, then, this chapter follows on from the analysis of how the ILGWU, by bringing political pressure to bear on municipal government to use the force of law (via zoning ordinances) to control the types of economic activities undertaken in particular places, played a significant role in shaping the built environment of New York City's Garment District as a means of protecting jobs. However, in what follows I take a slightly different approach by examining how workers struggled to shape the spread of new technologies and ways of organizing work across the landscape through their ability to negotiate certain work rules with their employers

Specifically, the chapter examines how the International Longshoremen's Association (ILA), the dockers' union in the U.S. East Coast longshoring industry, responded geographically to the threat of job loss occasioned by containerization.[1] As the introduction of new technology (specifically containerization) in the 1950s and 1960s promised to destroy its members' jobs, the ILA implemented a series of contract provisions aimed at preserving work at the waterfront. Arguably the most controversial of these were the "Rules on Containers," a set of restrictive work practices by which the union reserved for its members the right to pack ("stuff") and unpack ("strip") on the piers certain types of containers, thereby limiting the relocation of that work to cheaper inland freight-handling warehouses. The Rules are important for the present argument because they represent an effort by the union to shape the geography of work along East Coast waterfronts, to control the spatial extent of local labor markets, and, ultimately, to shape the economic geography of the entire longshoring industry as part of a strategy of ensuring that dockers had continued access to work. Indeed, consciously reshaping the geography of employment in the industry was a key element in the union's postcontainerization strategy to ensure that dockers' own livelihoods were maintained. Dockers' capacity to make space in a way that benefited them, rather than in ways that the employers or even other groups of workers (such as those who worked at inland warehouses) preferred, was crucial in shaping the evolution of the industry's economic geography during this period. As we shall see, whereas the "pure economics" of containerization may have favored the emergence of a geography of postcontainer employment in which all work was done at inland cargo terminals staffed by cheaper non-ILA labor, the *actual* geography of employment in which ILA dock labor maintained some work at the waterfront was determined by the outcome of intense political and legal struggles between competing interests. This is important, because it challenges traditional explanations that have seen the new economic logic that the employers unleashed (through introducing containerization) as driving the industry's spatial development, and instead places labor in a much more crucial and active role in shaping the economic geography of cargo handling.

By examining how union work rules can be significant determinants of what work gets done where and by whom, I seek to add to the argument I am pursuing in this volume in two ways. First, the present chapter shows how a different group of workers—this time, dockers—managed to control the location of work in their industry even though the economics of the industry was against them. The chapter, then, provides more empirical evidence for the conceptual argument I am trying to make. Second, following from the preceding chapter, this case provides another example of how workers were able to organize successfully around a series of very local labor market issues, in this case in the face of dramatic restructuring in the national and global shipping industry being brought about by containerization. Whereas containerization as a global force for change was transforming the shipping industry across the planet, East Coast dockers were able to confront locally the new competitive pressures that the nascent technology was unleashing worldwide. Again, this suggests that in some instances local praxis may indeed be quite effective in the face of extralocal global processes and pressures.

The chapter is organized as follows. The first section examines the impact that the introduction of containerization had on the economics of the industry and upon the labor process within it. The second section of the chapter examines how dockers and the ILA went about securing a number of work preservation agreements beginning in the 1950s. The third section analyzes the impact of these agreements upon the geography of work in waterfront areas and their hinterlands. The concluding section addresses some of the theoretical and political implications for conceptualizing the making of the geography of capitalism that the ILA's spatial negotiation of the new containerization technology raises.

THE IMPACT OF CONTAINERIZATION
ON WATERFRONT WORKERS

Prior to the 1950s, longshoring in the United States and elsewhere was a highly labor-intensive industry. In the late 1940s in New York, the U.S. East Coast's busiest and most important port, almost 50,000 dockers were registered with the ILA union, although only about half that number ever found work on anything like a regular basis (New York, 1952). Low levels of capital investment required moving most solid cargo through ports on a piece-by-piece "break bulk" basis. For exporting, loose cargo would first be delivered to the pier by truck, whereupon dockers would sort and check the goods. Once the cargo had been checked, dockers would place each individual piece onto pallets that were then lifted into the hold of a waiting ship by a combination of muscle power and small winches. For a ship bringing goods into port the process would operate in reverse, with dockers discharging cargo from the hold, whereupon it would be moved piece by piece to the local pier terminal. From there, cargo would either be loaded directly into trucks or hauled to local trucking terminals for reloading into over-the-road vehicles for delivery to its ultimate destination. Although specific work traditions varied in different U.S. ports (see Groom, 1965, for an account), the general practices of loading and unloading cargo were essentially the same throughout the industry.

Beginning in the 1950s, all of this started to change as the first tentative steps were taken toward developing an integrated intermodal system of freight transportation based upon the greater ease and speed of cargo handling made possible by containerization. Indeed, the introduction of containerization into the industry has been described by at least one commentator as "the single most important innovation in ocean transport since the steamship displaced the schooner" (Ross, 1970: 398). Although in the 1950s the idea of putting several small pieces of cargo into one larger box for ease of handling was by no means new—small reusable wooden containers (called "Conex" and "Dravo" boxes) had been used to transport military materiel during World War II—it was not until then that the commercial practice of containerization really began to take hold and, in turn, to threaten traditional longshoring work. Defined as "the utilizing, grouping or consolidating of multiple units into a larger container for more efficient movement" (Rath, 1973: 7), containerization itself is a relatively simple concept that involves packing cargo into an "enclosed, permanent, reusable, nondisposable, weathertight shipping conveyance fitted with a minimum of one door" (Tabak, 1970: 364). The types of commodities most frequently transported in containers are those of high value that attract relatively high shipping rates, those most susceptible to damage, and those that are most frequently pilfered. Such items include many perishable agricultural products, wine, liquor, pharmaceuticals, vehicles, and nonbulky machinery. Commodities of low value with relatively low shipping rates (for example, steel ingots) are generally only transported in containers if their weight, size, or other packaging problems require it. Other types of bulk cargo such as grain are moved in specialized ships (Van den Burg, 1975).[2]

The first company to operate a complete door-to-door distribution network using containers was the Pan Atlantic Steamship Corporation (which later became Sea-Land Services). Through the use of standardized containers, Pan Atlantic/Sea-Land sought to negate the need to physically rehandle each piece of cargo at traditional break-bulk points. In April 1956 the company initiated the first commercial application of containerization in the maritime industry in a run between New York and Puerto Rico. Despite the potential for huge cost savings that containerization offered the industry, this new system was not immediately adopted by other carriers. For years the federal government had fostered inertia in maritime innovation through its structure of subsidies that absorbed much of the cost of carriers' operating inefficiencies. Consequently, adoption of the new technology was initially somewhat slow, with containerization accounting for less than 3% of general cargo shipped through New York in 1966 (Rath, 1973; Ross, 1970). Significantly, neither Pan Atlantic/Sea-Land nor Matson Navigation (the first West Coast company to adopt containerization) received such subsidies, a factor that seems to have been crucial in their innovative activities. Although the continued availability of subsidies at first allowed the other U.S. cargo carriers to ignore these experiments with containerization, they were obliged to take up the challenge when, in 1966, Sea-Land signed agreements with more than 300 European trucking companies and began a weekly service to Europe. Sea-Land's lower rates and increased productivity compelled the already established lines on the trans-Atlantic route to convert existing vessels and to build specialized container ships in order to compete (Johnson & Garnett, 1971). As a result, just two years later 12% of cargo passing through New York was being carried in containers (Ross, 1970).

The Benefits of Containerization for Employers

The manner in which different employers were able to take advantage of the economic benefits offered by containerization has varied according to their position within the industry (see Johnson & Garnett, 1971, and Rath, 1973, for more on the economics of containerization). At least three groups of employers can be identified whose actions, while stemming from different specific interests, have converged to expand the use of containerization to the point where in the mid-1990s some 2.3 million containers were handled in the Port of New York alone. These three groups are the ocean carriers, manufacturers who use containers to transport their commodities, and the stevedoring companies contracted to hire dockers to work the carriers' ships.

Ocean Carriers

For ocean carriers, containerization has brought its greatest economies through a more efficient use of capital. Historically the labor-intensive nature of longshoring and break-bulk cargo handling meant that it could take several days for a ship to be loaded and/or unloaded. Because a ship only earns revenue through the act of transportation, this "dead" time represented nothing but overhead costs for carriers. The faster loading and un-loading that containerization allows has reduced that portion of a ship's life spent in port from (frequently) 60% to about 10% to 20%, bringing substan-tial savings on port costs and enablng ships to spend more time at sea trans-porting the goods for which the ocean carriers are paid (Chilcote, 1988). Furthermore, not only has containerization allowed carriers to cut their port costs drastically, but because faster turnaround times enable a ship to spend more time at sea, thereby increasing the cargo tonnage it can carry annually, it has also diminished the number of ships required to move any given vol-ume of cargo (Johnson & Garnett, 1971; Goldberg, 1968). Finally, the lower relative land transportation costs associated with containerized freight have also reversed the traditional economic logic that drove steamship compa-nies' distribution strategies. In the precontainer days, the high cost of using land transportation to collect and distribute cargo at a single port such as New York meant that steamship companies would instead schedule their ships to call at several ports on each side of an ocean, frequently taking up to several months to complete a round-trip journey. By negating the need for the costly loading and unloading of hundreds of individual pieces of freight at inland points—perhaps from truck to train to barge to truck again—containerization reversed this logic. The reduction of the relative cost of transporting cargo by land meant that ocean carriers could now draw business from larger hinterland areas. As a result, several of the major carri-ers began to restructure their shipping routes to call at a smaller number of more distantly spaced ports, thereby further minimizing their port costs and transit times (Chilcote, 1988) but also resulting in the potential loss of work in those ports increasingly bypassed.

Cargo Shippers

Traditionally, one of the largest expenses for shippers was the loss of in-come associated with the deterioration of cargo in transit (such as ripe fruit held up for several days at the port), losses due to damage and pilferage, and the high policy premiums charged by maritime insurance agents that such losses occasioned. Selna (1969) notes that dockworkers traditionally considered any booty acquired during their work as "fringe benefits" and that, for example, as much as 10% of the whiskey unloaded in New York was regularly "liberated." By packing their goods in sealed containers, ship-pers hoped to minimize the opportunities to steal individual items— although, as the Waterfront Commission of New York Harbor (1970) subse-quently observed, in several cases dockworkers merely substituted the theft of individual articles with the theft of entire containers filled with merchan-dise! Generally, however, containers have minimized many of the costs tra-ditionally associated with break-bulk cargo handling. By eliminating the need for special export packaging (traditionally a significant expense in putting goods into a foreign market), containerization enabled shippers to cut packaging costs by about 30% (Ross, 1970). Cost reductions have also been realized through the greater speed of processing cargo-handling documents and the concomitantly reduced time taken for cargo to clear customs (Selna, 1969). As with ocean carriers, manufacturers have also gained through the faster delivery times that containerization allows. With ships now capable of being in and out of ports in a matter of hours, the geography of product distribution has been radically transformed as many shippers have been able to use "land bridges" to hasten delivery schedules. Whereas goods moving from, say, Japan to Europe were previously sent across the Pacific and through the Panama Canal, the greater ease of cargo handling means that manufacturers can now send their goods to West Coast ports such as San Francisco or Vancouver, have them transported by train to New York, reloaded aboard ship, and sent on their way to European customers. This has typically cut delivery times from about 45 days to fewer than 20.

In addition to these benefits, containerization has allowed those shippers sending freight in quantities that are insufficient to fill entire containers—referred to as less-than-container loads (LCLs) or less-than-trailer loads (LTLs)—to make use of cargo consolidators. Consolidators arrange for small amounts of cargo to be delivered to their off-pier facilities, whereupon they are packed with other shippers' cargo in a single container. This provides savings to LCL shippers by allowing their goods' transportation under a single bill of lading at container rates substantially below break-bulk carrier rates. Because off-pier consolidators perform inshore much the same work as that done by dockers at the waterfront, the ILA has expended great energy to inhibit their operations. Indeed, as we shall see later, shippers' use of off-pier consolidators has been central to the legal struggles concerning the union's attempt to preserve work for its members through the Rules on Containers.

Stevedoring Companies

For the stevedoring companies that contract with the ocean carriers to load and unload their ships, the labor-intensive nature of break-bulk cargo handling had traditionally proved costly. The adoption of containerization has allowed stevedores to dramatically improve labor productivity. In 1954 over 35,000 dockers labored to move 13.2 million tons of general cargo through the Port of New York. By 1995 the impact of containerization had been so great that it took only 3,700 dockers to move 44.9 million tons of general cargo through the port, a 3118% increase in cargo handled per docker (Waterfront Commission of New York Harbor, various years). Rath (1973) cites figures indicating that containerization increased a typical cargo-loading rate in its simplest form from 15 tons per gang-hour to over 200 tons. Ross (1970) quotes industry representatives as indicating that, whereas loading 11,000 tons of cargo aboard a conventional precontainer break-bulk ship might typically take 126 dockers (six 21-member gangs) about 84 hours each (10,584 work hours), containerization allows the same amount of cargo to be loaded by 42 dockers working 13 hours each (a mere 546 work hours), amounting to a 95% reduction in total labor time. The result of these developments has been the drastic reduction of labor needs and the concomitant intensification of the work process. This has produced demands on the part of the employers for fewer and smaller gangs.

The Politics of Controlling the Labor Process

Clearly, there have been many economic advantages for container users. However, the introduction of containerization must also be seen as an attempt by employers to reestablish control over the labor process after a wave of rank-and-file wildcat strikes had shaken the industry during the late 1940s and early 1950s (Lamson, 1954). Although the potential for containerization had existed for years—containers are, after all, little more than standard-sized boxes—and various railroad companies had experimented with the use of standardized freight containers earlier in the twentieth century, it was, arguably, the growth of rank-and-file unrest after World War II that provided the greatest stimulus to employers to develop containerization as a means to tame waterfront labor, which showed signs of exerting itself after decades of quiescence. Employers hoped to use automation to attack the ILA politically by eliminating traditional restrictive work rules and practices (see Cockburn, 1983, for a similar example from the British printing industry).[3]

The New York Shipping Association (NYSA)—the largest employers' association in the industry, with a membership made up of some 85% of all employers in East Coast ports—had long maintained that union enforcement of customary work rules regarding staffing levels and job assignments constituted blatant "feather-bedding" practices that forced employers to hire many more dockers than they believed were strictly necessary for a particular task (for a study that contrasts the internal cohesiveness of West Coast employers with the internal divisions exhibited by the NYSA, see Kimeldorf, 1988: 77–78). Historically, the employers had figured it cheaper to concede such ILA practices than to risk provoking potentially costly strikes by fighting the union on this issue (Larrowe, 1955; Jensen, 1974). Furthermore, with the relatively low wages forced on dockers by a corrupt and compliant union hierarchy (in return often for various emoluments from the employers), this was a cost the employers could just about live with.[4] However, the NYSA greatly feared that any newfound rank-and-file militancy might upset this situation and increasingly sought to erase such work rules that might limit the introduction of containerization. To reap the full benefits of the new technology, the NYSA maintained, employers needed to enjoy "unfettered" rights "to inaugurate and regulate automation operations" free from union work restrictions (NYSA, 1959). Key to the NYSA's goal was its insistence that the written contract take precedence over the wide variety of customary rights and rules operating in the industry (Jensen, 1974: 142).[5] The subsequent history of containerization in New York and elsewhere has been indelibly marked by the operators' efforts to eliminate dockers' customs and practices and to replace them with the rule of the contract (this was a key NYSA demand in the 1968 contract negotiations, for instance [see *Journal of Commerce*, August 7, 1968]). In this manner the NYSA has sought both to establish standard rules allowing for the smoother introduction of the new technology but also to bring such unwritten arrangements within the realm of the legally enforceable contract wherein they could be regulated and, if need be, subsequently negotiated away.

Recognizing that employers introduce new technologies to further their political control over the labor process, however, is not to ignore the dialectic between workers' resistance and accommodation to new labor processes (Burawoy, 1985). Whether covertly or overtly, whether manifested, in the words of James Scott (1985), through the "big guns or small arms fire" of political struggle, the introduction of new technologies is mediated socially by worker opposition and acceptance, struggle and quiescence. Whereas the introduction of containers to the waterfront foreclosed on some possibilities for political struggle as old work practices and traditions were sometimes swept away, automation simultaneously sparked new conflicts over job displacement, job flexibility, and job security. As containerization augured massive labor savings for the employers, the union made the question of staffing levels and the interchangeability of work assignments a focus for labor–management discussions throughout the longshoring industry. Whereas the steamship companies and contracting stevedores stressed that containerization's maximum social benefit—presented publicly in terms of lower costs to the consumer but measured privately, as ever, by employer profit margins—could only be obtained through greater flexibility of work assignments and the shedding of excess labor, the ILA, fearing that containerization would idle thousands of waterfront workers and devastate waterfront communities that had relied for generations upon work on the piers, sought job security for those who remained in the industry while at the same time it tried to ensure that displaced dockers were provided for financially. The cost of automation, the union maintained, was not going to be borne by its members alone.[6] While accepting that they could not prevent the employers from adopting the new technology, beginning in the late 1950s ILA officials and rank-and-filers fought hard for protective contract clauses to minimize work dislocation and its impacts on waterfront workers, clauses that, as we shall see below, would dramatically shape the evolving geography of the longshore industry and the location of cargo-handling work.

Implications of the Changing Geographical Logic of Cargo Handling

The spread of containerization during the 1950s and subsequently is significant, then, for three reasons. First, it meant that many fewer dockers were required to do traditional waterfront work. By the mid-1990s, for example, only about 2,500 dockers still worked regularly in New York (see Phillips & Whiteside, 1985, for a comparison of containerization's impacts on employment levels in British ports). Consequently, during the next four decades job loss would be a continual union concern in contract negotiations with the employers throughout the industry. Second, the fact that the new technology was first adopted in New York meant that it was New York dockers who were the first to fight for and gain employment preservation agreements designed to minimize job displacement. However, union officials quickly realized that implementing restrictive job preservation agreements solely in New York would simply encourage employers to ship through other ports where similar protections did not exist. As a result, the union increasingly sought to develop a new geographical scale of contract bargaining by imposing on the industry a national agreement containing job preservation provisions that would replace the system of port-by-port negotiating that had historically characterized bargaining in the industry (we shall return to this geographical struggle in the next chapter).

Third, the new technology brought with it the possibility for the greater geographical mobility of work traditionally confined to the waterfront. Whereas prior to containerization each individual piece of cargo had been handled at the piers, now this was the case for only the containers themselves. This meant that the more labor-intensive work of cargo stuffing and stripping could now be done at cheaper locations inland, perhaps even at nonunion freight terminals. Not only were dockers thus facing job loss as a result of automation, but what work remained might also increasingly be relocated away from the waterfront. Accordingly, rank-and-file dockers and union officials began to push for contract terms that would preserve work *at the waterfront* in the face of the new technology. By so doing, the ILA sought to halt the growth of off-pier consolidating terminals that, it claimed, were taking work away from East Coast waterfronts. In securing certain work rules, the union hoped to force shippers and manufacturers to change their patterns of cargo distribution and warehouse utilization, thereby maintaining work at waterfronts up and down the East Coast.

WORK RULES AS GEOGRAPHICAL PRAXIS

Early Restrictive Practices

The ILA's first attempts to confront the employers' introduction of containerization and to challenge the potential growth of off-pier stuffing and stripping came in November 1958 when New York dockers forced arbitration after refusing to handle shipper-stuffed containers. Charging that carriers were actively encouraging the delivery of cargo in prepacked containers, the ILA argued that its contract allowed the union to boycott containers used by companies that had not engaged in container operations before October 1, 1956, the date on which the union's contract had gone into effect (Anon, 1973). This stipulation effectively exempted Sea-Land, the pioneer of containerization, because it had been using containers before this date (Ross, 1970). For several weeks many containers were held up in New York. ILA leaders argued that because it was the ocean carriers who owned and made the containers available to their customers, the NYSA was in a position to determine the conditions under which shippers, consolidators, truckers, and warehousers used them. The NYSA, however, maintained they were powerless to tell shippers how to send their goods and that cargo had to be handled in whatever fashion it arrived at the piers (Jensen, 1974). Negotiations continued for many weeks without agreement. As a result of this stalemate, the two parties entered the 1959 contract negotiations with diametrically opposed positions.

Initially the union demanded the right to stuff and strip *all* containers *at the pier*. This, union officials argued, would maintain traditional cargo-handling work jurisdictions, particularly between itself and the International Brotherhood of Teamsters, which represented many off-pier warehouse workers. The ILA argued that the convenience that containers provided shippers in easing freight transportation should not be translated into job losses for dockers. The NYSA, on the other hand, contended that in order to take full advantage of this new technology the movement of containers had to be allowed without any union hindrance (Federal Maritime Commission, 1973). Deadlock resulted and dockers idled the piers. The subsequent work stoppage ended only after arbitration and a "cooling off" period ordered by President Dwight D. Eisenhower under provisions of the 1947 Taft–Hartley Act.[7] Finally, in December 1959 the contract was settled with an agreement that the ILA would recognize any NYSA member's right "to use any and all type of containers without restriction or stripping by the union" (ILA/NYSA, 1959). In turn, the NYSA conceded that ILA labor would conduct "any work performed in connection with loading and discharging of containers for employer members of NYSA . . . in the Port of Greater New York." Additionally, the NYSA agreed to maintain the ILA's gang size at 20 (plus the gang boss) and to pay royalties on containers stripped and stuffed away from the pier by non-ILA labor, although it was only after further lengthy arbitration that the amounts of the royalties were decided upon (Anon, 1973).

Despite this settlement, the growth of off-pier consolidation facilities during the early 1960s greatly intensified conflict between the ILA and the employers' association, and strife in the form of wildcat "quickie" strikes continued, largely because of the two sides' differing interpretations regarding the precise nature of the provisions laid out in 1959. The ILA claimed that its agreement with the NYSA had only permitted the free movement of containers holding cargo owned by a *single* shipper or consignee (called "full shipper loads" [FSLs]) and that therefore the union still retained the right to stuff and strip at the pier containers holding less-than-container load (LCL) or less-than-trailer load (LTL) cargo—that is, cargo belonging to more than one shipper or consignee.[8] The NYSA, however, argued that the 1959 settlement, in addition to allowing FSLs to move through the port unhindered, permitted the free movement of LCLs/LTLs provided that royalty payments were made to the union for all such containers handled away from the pier by non-ILA labor. Bowing to union pressure, in 1962 the NYSA conceded to the ILA the right to work containers supplied by its members to consolidators for stuffing and stripping in the Port of Greater New York (Anon, 1973).[9] This extended the terms of the 1959 settlement—in which the ILA as opposed to Teamsters working at off-pier trucking terminals had won the exclusive right to stuff and strip containers for NYSA members—to cover carrier-owned and leased containers provided to consolidators who were themselves not necessarily members of the NYSA.[10] Such a concession, however, did not resolve issues to the union's satisfaction, particularly given the potential for much greater job losses augured by the introduction in 1967 of the first fully containerized ships specifically designed and built to carry large containers.[11]

By the late 1960s the struggle to implement container-handling rules had also become inextricably tied up with the ILA's effort to develop a uniform national contract to replace the port-by-port contracts that it had traditionally negotiated with the employers along the East Coast (see Chapter 5 for more on this struggle). Containerization had now spread well beyond New York, and dockers in other ports wanted protections against job displacement similar to those that New York waterfront workers were seeking to impose on the NYSA. At the same time, New York union officials worried that the implementation of any container handling rules solely in New York would encourage shippers and ocean carriers to turn to other nearby ports such as Philadelphia and Baltimore to avoid the restrictive provisions. Consequently, in July 1967 the union began demanding that ILA members again reserve the right to stuff and strip *all* carrier-owned and leased containers *at the pier* and that employers from Maine to Texas be forced to adopt as part of a national contract the container-handling rules negotiated in New York. When its contract expired in October 1968, the union called a strike from Maine to Texas in support of its claim (*Journal of Commerce*, October 1, 1968). In response, the NYSA returned to its 1959 position and demanded the ILA remove all restrictions on the free movement of all types of containers. The NYSA was particularly keen to preserve the unfettered movement of FSLs that the 1959 settlement had guaranteed and that were moving through the port in ever greater numbers. Other East Coast employers, meanwhile, were equally keen to avoid implementation of container handling rules in their ports altogether.

The "Rules on Containers"

After a national strike, in January 1969 the ILA finally forced the NYSA to adopt an agreement known as the "Rules on Containers," sometimes referred to as the "50-mile rule." The most crucial element of this agreement concerned the stuffing and stripping of consolidated full container loads. The Rules provided that any *consolidated* container coming from, or destined for, points less than 50 miles from a North Atlantic port that would normally be stuffed and stripped by employees of consolidators at off-pier warehouses would now have to be worked instead at the piers by ILA labor (see *Journal of Commerce*, January 14, 1969, for the specific terms of the agreement).[12] If such a container arrived at the pier prepacked, ILA members were entitled to strip it and restuff the contents into a different container. In such instances, the ocean carriers were responsible for ensuring that containers were not accepted for transportation or released to consignees until the required stuffing or stripping had been done. Most commentators accepted that the Rules would preserve for the union the work of stuffing and stripping only about 20% of containers moving through North Atlantic ports, with the remainder proceeding intact. The union was empowered to exact a $250 fine (later increased to $1,000) for each container scheduled to be stuffed or stripped at the pier but which the ocean carriers allowed to pass through the port in violation of the Rules. The carriers also had to pay a royalty for FSL (i.e., full nonconsolidated) containers moving over the piers. In addition to the penalties, the ILA demanded the inclusion of an "open-end" clause permitting the union to renegotiate the Rules and expand their coverage if it could show either that firms were shifting their operations beyond the 50-mile radius (see Figure 4.1) to evade the Rules or if the restrictive provisions were failing to preserve adequate work at the waterfront.[13] Should the latter prove to be the case, the union reserved the right to abandon the Rules and resume stuffing and stripping *all* containers at the waterfront. Seeing this as a potential setback to the 1959 agreement they had secured for FSLs, the NYSA had strenuously resisted the union's

FIGURE 4.1. Map showing extent of 50-mile zone for ports of New York City; Philadelphia, Pennsylvania; and Wilmington, Delaware.

demands for this clause. Only when the financial costs of the 1968–1969 strike became unbearable did they finally capitulate (Haynes, 1977).

Although agreement on the Rules was reached in New York on January 12, 1969, in their effort to force national contract negotiations on the issue of containerization and job preservation ILA leaders refused to submit the agreement for ratification by New York dockers until ports from Maine to Texas had accepted contracts with the same container clauses as New York.[14] The employers' associations in these ports refused, however, to consider including the container-handling rules as a part of a uniform national contract, and so the strike dragged on. Only on April 14, 1969, was agreement finally reached in the last holdout ports in Texas that the Rules would be included as part of local contracts along the coast (rather than as part of a national agreement). The strike had lasted 57 days in New York (the lengthiest in the port's history until that date) and had cost local employers nearly $75 million in lost business (Board of Inquiry, 1968). It had lasted over 100 days in some other Atlantic and Gulf Coast ports.

The 1968–1969 strike's resolution was significant for several reasons. First, whereas the union was unsuccessful in its effort to include the Rules in a national "master" agreement, it did force North Atlantic, South Florida, and West Gulf ports' employers' associations to adopt as local contract items the same container-handling rules as New York had implemented, though the ILA was unable to secure them in the remaining ports in the South Atlantic and the East Gulf. Many dockers, then, now enjoyed some measure of protection against containerization, although because of the union's failure to secure the Rules as part of a national contract not all enjoyed uniform protection. Second, although the North Atlantic version of the Rules was not written into all local contracts from Maine to Texas, the strike's settlement did nevertheless mark the first time that container-handling practices developed in one region represented by the ILA were implemented in other ports. This was no mean feat in the face of such entrenched employer opposition. Third, the 50-mile radius had clearly been chosen on the basis of conditions prevailing in New York, particularly the pattern of LTL movements in the port's hinterland, but it formed the standard for other ports. The use of New York as a template for contracts elsewhere, however, exacerbated other North Atlantic ports' employers' growing concerns that their own contract conditions were too frequently shaped by the agreements negotiated by the union's national leadership and employers in New York. This concern would soon culminate in the formation of a new multiport bargaining agent for the North Atlantic ports from Maine to Virginia, the Council of North Atlantic Steamship Associations (CONASA) (this development is covered in more detail in Chapter 5).

The result of the strike, then, was that dockers in many East Coast ports had forced their respective employers to include the Rules in one form or another in their local contracts. Through such success the union hoped that work would be preserved at waterfronts up and down the coast. Yet, conflicts between the ILA, ocean carriers, consolidators, and shippers over containerization continued. The union claimed that carriers were seeking to evade the Rules both by supplying containers to off-pier consolidators and by structuring their tariffs to favor off-pier consolidators (International Association of Non-Vessel Operating Common Carriers, 1981). Indeed, since it was the carriers who had to bear the cost of any stripping and restuffing done at the pier, they had little incentive to enforce the Rules stringently.[15] Furthermore, the fact that South Atlantic and East Gulf employers had not adopted the Rules deeply concerned ILA negotiators. Fearing that consolidators in these regions would be able to evade the Rules' work preservation provisions, in 1971 the union began a two-month-long national strike to force port employers' associations to accept them as a national contract item. This strike's success led to the Rules' inclusion in the 1971–1974 master agreement negotiated between the ILA and the newly constituted CONASA. (This agreement, though negotiated for the North Atlantic ports from Maine to Virginia, also served as the basis for agreements in most ports from North Carolina to Texas, thereby effectively making them nationally applicable.)

The Dublin Agreement

Despite their inclusion in the ILA's master contract agreement with the CONASA, the union continued to claim the Rules were failing to adequately protect longshoring jobs from the effects of containerization. In New York the ILA's principal complaint was directed against several ocean carriers which, the union claimed, were persisting in providing containers to off-pier consolidators in violation of the Rules (Anon, 1973: 32–34). In other ports the controversy centered on "short-stopping" FSLs, a practice whereby truckers would pick up FSL containers at the pier and haul them to local warehouses or trucking stations (*International Longshoremen's Association v. National Labor Relations Board*, 1979). Upon arrival, the freight would be stripped and reloaded to meet over-the-road trucking weight and safety regulations or delivery needs unrelated to maritime transportation requirements. The union claimed, however, that short-stopping contravened the Rules on Containers, which stated that *any* stuffing and stripping conducted within 50 miles of the waterfront by workers other than employees of the FSLs' beneficial owners or consignees had to be done at the waterfront by ILA members. Invoking its right to abandon the Rules, the ILA refused to handle all containers stuffed or stripped anywhere inland unless the ocean carriers who owned and leased them to their customers ensured that these perceived violations were ended. Anticipating a costly strike over the issue, the CONASA met with ILA representatives in Dublin, Ireland, in January 1973. The resulting agreement, the CONASA-ILA Container Committee Interpretive Bulletin No. 1 (known as the "Dublin Agreement"), clarified two important points that had been the source of greatest conflict between the union and the employers since the Rules' initial adoption (see Federal Maritime Commission, 1987, for a full listing of the terms of agreement).

First, the Dublin Agreement declared that the Rules applied to *all* containers *owned or leased* by NYSA members that were to be stuffed and stripped within 50 miles of any ILA-controlled port unless that work would normally have been done by employees of the cargo's beneficial owner.[16] Under the 1968 and 1971 contracts, FSLs had been exempted from the Rules, even if destined for, or coming from, points less than 50 miles from the port. Now, for the first time, FSL containers scheduled to be short-

stopped were covered by the ILA's contract. This was a significant victory
for the ILA because it allowed the union to regulate short-stopping by giving
dockers authority to treat all short-stopped FSL containers as consolidated
loads that they then had the right to stuff and strip at the pier. Such authority, the union averred, would remove a potential means by which shippers
could evade the Rules, namely, hauling containers over the piers and working them perhaps only a few hundred yards from the waterfront. Because
some warehousers had traditionally loaded and unloaded stored cargo for
reasons unrelated to marine transportation, the ILA agreed to exempt those
FSL containers that were to be warehoused for at least 30 days within the
50-mile limit. The 30-day provision represented the union's attempt to dis-
tinguish between what it called traditional bona fide warehousing functions
and warehouses that were simply used as drop-points where Teamsters immediately stripped containers and loaded their contents into trucks (for
more on such efforts to delineate geographical lines of work jurisdiction,
see Herod, 1998c).[17] Containers coming from, or destined for, points farther
than 50 miles from ILA-organized ports were still exempt from ILA restrictions and were allowed to pass over the piers intact.

   Second, the agreement allowed the ILA to maintain a list of consolidating firms and terminals that operated in violation of the Rules and to whom
ocean carriers were to be prohibited from supplying their containers.[18] As a
result, such firms would no longer be able to use containers provided by
carriers but would be forced instead to purchase or lease containers from
elsewhere if they wished to continue delivering cargo to the piers. The
added cost of leasing a container (as much as $300 for a 20-foot container
on a one-way North Atlantic crossing), the union believed, would greatly reduce consolidators' competitiveness relative to "qualified" shippers and consignees, to whom the carriers provided containers free of charge (*New York
Shipping Association v. Federal Maritime Commission*, 1988: fn.3). This, the
union hoped, would encourage manufacturers to send all LCL/LTL cargo to
waterfront facilities in uncontainerized form, in the same manner in which
break-bulk freight had traditionally been sent.[19]

Challenges to the ILA's Geographic Vision

Making landscapes is a process run through with struggle, for the production of space is a highly contested social activity. The ILA's efforts to preserve work for its members were clearly struggles over the genesis of the
new economic geography of the longshoring industry in the age of automation. By enforcing the Rules on Containers, the union intended to carve out
both a *space* and a *scale* (defined by the 50-mile radius) within which it
could regulate container-handling work. The Rules were the union's attempt
to impose a geographical solution to the problems of job displacement that

its members faced as a result of containerization. For the ILA, they were a means to mediate geographically the new technology's impacts and to control the geographical restructuring of traditional work patterns which, the union claimed, containerization and the growth of off-pier consolidating terminals were causing. However, while the ILA struggled to implement one vision of the new geography of cargo handling for the industry, off-pier consolidators and their Teamster warehouse employees sought to create a rival and quite different geography of work. Whereas the ILA maintained that the Rules were a legitimate attempt to preserve work for its members, opponents argued they were nothing but a thinly disguised and unlawful attempt to acquire work to which the union was not entitled. Citing cases in which shippers had sent stuffed, nonexempt containers to the pier, only for ILA dockers to strip and restuff the cargo into a different container at a cost of several hundred dollars, opponents also argued that the Rules were economically prohibitive.[20] Further, they complained that in such instances cargo delivery was frequently delayed several weeks while, because of loose packing, ILA dockers invariably restuffed cargo into a greater number of containers than had been required to bring it to the pier, resulting in higher transportation costs (*Container News*, 1969; Ullman, 1983).

The crux of the argument over job preservation versus job acquisition concerned the reconstruction of rival work histories and their spatial inscription in different geographies of work. Two principal traditions were articulated by opponents and supporters of the Rules to lay claim to the work of stuffing and stripping containers away from the pier. These derived their origin both from contested notions concerning the functional nature of a container and from a basic disagreement over the historical and geographical development of the consolidating industry. Whereas the ILA and various port employers' associations maintained that a container was the functional equivalent of a ship's hold and thus that dockers were entitled to work them, the Teamsters and the American Trucking Associations (whose members are state trade associations representing interstate trucking interests) argued that a container hooked onto a truck's trailer was instead the equivalent of an ordinary tractor-trailer tandem and that, consequently, truckers and warehouse loaders had jurisdiction over them.[21] Furthermore, trucking interests argued that, because there was usually insufficient export cargo for most long-haul trucking operations to fill their trailers on any given trip to waterfront areas, historically they had brought a mixture of export and domestic freight to ports. Upon arrival, the domestic freight was distributed to local customers and the export freight was consolidated into containers at trucking terminals ready for loading aboard ship (Lytel, 1969). Allowing ILA dockers to do this work, critics suggested, would rob Teamsters of their traditional jurisdiction over these container-stuffing jobs, while off-pier trucking terminal operators would lose business to waterfront facilities.

The Rules' defenders, while conceding that the consolidating business per se predated containerization, sought to show that its *recent growth* had been occasioned by the use of containers. It was the impact of this growth on dockers' waterfront employment opportunities that defenders sought to reverse with the Rules. The ILA stressed this was not merely a jurisdictional squabble concerning whether itself or the Teamsters were entitled to work the cargo in question. Rather, ILA officials claimed, it was fundamentally about controlling the geographic location of container-handling work and restricting it to the waterfront. Even if off-pier warehouses within the 50-mile limit were staffed *by ILA members*, the ILA averred, the Rules required the work to be done instead at waterfront terminals (ILA, 1980; see also Gleason, 1973: 51). For the ILA, the issue was preserving work for its members *at the waterfront*, even to the extent of denying it to ILA members who might be working as warehouse employees farther inland.[22] Fearing that the Rules' rescission would lead to further costly waterfront strikes, the NYSA supported the union's claim.

Consolidators countered ILA claims by arguing that, because many of them had operated before the advent of containerization, they had a tradition independent of longshoring work (Ullman, 1983). Together with trucking and warehouse interests, and the unions representing workers in these industries, freight consolidators took the position that the Rules were simply an attempt by the ILA to acquire work historically done at off-pier facilities. For opponents of the Rules, it was not the contemporary growth of consolidating that should define the argument, but, rather, its long history. Although many individual consolidators' businesses had indeed been founded on the basis of containerization, these groups claimed that the consolidating business itself predated containers and that truckers and warehousers had historically handled cargo at off-pier facilities. In fact, the very structure of the consolidating industry, they maintained, meant that it was inefficient for them to operate at crowded waterfront facilities (Ullman, 1968). If consolidators were forced to locate at spatially cramped waterfronts, traffic congestion would greatly increase because containers stuffed at one pier might have to be trucked to a ship docked at a different pier elsewhere in the port. By locating away from the waterfront, consolidators maintained, they could dispatch containers directly to specific piers for loading aboard ship and avoid much of this congestion. Seeking to undermine the legitimacy of the ILA's claims, several consolidators recounted instances in which shippers and carriers had paid union members to allow containers that were subject to the Rules nevertheless to pass over the piers intact (Jacobs, 1973). If dockers had not even carried out the pierside stuffing and stripping to which they claimed entitlement under the Rules but had instead allowed these containers to move through the port unopened, how, opponents asked, could the ILA seek to articulate a tradition of working consolidated containers at the waterfront? It appeared, critics of the Rules argued, that the ILA was merely seeking to claim work to which, based on precontainer lines of demarcation between dockers, truckers, and warehousers, it simply was not entitled.

For the ILA, efforts to articulate such a tradition throughout the East Coast as part of a uniform national contract that laid claim to consolidating work were complicated considerably by the great variety of local work practices that developed in different ports prior to containerization.[23] (As we shall see in the next chapter, the problem of attempting to create national contracts in the context of varied local conditions has been a constant in the union's efforts to address the issue of containerization.) This was exacerbated by the fact that the geographical restructuring of the waterfront's industrial structure stimulated by containerization further complicated attempts to trace discrete traditions. Labor-intensive break-bulk methods of loading and unloading ships had traditionally required the storage of large quantities of cargo at waterfront facilities, but the speedup brought about by containerization now made the storage of many goods at dockside warehouse sheds practically obsolete (Chilcote, 1988). The elimination of large quantities of waterfront cargo work, together with the greater geographical mobility of that which remained, stimulated intense conflicts over the extent to which traditional waterfront work had been destroyed in situ relative to its loss caused by the movement to off-pier locations.

These, then, were the debates swirling around the contentious issue of work jurisdiction in the face of containerization. As might be expected, sooner or later the issue would come to be settled in court. In 1973 two New Jersey consolidators (Consolidated Express, Inc., and Twin Express, Inc.), who had been denied access to containers by several carriers, filed separate unfair labor practice charges with the National Labor Relations Board's (NLRB) 22nd Region (Newark, New Jersey) in an effort to have the Rules in New York struck down.[24] The consolidators claimed that the ILA's refusal to allow carriers to provide them with containers was a "hot cargo" agreement operating in violation of the Taft–Hartley Act's prohibition on secondary boycotts against third-party employers.[25] Initially an administrative law judge (ALJ) for the Board ruled that the ILA had not engaged in a secondary boycott and was using the Rules as a valid form of work preservation (*International Longshoremen's Association*, 221 NLRB 956 [1975]).[26] In response, the consolidators appealed the decision to the full Board, which overruled the ALJ and determined that the ILA was indeed seeking not to *preserve* traditional work but, rather, to *acquire* work from others, namely, Teamsters and consolidators.[27] After much subsequent legal wrangling the Rules became the subject of two U.S. Supreme Court decisions.

In its first decision on the matter in 1980, the Supreme Court determined that the NLRB had erred as a matter of law in its definition of the work in controversy when it ruled that the ILA was seeking to unlawfully obtain work that belonged to other workers. The Court argued that, by focusing on the off-pier stripping and stuffing conducted by truckers and consolidators *after* containerization's introduction, rather than on the work traditionally performed by the ILA, the NLRB had failed to "examine the relationship between the work as it existed before [containerization] and as the agreement proposes to preserve it."[28] The Court overturned the NLRB's decision and ordered the Board to reexamine whether the Rules constituted a legitimate means for the union to preserve work in the face of technological displacement. In 1985 the Supreme Court was asked to decide whether the NLRB had been correct to determine that the Rules, when applied to containers used by short-stopping truckers and traditional bona fide warehousers, represented an illegal secondary boycott under the National Labor Relations Act. The NLRB had earlier adjudicated that, because the Rules sought to preserve longshoring work which had been eliminated by containerization, they had an illegal work acquisition objective. The Court, however, ruled that agreements negotiated to preserve work that duplicated that eliminated by automation did not constitute unlawful work acquisition if the union's *primary* purpose was to protect its members' jobs rather than to enforce a secondary boycott.[29] Finding the NLRB's actions inconsistent with its 1980 decision, the Court held that the Board had failed to concentrate solely on the matter in question (namely, the Rules as a form of work preservation) but had instead focused on their *effects* upon short-stopping truckers and warehousers. The Court dismissed these effects, no matter how severe, as irrelevant to the issue of whether the Rules themselves constituted an illegal secondary boycott and reaffirmed the ILA's right under federal labor law to enforce them.

Although the Rules' validity had been secured under federal labor law, they still had to face scrutiny under maritime law. In particular, opponents argued that, because the Rules resulted from collective bargaining agreements rather than from the nature of container transportation needs, incorporating them into transportation tariffs violated the 1916 Shipping Act's proscriptions against inhibiting competition.[30] In particular, they maintained that the Rules unfairly discriminated against shippers and consolidators who were not party to the ILA collective bargaining negotiations but who were, nevertheless, obliged to abide by the Rules because they used containers owned or leased by the carriers. Two elements of the Rules' provisions were emphasized as evidence of discrimination. First, shippers located within the 50-mile limit were treated differently than shippers sending the same items from outside the limit because the latter's containers were not subject to stripping and stuffing by ILA labor at the pier. Second, whereas exporters packing cargo at their own warehouses were not subjected to the Rules— private warehouses had been exempted—exporters using public warehouses were obliged to follow their prescriptions. After a series of hearings on the matter, the Federal Maritime Commission (FMC) determined that under maritime law tariffs structured to include the Rules were indeed discriminatory and therefore unlawful, regardless of the fact that they may have resulted from an attempt to incorporate a lawful work preservation rule under labor law (Federal Maritime Commission, 1978). Although the ILA appealed this ruling, ultimately the union lost its case and, following three decades of litigation, in February 1989 the Rules were finally abolished.[31]

Although my purpose here has been to show how the ILA used the Rules as a form of "spatial sabotage" in the face of new technology, the Rules' abolition in 1989 was certainly not the end of the union's attempt to control the location of work so as to mitigate the impacts of containerization on its membership. Within two months of the FMC's decision, the ILA had negotiated a new agreement to preserve work in those ports—principally North Atlantic ports such as New York, Baltimore, and Philadelphia—most affected by the Rules' negation. Under this new compact (called the Container Freight Station Program [CFSP]) employers representing all the major ports along the Atlantic and Gulf ports agreed to create a number of container freight stations in which cargo-handling work would be done by ILA members. Through a combination of liberalized ILA work rules and the subsidizing of dockers' wages by employers, the program was designed to attract to the waterfront jobs that had moved to inland warehouses. As with the Rules on Containers, so, too, with the CFSP, the ILA has sought to manipulate the geography of local labor markets to keep work from migrating inland and away from the waterfront. In so doing, the union has continued to shape the local economic landscape (for more on the CFSP, see *Journal of Commerce*, June 27, 1989).

THE RULES AS DOCKERS' SPATIAL FIX

Whatever else they may have been, legal conflicts over the Rules' enforcement were fundamentally struggles over the production of new industrial geographies in response to the restructuring of traditional patterns of employment conditioned by technological innovation. Although they were ultimately ruled illegal under maritime law, during the period of their implementation the Rules nevertheless had significant impacts upon the evolution of both waterfront and hinterland industrial landscapes. Certainly the history of litigation—suit and countersuit, injunction enforcement and dismissal—means that in some ports the Rules were enforced intermittently during the two decades after their inception, thereby making many of their impacts on consolidators and others also intermittent. In several instances, for example, consolidators developed strategies either to conform to, or to evade, the restrictive provisions of the ILA's work agreements but put these into action only during periods when the Rules were being strictly enforced, merely to abandon them upon their enjoinment by a new court injunction. Whereas such responses may thus have had ephemeral (though nevertheless significant) impacts on the geography of work, others have had more lasting impacts on the landscape, resulting in the permanent shutdown of off-pier facilities, changed work locations, and cancelled plans for business expansion.

Because of the limitations of using large-scale aggregate data to examine the intricate geographical responses to, and impacts of, the Rules, in what follows I have adopted an "intensive" (see Sayer, 1984) case study approach both to illustrate how the Rules directly impacted the waterfront landscape but also to examine how some consolidators sought to avoid their provisions, avoidances which themselves led to the creation of new geographies of work. Although under its contract the ILA was entitled to extend the Rules to include consolidators who moved to avoid the handling restrictions, in practice consolidators were often able to keep one step ahead of union investigators. Indeed, union investigators, truckers, and consolidators played a game of cat and mouse that was extremely sensitive to local waterfront geographies, a game that often saw dockers and ILA officials following trucks out of port areas to see to which terminals they were hauling containers, while simultaneously truckers sought to lose the ILA investigators in traffic or in the labyrinthine backstreets near the piers. Four sets of case studies are examined to demonstrate how the ILA's political practice has shaped the production of the contemporary economic geography of the longshoring industry and the industrial landscapes of port waterfronts and hinterlands all along the East Coast. Specifically, these case studies analyze the termination of operations conducted in violation of the Rules, consolidators' cancellation of plans for business expansion, hauling freight beyond the reaches of the 50-mile provision and/or rerouting freight to nonunion ports, and the geographical reorganization of distribution systems by consolidators' customers.

Termination of Operations

In 1969 Mahon's Express, a New Jersey trucking business, operated three off-pier consolidating terminals in Newark, New Jersey, these being located at Jabez Street, Commercial Street, and the Port Authority Truck Terminal of Newark on Delancy Street (Mahon, 1984). The company had been founded in 1905 and was a family-owned corporation serving the New Jersey, New York, and eastern Pennsylvania hinterland. At the time, Mahon's employed about 100 people, most of whom were members of the International Brotherhood of Teamsters Local 478, and counted F.W. Woolworth and S.S.

Kresge (predecessor to K-Mart) as two of its oldest customers. Among other services, Mahon's acted as a consolidator for Woolworth and Kresge, containerizing freight at its off-pier warehouses. After stuffing, Mahon's would transport these containers to Port Newark for shipment to Woolworth and Kresge retail stores in Puerto Rico, the Virgin Islands, and St. Thomas.

Until 1971, the ILA had not attempted to interfere with Mahon's operation. In that year, however, a number of containers that had been stuffed at Mahon's terminal facilities were, upon arrival at the pier, stripped by ILA labor and their contents restuffed into other containers. On several subsequent occasions ILA dockers rehandled containers stuffed by Mahon's in this manner, although only on a sporadic basis. In 1975, however, Puerto Rican Marine Management (PRMM), an ocean carrier, refused to supply Mahon's with containers for its off-pier consolidating terminals because its facilities were operating in violation of the Rules on Containers, which the carriers were now enforcing more stringently as a result of the Dublin Agreement. The carrier informed Mahon's that the company would only be supplied containers if it brought them to PRMM's waterfront terminal for stripping and restuffing by ILA labor. Each year during the late 1960s and early 1970s, prior to the Rules' more strict enforcement, Mahon's had consolidated and delivered to the piers approximately 2,000 containers. With the stricter application of the Rules, however, this number fell to fewer than 400. As a direct result of this decrease in business, Mahon's ceased its operations at both the Commercial Street and the Port Authority Truck Terminal facilities (Mahon, 1984).

Although the Rules' provisions concerning pierside stuffing and stripping had been primarily directed at consolidators, ocean carriers were also forced to close off-pier stuffing and stripping stations. In the early 1960s, even before the Rules on Containers' official adoption, the union had forced Sea-Land to close an off-pier terminal in Jacksonville, Florida, and transfer that work to a dockside facility to avoid container rehandling by the union (*International Longshoremen's Association Local 1408*, 245 NLRB 1320 [1979]). In the New York region the Rules' implementation also forced carriers to shut down their own off-pier consolidating terminals. For a number of years, Seatrain had conducted LTL consolidating operations at off-pier facilities located in Secaucus and North Bergen, New Jersey. These facilities received small export LTL shipments for stuffing into Seatrain containers that were then sent to the carrier's main terminal in nearby Weehawken for loading aboard ship. The ILA had long complained that these facilities were siphoning work away from the waterfront. In response to union threats to enforce the contract's stripping and restuffing provisions against containers coming to the waterfront in violation of the Rules, in March 1973 the company closed its two off-pier warehouses. Seatrain officials explained that their decision to close these warehouses had been shaped by the more stringent enforcement of the Rules that the just-negotiated Dublin Agreement would bring (Novacek, 1973).[32] Realizing that it would no longer be able to operate these warehouses without facing huge ILA fines that the Dublin Agreement now made increasingly likely, Seatrain subsequently accepted LTLs only at its waterfront facilities.

The ILA had also been concerned about one of Sea-Land's off-pier stuffing and stripping facilities located in Manhattan. Like Seatrain's Secaucus and North Bergen terminals, this, too, received small quantities of LTL cargo for consolidating before loading aboard ship. In preliminary negotiations prior to the signing of the Dublin Agreement, the ILA had specifically insisted that this particular Sea-Land terminal be shut down because its operation was not in full accordance with the Rules (NYSA, 1972). As a result of such pressure, Sea-Land officials closed the terminal in March 1973 and began accepting LTL cargo only at the company's waterfront facility in Port Newark for stuffing and stripping by ILA members (Dickman, 1973; Anon, 1973).

Lost Business and Cancelled Plans for Expansion

Perhaps one of the most difficult tasks in evaluating the impacts of any set of restrictive work practices is to attempt to determine what would have happened had they *not* been implemented. Most contemporary social science, certainly that which draws inspiration from a positivist tradition, is primarily concerned with measuring events that have already occurred as the result of particular decisions or sets of circumstances. In the current context, for instance, it would generally accept the recording of the closure of specific warehouses as a valid form of evidence by which to measure the impacts on the landscape of the Rules' enforcement. Given this concern to measure and catalogue, however, positivist social science by its very nature excludes from its purview consideration of events that, equally, did *not* occur because of a particular set of decisions or circumstances. Yet, examination of how events are prevented from taking place can be as revealing, if not more so, of the power relations that shape the landscape as can examination of those that did occur. This is not meant to suggest that researchers should engage in mere "counterfactual" speculative flights of fancy. Rather, it is a recognition that concrete examples of events that did not occur (such as the failure to open off-pier warehouses) are also valid forms of evidence. Indeed, evidence from documents and testimony shows that at least two companies shelved plans to expand their operations as a direct result of the loss of business caused by the application of the ILA's restrictive work practices.

The first of these two companies whose expansion plans were affected

by the Rules—Twin Express—began operation in 1967. The company consolidated LTL freight (using members of Teamsters Local 707) at its off-pier
Manhattan terminal for shipment to company facilities in Puerto Rico, work
claimed by the ILA (*International Longshoremen's Association*, 221 NLRB
956 [1975]). Although Twin itself did not belong to the NYSA, it utilized containers supplied by two members, Sea-Land and Transamerican Trailer
Transport (TTT). Less frequently the company used Seatrain, also a NYSA
member. For a brief period in 1968 and again in 1971 the ILA stripped and
restuffed at the waterfront those containers coming from Twin's terminal. In
early 1973, however, Twin was listed by the NYSA–ILA Container Committee as one of 14 consolidators operating in violation of the Dublin Agreement. The ILA coerced Sea-Land and Seatrain to cease providing containers
to Twin because of the union's labor dispute with the company over off-
pier stripping and stuffing. At this time TTT, while not providing Twin with
its own containers, did help the company lease containers from local rail-
roads, although to conform to its agreement with the ILA the carrier then insisted upon stripping these containers at the pier. Whereas Twin had
steadily expanded its operation (both in volume of business and the number of containers shipped) since the company's inception in 1967, the loss
of business associated with the carriers' refusal to supply containers reversed this trend. As a result, Twin cancelled plans to lease a new terminal
in Maspeth, Long Island, and was forced to remain in its obsolete Manhattan
facility (Sanjuro, no date).

A second New Jersey consolidator, Jayne's Motor Freight, experienced
similar problems as a result of the Rules' provisions. Jayne's had been in existence since shortly after World War II and operated terminals at Elizabeth
and Vincentown, New Jersey, both of which were located inland from the
coast but within the 50-mile radius of ILA-organized ports (New York and
Philadelphia, respectively). Typically, importers had arranged for Jayne's to
pick up full truckloads of import cargo and haul them to the company's terminals for deconsolidation. Upon unloading this cargo, Jayne's employees
would either store it for subsequent distribution or would immediately deliver it to locations designated by their customers. Part of the company's
business also consisted of making small local LTL cargo pickups for consolidation at its terminals. Once consolidated into full loads, the freight would
be sent to the pier for transport to Puerto Rico. The refusal of the ILA during
periods of the Rules' enforcement to allow these containers to be loaded
aboard ship without stripping and restuffing at the pier led a number of
Jayne's customers to cease using the company because of the added cost of
shipment and delays caused by this rehandling. As a result, Jayne's lost
much business, leading the company ultimately to cancel plans for the expansion of its consolidation services (Hagemann, 1984).

Rerouting Freight/Hauling beyond the 50-Mile Radius

Although illegal under the provisions of the ILA's collective bargaining agreement, consolidators developed several techniques to evade the Rules' provisions during periods of their enforcement. A strategy used by a number of consolidators on export shipments, mostly to Puerto Rico, was to reroute freight to ports employing non-ILA labor for transport. One such consolidator was Dolphin Forwarding. Founded in 1964, Dolphin provided service between New York and Puerto Rico and the U.S. Virgin Islands. Typically, it received goods at one of several off-pier facilities located within 50 miles of New York. Once manufacturers delivered the cargo to Dolphin's warehouses, it was stuffed into containers either by nonunion employees or by Teamsters. These containers were obtained from ocean carriers, from container pools at local railroad yards, or directly from leasing companies. The company was not subjected to the Rules until late 1974 and early 1975, when steamship companies began refusing Dolphin access to booking space on ships leaving New York and Newark. In January and September of 1975 Dolphin was also fined a total of $11,000 by a Puerto Rican carrier for alleged Rules' violations, fines that Dolphin refused to pay (Lee, 1980). As a result of this refusal by carriers to provide booking space for its containers, Dolphin chose to transport its containers overland (at an additional per-container cost of some $1,400) to Jacksonville, Florida, from where they were shipped to Puerto Rico by a steamship company that did not employ ILA labor and was therefore not subject to the Rules.[33] A measure of the Rules' significance for shaping the actions of shippers—and hence the geography of the industry—is the fact that Dolphin returned immediately to shipping containers through New York during times when the Rules were enjoined from operation by court order.

A second company that used the same approach as Dolphin was Mahon's Express (Mahon, 1984). During the periods when the ILA refused to handle containers consolidated at its terminal, Mahon's stuffed export cargo into railroad truck-train trailers for delivery by rail to Jacksonville. Upon arrival, the intact trailers would be loaded onto a barge for ocean transport to Puerto Rico, after which they were picked up at the pier for delivery to Mahon's customers' stores. Whereas some consolidators shipped through Jacksonville, others turned toward Canada to evade the Rules (Ullman, 1983).

While the practice of rerouting cargo through Jacksonville mostly applied to containers destined for export, to avoid ILA handling of import cargo some consolidators hauled containers to facilities beyond the 50-mile radius for stripping. The ILA's contract provided that, if containers were shown to have been hauled beyond the 50-mile limit for the express purpose of evading their work preservation agreement, the ILA was entitled to consider such facilities in violation of the Rules and could then insist upon doing the work. This was a difficult provision to enforce, however, particularly in cases where, to evade the Rules, consolidators and manufacturers hauled containers that were due to be stuffed and stripped at the piers to warehouses located perhaps several hundred miles outside the 50-mile limit—an apposite example of how the unintended consequences of actions subvert desired goals, in this instance the ILA's effort to bring work back to the waterfront. In March 1974, for instance, the McLean Trucking Company was found to be diverting containers obtained from the Dart Containers Lines (a member of the Hampton Roads Shipping Association) from McLean's Virginia warehouse to points outside the 50-mile limit for stripping prior to the movement of the cargo to its owner's place of business in Memphis, Tennessee (ILA, 1977). Rather than divert individual containers to points beyond the 50-mile limit, still other consolidators removed entire operations. United Freightways, for example, closed its consolidating operation in New York City and moved its stuffing and stripping facilities beyond the 50-mile radius (ILA, 1980).

Reorganizing Distribution Systems

The fourth set of vignettes concerning the Rules' impact on the geography of work examines how manufacturers and shippers reorganized their distribution systems.

D. D. Jones Transfer and Warehouse Company was a local hauler and warehouser operating four off-pier distribution warehouse facilities in the port of Hampton Roads, Virginia (McNeil, 1983). Upon arrival at these warehouses cargo was broken down and stored for periods of time ranging from one day to one year before delivery to its consignees. The warehouses also handled export cargo, which would either be picked up by Jones's employees or brought to them by the freight's owners. If several small lots were delivered, they would be consolidated into one shipment at the warehouses before being sent to the pier. The company operated as a public distribution warehouse primarily serving areas within 250–300 miles of its facilities. J.C. Penney and General Electric provided some of its largest accounts.

Unlike freight destined for private warehouses (i.e., those operated by the company owning the goods stored), freight headed for public warehouses came within the ambit of the Rules on Containers unless it fell under Dublin Agreement's 30-day proviso. The fear that valuable capital and goods would be tied up at Jones's warehouses by enforcement of the 30-day provision led many of the company's customers to reorganize their distribution systems in at least two fundamental ways. First, several customers ceased using Jones as their port area agent and distributor and instead had other trucking companies haul the containers carrying their goods to container facilities beyond the 50-mile limit. Second, some (such as the paint manufacturer Sherwin Williams) no longer used Jones's public warehouses but established their own private warehouses, which were not subject to the Rules, in the process restructuring the geography of warehouse work.

Mahon's Express of New Jersey had a similar experience with some of its customers (Mahon, 1984). Under the Rules, manufacturers who would normally have used off-pier consolidators located within the 50-mile radius were required to send their shipments to the piers in break-bulk fashion. However, due to the added expense involved in packing and transporting individual pieces of freight, together with the mountains of costly paperwork that would be needed to document shipments to dozens of different stores in Puerto Rico, this requirement was considered prohibitive by many manufacturers. To avoid these costs, during the periods of the Rules' enforcement several of Mahon's customers reorganized their distribution patterns by consolidating shipments at their own private facilities rather than sending LTLs to Mahon's terminals.

DOCKERS, THE ILA, AND THE PRODUCTION OF SPACE AND SCALE

Below I want to draw out some larger points from the vignettes presented above that help move forward the arguments about workers' spatial praxis and the geography of capitalism that I am seeking to develop. As we have seen, the ILA's struggle to implement the Rules on Containers was fundamentally a struggle to control the location of container-handling work and to restructure the industrial geography of longshoring to favor dockers at the waterfront. Through its regulation of certain work practices, the union was instrumental in shaping the industry's new economic geography in two ways. First, through their direct manipulation of the nascent geography of work on the waterfront in the wake of technological innovation, the union and dockers who struck on numerous occasions throughout a 30-year period to support the Rules forced consolidators and others to change their ways of doing business and the locations at which that work was done. Second, the union and its members shaped the economic geography of the cargo-handling industry indirectly through the responses to the ILA's regulation of the 50-mile zones undertaken by consolidators, other segments of capital, and even other workers—responses such as seeking to evade the Rules through a variety of geographical subterfuges. Although the vignettes presented above are not meant to be comprehensive accounts of all of the Rules' impacts on the industrial landscape of East Coast waterfronts and their hinterlands, they do serve as illustrations of at least some of the ways in which the union's implementation of restrictive work practices shaped the production of new economic and social geographies in response to the destruction of the waterfront's traditional industrial structure by containerization. Thus, while particular companies have been highlighted, their experiences are suggestive of the myriad ways in which the ILA's actions have contributed to the building of the contemporary industrial landscape. Furthermore, it should also be noted that the issues addressed here are not unique to the East Coast of the United States. West Coast dockers, represented by the International Longshore and Warehouse Union, also negotiated their own version of the 50-mile rule, while British dockers implemented a similar set of work rules in the 1970s, though in the British case the cutoff distance was only 5 miles.[34]

The multitude of such political struggles to enforce, evade, and otherwise geographically negotiate the Rules highlights the fundamentally contested nature of the production of space and, particularly, labor's role in the making of economic geographies. Although in the case of the Rules it is, perhaps, easier to see the ILA's influence on the economic geography of their industry because we are dealing with the very visible physical movement across the landscape of work and commodities, the study outlined above is also, I would argue, suggestive of the ways in which workers shape industrial landscapes even when no physical movement of work or commodities is involved—such as through securing work rules that limit employers' introduction and/or use of new technology and that thereby affect the geography of production, profitability, (un)employment, and the like. However, the union's struggle was not just one that concerned the geographical *location* of work, but it was also one concerned with the geographical *scale* at which dockers could shape the industry's new economic landscape. This concern with scale played out in two ways. By restricting certain types of work to waterfront piers, the Rules effectively shrank the spatial extent of labor markets that previously had stretched up to 50 miles inland from ILA-organized ports. In essence, the Rules "scaled down" the labor markets for consolidating work, bringing that work from port hinterlands to the ports themselves. But the union's work restrictions also had regional and national purchase. The fact that 50-mile zones were created up and down the East Coast shows how the union was able to link local struggles and labor markets with regional and national concerns, struggles, and labor markets. By forcing employers from Maine to Texas ultimately to adopt the Rules first as local contract conditions and later as part of a national contract, the ILA not only shaped local labor markets in each of the 30-odd ports in which it organized dockers but also shaped the economic geography of the industry nationally. The 50-mile zone around each port represented the ILA's success in carving out *locally* a space and a scale of control over the labor process and the location of work, but its ability to implement the Rules *nationally* was indicative also of its success in expanding its efforts to the national stage, an issue we shall examine in more detail in the next chapter.

The account presented above highlights three more general points that are important to bear in mind when seeking to understand how workers' spatial praxis may shape the geography of capitalism. First, the elision of apparent interests between workers and the employers on both sides of the conflict concerning the disputed work and the Rules exemplifies the "locality dependence" (Cox & Mair, 1988) that we have already touched upon in Chapter 3, in which social actors' spatial entrapment in particular locations relative to more mobile capital often leads businesses and workers in these places to come together to form geographically defined, cross-class, growth coalitions designed to foster investment in their own communities. In this instance, the ILA and dockers' employers—particularly the stevedores who contracted with the ocean carriers to load and unload ships and who would be put out of business if this work migrated inland—came together to ensure that cargo-handling jobs remained at the waterfront, whereas the Teamsters joined with various trucking and consolidating interests to prevent the destruction of the off-pier freight consolidating business. Employers and workers in these industries may have struggled with one another over many other things, but when it came to the issue of the geographical location of work in their respective industries, they came together to defend particular (and different) localities against job loss. Clearly, inter- and intraclass politics cannot be understood devoid of their geographical context. Likewise, geography can be a powerful influence on political praxis, as spatial interests frequently win out over class ones.

Second, struggles to remake the geography of the cargo-handling industry in the postcontainerization period show the complexities of workers' production of economic landscapes on the ground. That working people produce geographies through their social practices does not mean that they do so only in opposition to capital. Working people do not always organize simply along class lines. For example, as was evident in this situation, jurisdictional disputes among labor unions or within unions (as in the ILA national leadership's decision to favor its waterfront workers over members of its own inland warehouse division) are often about which group of workers will have access to certain jobs, and may or may not have anything to do with the activities of capital. The ILA's attempts to inscribe spatially its vision for the new waterfront industrial structure affected many different interests, both employers and other groups of workers. Its actions were inimical not only to the interests of off-pier consolidators and trucking operators but also to its own inland members and to Teamsters and warehouse workers who would lose access to this work. In theorizing the making of the geography of capitalism—and workers' roles in this—we should not, then, assume that all workers have the same spatial interests. Rather, we must adopt approaches that recognize that, depending upon where they are located, workers may have very different visions of how landscapes should be made. At the same time, we should also recognize that, while the spatial practices of certain groups of workers may have advantages for certain segments of capital (e.g., stevedores, in this case), this does not deny the fact that workers are engaging in such spatial praxis for purposes of achieving their own geographic goals.

Finally, in taking this story back to where the chapter began, it is necessary to think of struggles over the introduction of new technologies and ways of organizing production not simply in historical terms (i.e., how such struggles play out over time) but also in spatial terms. While there has been a fair amount of work that has examined how new technologies may diffuse across the landscape—Hägerstrand's (1967) contribution is, arguably, the benchmark publication in this literature—we might also think about how those who will be negatively impacted in this process may manipulate the geography of capitalism through "spatial sabotage" to minimize those negative effects or even, perhaps, to shape spatial relations so that the benefits of introducing new technologies will not be reaped by employers, who may themselves subsequently decide that the cost of implementing such technologies is greater than the anticipated returns from doing so. Through such "spatial sabotage" workers will significantly shape how the geography of capitalism unfolds in particular places at particular times.

CHAPTER 5

Scales of Struggle

Labor's Rescaling of Contract Bargaining in the U.S. East Coast Longshoring Industry

The industry's oldest and deepest conflicts center in customs and practices, in work rules and cargo handling methods, habits that vary widely from port to port, creating multiple contradictions and needless fragmentation. . . . A national agreement, identical in all major provisions, would gradually erode industrial anarchy, would curb or halt wildcat strikes, would tend to equalize profits, and would slice through and ultimately resolve the welter of conflicting special conditions that now exist in every port.

> —ANTHONY SCOTTO,
> President of Local 1814 (Brooklyn)
> and International Vice President
> of the International Longshoremen's Association

It has been our position now that all the companies that we have any control over should sit down around the table with us . . . and make a rate for the entire United States. . . . I believe it takes just as much strength to lift 100 pounds of sugar in Tacoma . . . as it does in Baltimore. [A docker] should get the same money for doing it.

> —THOMAS "TEDDY" GLEASON,
> General Organizer (later International President)
> of the International Longshoremen's Association.[1]

Historically, one of the key goals of unions has been to take wages and working conditions out of competition. Unions' failure to do so invariably leaves them vulnerable to employers who are either multilocational or who are geographically footloose and not tied to any particular location and who can then pit workers in different regions against one another (a practice most commonly referred to as "whipsawing") on the basis of what may be

102

quite significant geographical variations in wages and conditions. If they are to prevent employers from using such spatial variations as a source of economic and political power, labor unions will invariably have to address any regional or local differences in work practices, wage rates, costs of living, unemployment rates, and a whole host of other social phenomena that may exist. To do so, workers and their organizations must frequently consider developing networks and bargaining mechanisms that will allow them to equalize conditions among different places. Put another way, they must think about building scales of mobilization and organization that link together places that may be many hundreds or thousands of miles apart. This involves their developing, among other things, an appreciation of how wages and working conditions vary across the landscape, how places are linked together geographically (e.g., are they close together in space or far apart, and what difference does this make?), and how efforts to take wages and conditions of work out of competition may play out in different ways in different places—for example, efforts to impose a single national wage rate might have to address how labor markets in one part of a country may be booming while simultaneously those in another part are in recession. Whatever else they may be, then, unions' abilities to develop regional, national, or even international scales of organization and collective bargaining so as to minimize employers' opportunities to whipsaw workers are fundamentally geographic practices that must take into account the myriad spatial variations existing in the economic landscape in which they are implemented. It is to an example of just such a struggle to build a scale of collective bargaining that minimized differences in wages and working conditions among places that I turn in this chapter.

Examining how workers build new scales of collective bargaining that operate at the regional or national scale raises a somewhat different set of conceptual questions than those addressed in the preceding two chapters. So far I have explored how efforts to confront job losses brought about by global processes of economic change may center upon quite local issues and how locally organized responses may be quite effective—under certain conditions—in countering the impacts of such processes. Although in the preceding chapter I indicated that in the case of the East Coast longshoring industry ILA officials and the dockers who supported them were finally successful in getting the Rules on Containers written into their national contract, the focus of this volume so far has largely been upon how the unions examined organized locally to defend particular spaces—the Manhattan Garment District, East Coast waterfront communities—in the face of economic restructuring. Furthermore, although in both cases these unions were embroiled in struggles to inscribe in the landscape scales of opposition to the extralocal processes that were impacting their various industries, in neither chapter did I really explore the process of constructing what Kevin Cox

(1998a) has called "spaces of engagement" but which I would prefer here to call "scales of engagement." Rather, the focal point was on the construction of singular scales of social action, delineated respectively by the special garment manufacturing preservation zone and by the individual 50-mile restricted areas placed around East Coast ports.

In this chapter, in contrast, I want to extend the argument about workers' spatial praxis and the politics of scale by showing how a group of workers—again, East Coast dockers—and their union went about building *extra*local scales of organization as part of their efforts to defend their economic and political interests as technological innovation transformed their industry. Specifically, I show how they articulated the relationship between these scales in different ways at different times as a central part of their struggle to minimize the negative impacts upon them of containerization and to secure their own vision as to what the future geography of the industry should look like. As I indicated in Chapter 4, while the introduction of containerization had important implications for the location of work *within* each port's local hinterland area and stimulated intense conflicts up and down the East Coast of the United States, it has also dramatically changed the relationships *among* ports along the coast. Faster overland distribution of goods—the result primarily of a number of developments in the field of transportation, including the building of a national interstate highway system and the growth of intermodalism facilitated by containerization—has reduced manufacturers' dependence on using ports located close to particular markets and instead has given them the opportunity to ship through cheaper, more distant, ports. As a result, during the past four decades ports with previously fairly geographically discrete hinterlands have increasingly felt the pressures of competition as local market boundaries have been eroded and the national economy of the United States has become more spatially integrated.

Such increased spatial integration of the economy has forced the ILA to confront two issues in particular during the post-World War II period: depressed wage levels in the South and the potential for steamship operators to more easily divert cargo should any port be tied up through strike action. Consequently, as intermodalism developed during the 1950s and 1960s, eliminating wage differentials among ports and presenting a unified front to the employers came to assume an increasingly prominent position in the union's political agenda (Gleason, 1980a). To attain these goals the union adopted a strategy designed explicitly to replace the traditional system of port-by-port bargaining that prevailed until the 1950s with a national "master" contract extending from Maine to Texas. In essence, the ILA sought nothing less than to construct a new geographic scale of bargaining and labor relations in the industry. In so doing, I argue, the union and its members played a central role in refashioning the economic and political geography of the longshoring industry, particularly by forcing the employers to redefine their own scales of political and economic organization with one another and with the union.

There are five elements of the union's struggles to define a new geographical scale of contract bargaining with the employers that I examine in this chapter. First, after initially opposing the expansion of its contractual obligations to include other ports along the coast, in 1957 the NYSA, the most powerful East Coast employers' association, was forced to agree to negotiate over five issues on a single basis for ports from Portland, Maine, to Hampton Roads, Virginia. This five-item "master contract" was usually mimicked in other ports in the South Atlantic and Gulf. Second, having secured a regional master contract covering basic monetary issues, the union then attempted to force the steamship operators to adopt nationally a series of work preservation and income maintenance measures (including the Rules on Containers) that would protect dockers from the effects of containerization. Third, in response to the union's efforts, in 1970 the employers formed a new multiport association (the Council of North Atlantic Steamship Associations) for the purpose of bargaining with the ILA's North Atlantic region on issues specifically related to containerization.

Fourth, in 1977 a Job Security Program adopted in 34 ports from Maine to Texas effectively became the first legally binding agreement negotiated between the ILA and the industry to cover the entire East Coast. Finally, despite the ILA's success in expanding the master contract, in the 1980s powerful forces emerged that threatened to break apart the master contract and perhaps even the union itself. In 1986, amid economic restructuring in the maritime industry and intensified political attacks on labor, 11 West Gulf locals took the unprecedented step of preempting master contract negotiations by voting to slash wages, cut gang sizes, freeze some benefits, and scrap a number of job preservation agreements in an attempt to halt the spread of nonunion longshoring operations. The emergence of a concessionary bargaining movement in the Gulf exposed a growing geographical schism between different segments of the union, a line of fracture that could inhibit moves toward the further development of a coordinated system of national bargaining.

My purpose in recounting the story of how one group of workers went about constructing new scales of labor relations in their industry to respond to technological innovation is threefold. First, by providing a detailed empirical analysis of struggles over the scale of bargaining, I locate workers' struggles at the core of efforts to understand the transformation of the economic geography of one particular industry (longshoring) in the post-World War II period. The implication, of course, is that a more sensitive reading of labor's activities may encourage a similar evaluation of the geographic development of other industries. (In this vein, Page [1998] has shown, for example, how struggles between rival union factions were central to the spatial evolution of the meatpacking industry in the United States.) Second, while in the preceding two chapters I illustrated how workers could organize at the local scale as a way of challenging nonlocal processes and actors, in this chapter I show how workers may seek to transcend the local by reconfiguring the geographical scales at which they interrelate with one another and the world around them. In so doing, I want to demonstrate how producing new geographical scales of social interaction can be a crucially important part of political struggles to shape the geography of capitalism in some ways and not in others. More specifically, by examining how the ILA preferred, once it had secured a national scale of bargaining, to negotiate some matters nationally while continuing to negotiate others locally (i.e., on a port-by-port basis), I examine how social actors may articulate the relationship between the various scales of their organization differently to achieve different goals at different times or in different places.

Third, I want to emphasize both that the production of scale by these workers should not be seen as having some kind of internal or inherent logic of its own and that the production of scale can vitally hinge upon sometimes almost imperceptible changes in the everyday political or economic environment in which actors find themselves. Depending upon the circumstances, workers may have an interest in expanding the scale of bargaining at particular historical junctures in particular places, yet in reducing it at other times. For example, as we will see below, whereas high-wage dockers in New York saw their own immediate interests best served by national bargaining so that they did not run the risk of losing work to lower-waged workers, during the 1980s West Gulf coast dockers competing locally with lower-waged nonunion workers increasingly sought to remain competitive by bargaining locally. Equally, high-cost employers may prefer a national agreement so that their competitors elsewhere do not enjoy the benefits of lower wages, whereas low-cost producers may prefer local agreements precisely to take advantage of their status as low-wage employers. To put it bluntly, context is important for the making of geographic scale and landscapes.

TECHNOLOGICAL INNOVATION AND THE ILA'S PRODUCTION OF A NEW SCALE OF BARGAINING[2]

Prior to World War II contract bargaining in the East Coast longshoring industry was principally conducted on a port-by-port basis, with employers and the ILA locals in each port negotiating separate contracts. The earliest moves toward the equalization of conditions between ports had been initiated during World War I when the ILA proposed a uniform wage scale for three broad geographic units—the North Atlantic, the South Atlantic, and the Gulf Coasts. The union saw the equalization of wage rates within these regions as a means both to fulfill federal government requirements to facilitate worker migration between ports to satisfy wartime demands for longshoring labor and to secure (at least partially) the union's goal of minimizing differences between ports with regard to a number of basic contract provisions (Department of Labor, 1980). In 1918 the union's proposal was implemented by the federal National Adjustment Commission, which established uniform rates in each of the three districts. However, this early gain proved to be short-lived, and as the exigencies of wartime passed and as recession hit the economy in 1919 many South Atlantic and Gulf ports soon returned to the prewar situation in which employers principally determined wages and working conditions on a port-by-port basis. Only in the North Atlantic did any consistent form of pattern bargaining remain, with employers in ports from Maine to Virginia adopting on a voluntary basis the basic wage and benefit provisions worked out in New York. Although no single legally enforceable "master" contract covered this section of the coast, gains won in New York were generally included in the individual contracts negotiated between the union and the employers in each port throughout the region.

Between 1919 and the early 1950s, then, a system of pattern bargaining existed in the industry in which some ports' employers voluntarily adopted the basic wage and benefits provisions worked out in their region's largest port (in the case of North Atlantic employers this was New York). However, the development during the 1950s of new labor-saving cargo-handling technologies, particularly containerization, together with political conflict within the union and broader labor movement, provided a powerful impetus for the ILA to seek anew to develop a national contract for the industry. Containerization did not affect all ports simultaneously. Rather, the new technology was introduced into the industry in a quite geographically uneven manner. It was dockers in New York, where labor costs were highest and where the ILA was strongest politically, who were first to feel the chill prospect of job loss. Equally, New York dockers were the first to fight for and win new restrictive cargo-handling rules and benefits in the face of automation. This had a number of implications for the union's struggle to expand the geographical scale of bargaining. First, International officials in New York feared that implementing such restrictions and benefits solely in New York would provide steamship operators with an incentive to divert cargo to ports where they did not apply, further diminishing waterfront jobs in New York (Gleason, 1980a). Second, union officials and rank-and-filers in other ports quickly came to see the writing on the wall concerning the potential for job losses caused by automation and soon began to hold out for the protections and benefits enjoyed by their New York counterparts, de-

mands strenuously resisted by their respective employers. As containerization spread throughout the industry, the ILA's strategy to expand the
geographical scale at which bargaining took place increasingly became in-
separable from the issue of technological innovation and job security. As the
Board of Inquiry set up to investigate the 1968 industrywide strike would
later note, "The two critical issues [in the industry] relate to union-wide collective bargaining and the impact . . . of containerization."

In seeking to develop such a national contract the ILA has had to address a great diversity of commercial activity and labor conditions along the
coast (e.g., see Groom, 1965), activities and conditions that constitute the
material basis for a remarkable heterogeneity of political and economic interests, both between the ILA's affiliated local unions and the various employers' groups. These regionally defined interests have shaped the political
struggles over the union's efforts to expand the geographical scale of
bargaining—together with employers' responses to such expansion—in several ways. The variety of conditions has posed immense problems for the
union's attempt to develop a unified national contract. For example, which
work rules would form the template for any future national agreement?
Given New York's dominance within the International union, it would most
probably be that port's practices. But would this cause resentment among
other ports' dockers whose traditions and customs would be sacrificed for
the sake of geographical uniformity? Certainly New York's dockers enjoyed
generally the most favorable conditions and wages in the industry, but this
did not mean that dockers in other ports were necessarily willing to give up
their own treasured customs or to allow union officials in New York to determine local work practices and conditions. In the Gulf, for instance, many
dockers still harbored resentment toward New York leaders for the manner
in which International president Joe Ryan had deserted them during a particularly bitter strike in 1935, abandoning the South to secure a favorable
contract in the North Atlantic.[3] Would these dockers reject New York's pattern and thus uniformity, or would they accept the port's work rules as the
basis for any future national agreement?

Equally, the multitude of conditions has also meant that the various employers' associations have different material interests around which they
have organized, interests that have shaped their attitudes toward the ILA's
efforts to impose on the industry a national agreement based on conditions
in New York. Thus, would southern carriers of mainly agricultural products
take their lead from New York steamship operators whose business relied
more heavily on consumer and industrial products that were more likely to
be containerized? Would their common interests as employers lead them to
seek unity in the face of union demands? Were southern operators willing to
tie themselves legally into the higher wage and benefit settlements negotiated in New York to preserve such unity? Or, instead, would divergent eco-

nomic interests combine with traditional enmity toward Yankee capital to split the employers along regional lines?[4] Furthermore, whereas the NYSA had historically opposed multiport bargaining for fear that disputes in southern ports might tie up New York, if forced to concede increasingly generous contracts might not the NYSA itself have an incentive to encourage national bargaining as a means of similarly burdening southern operators with equally expensive contracts, thereby eliminating any competitive advantages they might accrue from negotiating lower wage rates locally? In the final analysis, such questions would only be answered by political struggle.

The ILA Pushes for a Regional Master Contract

The early 1950s were a time of much political conflict in the East Coast longshoring industry. In August 1953 the American Federation of Labor (AFL) expelled the ILA for having engaged in a host of nefarious and undemocratic practices.[5] Soon after it had disowned the ILA (which now began calling itself the ILA-Independent [ILA-IND]), the AFL sponsored in its place a new union known as the AFL-ILA. These events began a period of intense factional conflict on piers throughout the East Coast. The AFL-ILA was keen to unseat the old ILA-IND as the dockers' negotiating agent with the employers and set about trying to encourage waterfront workers to leave the old union and support the new one in NLRB elections that would have to be held to determine which union was favored by a majority of dockers (this is covered in more detail in Herod, 1997d).[6] Nowhere was this interunion rivalry more animated or more bloody than in New York, the traditional heart of the old ILA, where, on several occasions, rivalries erupted into pitched battles along the waterfront. To counter the challenge from the new AFL union, the ILA-IND began pressuring the employers' associations from Maine to Texas to accept national bargaining because this, the old guard hoped, would allow it to use its national organization to isolate the AFL-ILA, which was largely concentrated in New York. The employers, however, vehemently opposed this move toward national bargaining for fear they would lose the ability to negotiate contracts suited to local conditions in their ports.

Although the AFL-ILA's efforts to unseat the ILA-IND as the bargaining agent for New York's dockers narrowly met defeat in two NLRB representation elections held in December 1953 and May 1954, thereby placating at least temporarily the ILA-IND, in October 1956 the AFL-ILA (now calling itself the International Brotherhood of Longshoremen [IBL]) once again challenged the old union in New York.[7] The IBL's challenge served as a catalyst for the Independent's renewed efforts to implement national bargaining. Proclaiming that "we want national bargaining, the membership wants it, and we intend to get it, even if we have to fight all the way for it," ILA-IND

President William Bradley ordered dockers from Maine to Texas to strike at the expiration of their contracts (quoted in *Journal of Commerce*, October 19, 1956). Throughout the 1956 contract negotiations and industrywide strike, the ILA-IND continued to press its claim for a national contract on the basis of two issues, one old and one new. First, International officials insisted that eliminating wage differentials between ports and presenting a geographically unified front to the employers had been the union's long-standing policy. For the union, the existence of separate contracts in each port meant that during strikes the steamship companies had historically been able to divert ships from struck ports to those whose contracts had already been settled, "thereby compell[ing] longshoremen, members of the ILA, to act as strikebreakers against fellow longshoremen."[8] A national agreement would prevent this, for a strike in one port (at least over basic contractual items included in a master contract) would now mean a strike in every port.

Second, during the 1956 contract negotiations the NYSA pushed for a three-year contract rather than for the traditional two-year agreement. By extending the length of the agreement NYSA officials hoped to reduce the frequency of strikes that had plagued New York in the immediate postwar years (Lamson, 1954; Jensen, 1974). However, high-level ILA-IND officials feared that implementing a three-year contract that applied solely in New York would place them in a potentially vulnerable situation should the IBL decide once again to seek to represent waterfront workers in other Atlantic and Gulf ports. Although New York ILA-IND locals would be legally obliged to abide by their contracts for an additional year, the AFL union would be free to offer dockers elsewhere the promise of higher wage and benefit levels than those scheduled to take effect during the third year of the Independent's contract, thereby perhaps enticing them to affiliate with the new union.[9] Pointing out that NYSA members controlled some 85% of the East Coast shipping business, Independent officials insisted that New York employers were in a position to speak for the whole industry on contractual matters. If only they would agree to an industrywide contract, Bradley told them, the danger of the IBL winning a national election would be greatly reduced and the ILA-IND would be able to consider the NYSA's request for a three-year contract (ILA-IND, 1956). Otherwise, he hinted, the union could not possibly consider extending the contract period and the NYSA could look forward to a lengthy strike.

Fearing that a strike over such issues could drag on indefinitely, in January 1957 the NYSA finally agreed to a modified form of multiport bargaining from Maine to Virginia. The ILA-IND, however, still pushing for an agreement covering the entire East Coast, indicated final acceptance would depend upon which provisions were included in such a contract. Furthermore, whereas the NYSA presented a settlement package that would cover the North Atlantic, other North Atlantic employers' associations in the "outports" (i.e., the North Atlantic ports other than New York) continued to offer several different contract proposals.[10] Only after another month of dispute did Bradley announce that a master contract acceptable to the ILA-IND had at last been reached that would cover all North Atlantic ports. Under the terms of the newly negotiated agreement, the NYSA was authorized by employer associations in Boston, Philadelphia, Baltimore, and Hampton Roads to negotiate on their behalf with the ILA-IND on five contract terms: wages, hours of work, length of the contract, and employer contributions to welfare and pension funds.[11] Once the terms of the master contract and New York's local conditions had been agreed upon by the NYSA and the union, each individual outport would then negotiate on a local level over such items as holidays, vacations, working conditions, gang sizes, and other benefits.

The regional master contract, then, was finally won by the ILA-IND over and above the entrenched opposition of the NYSA and the outport employers. It represented a significant change in the system of contract bargaining in the industry because, for the first time, the union had secured a legally enforceable multiport agreement. In a very real sense the union had constructed a new geographical scale of bargaining, one which was in marked contrast to the port-by-port negotiations that had historically been the norm for the industry. The development and implementation of the master contract is significant for at least two reasons. First, it was clearly a victory in the union's campaign to equalize certain conditions between ports. Although the agreement only covered the North Atlantic ports rather than the whole East Coast, it nevertheless meant that dockers in each member port could expect uniformity in the five contract issues and no longer needed to worry about undercutting one another in terms of wages and other issues that were now to be centrally agreed upon. Second, because dockers in these ports were now party to the same master agreement, they could henceforth legally strike in support of one another's goals without fear of legal challenges under the federal Taft–Hartley Act, which prohibits secondary boycotts.[12] Employers would now no longer be so easily able to use dockers in one port to break strikes in others. By forcing the adoption of a regional scale of bargaining, union officials and rank-and-file dockers had eliminated a powerful weapon in the employers' arsenal.

From Regional to National Contract?

The ink was barely dry on the master agreement when ILA-IND officials again raised the thorny issue of national bargaining. Seeking to build on their recent success, union officials demanded the 1959 contract negotiations include all Atlantic and Gulf ports—a demand that southern employers adamantly rejected. But by 1959 the conditions that had shaped bargaining during the mid-1950s had changed in at least two fundamental ways. First, the IBL was on the verge of disbanding itself and, indeed, in October voted to merge with the old ILA. For its part, the ILA-IND had been admitted to the newly formed American Federation of Labor–Congress of Industrial Organizations (AFL-CIO) in August 1959.[13] Consequently, the threat of dual unionism had finally been laid to rest. Second, technological innovation had begun to impact the industry. In particular, containers were making their presence felt in New York, the union's stronghold. Although still only a handful of steamship operators were using them, there could be little doubt in the minds of many dockers that unchecked containerization would have dramatic employment implications throughout the industry.

As we saw in Chapter 4, by the early 1960s the impacts of automation had come to dominate labor–management relations on the New York waterfront and, given the NYSA's role in master contract negotiations, elsewhere too. During the 1964 contract talks union officials called a national strike in support of their efforts to negotiate some form of guaranteed income for those workers in New York who would be displaced by containerization. The NYSA, on the other hand, opposed such a measure and pushed instead for greater work flexibility on the piers as a means of facilitating the more widespread use of the new technology and of reducing the gang size, which in New York then stood at 20 dockers plus a gang boss for general cargo. Negotiations continually faltered on the issue of staffing levels and job security. Although by November both the NYSA and the union had agreed in principle to gang-size reductions and greater flexibility of work assignments in return for a guaranteed annual income (GAI) that dockers would receive even if there were no work available, the central point of conflict remained the formula by which such job and income security would be provided.[14] This impasse not only held up talks in the North Atlantic—the master contract could not be signed until local conditions in New York had been settled—but, because the New York settlement had traditionally provided the broad basis for agreement in other ports, it also paralyzed contract talks in the South Atlantic and Gulf ports, whose union leaders usually looked to New York for guidance on bargaining issues such as wages and some benefits.

Such deadlock was particularly frustrating for southern steamship operators, who saw their own negotiations held up over issues that were, they felt, relevant only to New York. For them, the NYSA's concern with gang sizes and labor flexibility, together with the ILA's desire to negotiate compensatory benefits for those displaced by containerization, seemed unrelated to problems faced by southern employers for two reasons. First, in southern ports agricultural products made up a much larger proportion of total cargo shipped than was the case in the North Atlantic. This was an important difference because southern operators anticipated that agricultural products would continue to be shipped primarily in bulk rather than in containers. For them, the "exorbitant" contract package demanded by the ILA in New York seemed unnecessary to protect southern dockers from a technology that, the employers argued, was hardly likely to have anywhere near the displacement effects expected in New York. Second, reducing gang sizes simply was not an issue for most southern employers in the way that it was for the NYSA: in the East Gulf the gang size for general cargo already stood at 18 (3 fewer than in New York), whereas West Gulf and South Atlantic ports had no minimum gang sizes at all.

With talks stymied in New York over matters that were seemingly inconsequential to them, southern operators became increasingly resentful of the union's policy of trying to impose the New York settlement as the pattern to be followed throughout the industry. Members of the New Orleans Steamship Association (NOSA) complained they had been on the verge of signing a contract locally when ILA officials in New York ordered dockers from Maine to Texas to walk out. Arguing that it was "indefensible" for their ships to be tied up over an issue affecting only New York, one NOSA official commented that "maybe this is the time to divorce ourselves from the New York negotiators" (*Journal of Commerce*, December 18, 1964). If union policy was to seek common contract goals along the coast, southern shipping interests made it clear that they were equally determined to prevent this coalescence of contract conditions. Yet, if dockers in the southern ports could not force shippers to concede the North Atlantic wage and benefit terms, the union's hopes of building a national contract identical in most provisions would be dashed.

In mid-December 1964, after a short strike and imposition of a presidentially mandated 80-day Taft–Hartley "cooling-off" period, the ILA and the NYSA finally agreed to a guaranteed annual income in New York modeled after the union's formula. The GAI's inception would prove to be a significant moment in the union's struggle to develop a national contract, for not only did it provide a measure of protection against wage loss brought about by the new technology but also ILA officials in New York soon began to worry that the added expense of funding the GAI would encourage operators to divert cargo to other ports where the provisions did not apply.[15] Only by extending the GAI nationally, New York union officials argued, could this threat be averted. Equally, dockers in the North Atlantic outports eyed the GAI with considerable envy, while southern dockers were eager to secure (larger) minimum gang sizes patterned after New York. These dockers' struggles to match New York dockers' job preservation and income maintenance agreements soon became central to local negotiations throughout the East Coast and increasingly provided one of the key issues around which the effort to impose uniform contract provisions in the industry would revolve.

While the gang-size issue preoccupied the South, dockers' desire to secure GAIs similar to New York's dominated talks in the North Atlantic. The
GAI was not a master contract issue but had instead been negotiated as part
of New York's local conditions. ILA locals wishing to secure a similar provision would thus have to fight for it on a port-by-port basis. The variety of labor and commercial conditions throughout the North Atlantic, however,
made this a problematic proposition for such locals. In particular, the
amount of cargo moving through other North Atlantic ports (and hence the
total number of hours worked in each port) was significantly less than that
moving through New York. North Atlantic outport employers maintained
that they simply did not have the volume of work to offer GAIs patterned
after New York's. Nevertheless, after a three-month work stoppage ILA locals finally forced employers in Boston and Philadelphia to concede GAIs,
while dockers in the South Atlantic and West Gulf ports succeeded in pressuring their employers to accept minimum gang sizes.[16] Furthermore, all locals from Maine to Texas successfully exacted the same monetary package
(based on New York's formula) from their respective employers in the face
of entrenched opposition.

By enforcing the 1964 strike industrywide the ILA had sought to demonstrate its claim to national bargaining. Moreover, by focusing on the issues
of the GAI in the North Atlantic and minimum gang sizes in the South, the
union had expanded the geographic scale of its fight to mediate the future
impacts of containerization. Although they had not yet achieved equivalent
provisions throughout the industry, in a very real sense rank-and-file dockers and union officials had begun to build, piece by piece, a new scale of
contract bargaining that, they hoped, would ultimately establish a national
system of protections against the job losses augured by containerization.

The Employers Reorganize

As the use of containerization began to spread throughout the industry in
the late 1960s, the International ILA leadership became ever more concerned to protect its members from the job displacement effects of the new
technology. During the 1968 talks the union's major demand was to increase
the number of master contract items to 12, most significant of which were
the inclusion of the GAI and the Rules on Containers. For the union, including these items in the master contract would ensure that all dockers enjoyed
the same protections against job loss regardless of where they worked.
However, after 10 years of the regional master contract North Atlantic outport employers had become increasingly resentful of the costly agreements
that the NYSA was negotiating on their behalf. Concerned that the master
contract was too reflective of the NYSA's needs and that it paid insufficient

attention to their own concerns, in 1967 the outport employers' associations indicated they would abandon any master contract that included either a GAI or the Rules on Containers.

Despite immense pressure, ultimately the ILA was unable to overcome the outport employers' opposition in 1968 to including either the GAI or the container-handling rules as a master item in the contract. However, the union did succeed in wringing a substantial monetary settlement from the NYSA and agreement to include the Rules on Containers as local items in the North Atlantic and West Gulf regions (see Chapter 4). Yet, this further aggravated the outport employers, who complained bitterly that the NYSA was trying to buy off the union's opposition to containerization with ever more generous wage and benefit agreements, a strategy that, they argued, meant that the outports were effectively subsidizing the costs of containerization in New York and thereby encouraging its quicker adoption by NYSA members whose larger container facilities were, in turn, siphoning business away from the outports. Although pressure from the union ultimately forced all North Atlantic (and southern) ports to accept the New York monetary package, the strains between the NYSA and the outport employers were becoming increasingly visible. The union's determination to include the GAI and the Rules on Containers in the master contract, together with the size of the monetary settlement, had generated intense divisions among the employers, divisions that were explicitly geographical in nature. These divisions now erupted into a call on the part of the outport employers to reorganize the way in which the master contract was negotiated.

Emerging from what many saw as the economic catastrophe of the NYSA's 1968 master contract, outport employers determined to have greater say in future negotiations. By so doing they hoped to prevent the NYSA from agreeing to include container-handling rules and the GAI as master contract terms. Whereas New York employers might actually derive some significant economic benefit from having these items included, thereby ensuring they were not placed at an economic disadvantage vis-à-vis ports that had not been forced to adopt the terms, for outport employers the cost of maintaining the GAI at New York levels would be prohibitive and might even force some ports to close down altogether. As a result of their dissatisfaction, in October 1970 the outport employers formed a new multi-employer bargaining group, the Council of North Atlantic Steamship Associations (CONASA). Made up of the five employer groups that had customarily subscribed to the master contract (including the NYSA), together with the Rhode Island Shipping Association, which had not, the structure of the CONASA would give the outports a greater voice in the master contract negotiations.[17] This was underscored by their intention to negotiate local issues at the same time that they negotiated the master contract rather than waiting for New York to settle before discussing local terms, as had been the traditional pattern—a development that would have significant implications for the 1977 negotiations.

North Atlantic employers proposed to bargain with the union for the 1971 contract through the CONASA. Union officials indicated they were agreeable, provided that the CONASA include three additional issues in the customary five-item master contract, these being the GAI, the Rules on Containers, and bargaining on a new phenomenon known as "Lighter Aboard Ship" or "LASH."[18] CONASA officials refused to consider including the GAI, arguing that to do so would unduly burden smaller ports, although they did ultimately agree to the other two additional items. Whereas the CONASA talks proceeded relatively smoothly, in the South the GAI issue proved to be a major stumbling block. The union's failure to have the GAI included as a master contract item meant that dockers would have to fight for it on a port-by-port basis as a local contract issue. This they eagerly set about doing. After several weeks of negotiations dockers in West Gulf ports finally forced their employers to fund GAIs at 2,080 hours per year, the same as in New York, while dockers in the South Atlantic extracted a uniform 1,000-hour GAI from the newly formed South Atlantic Employers' Negotiating Committee (SAENC), which bargained for the ports from North Carolina to Tampa (with the exception of operators in Miami, Port Everglades, and Cape Canaveral, who had negotiated separately).[19]

The formation of the CONASA and the SAENC is weighty testimony to the union's power to remake the scale of bargaining in the industry. In many ways the new bargaining groups had been in the making at least since 1957, when the master contract first replaced port-by-port negotiations. The union's desire to negotiate job preservation and income security schemes had clearly grown out of concerns about job losses in New York. Yet, dockers had successfully foisted GAIs and container-handling rules on the still predominantly break-bulk outports. The ILA's insistence upon applying such provisions uniformly during a period of fundamental restructuring in the industry had forced North Atlantic employers to develop their own regional multiport organization in response. Furthermore, the 1971 settlement proved to be something of a watershed in the industry because for the first time ILA locals in virtually every port from Maine to Texas voted at the same time on both the master and local contracts. The CONASA agreement covering the (now) seven master items served as the monetary package in all East Coast ports in which the ILA represented workers. In addition, West Gulf dockers had negotiated a GAI at the same level as New York, while in the North Atlantic dockers in both Baltimore and Hampton Roads had succeeded in increasing their own GAIs. Throughout the South Atlantic a uniform GAI, albeit smaller than in New York, was now in place for the first time in the union's history, as was the case for several East Gulf ports. Finally, dockers from Maine to Texas now enjoyed the same protections against job loss offered by the restrictive Rules on Containers.

In sum, the 1971 settlement represented a significant union achievement in the historical geography of contract negotiations in the industry. Never before had the ILA enjoyed such a uniform agreement. Additionally, the ILA's insistence on developing a national contract or, at the very least, several regional contracts had forced the employers to respond by establishing a number of new regionally organized negotiating blocs, most particularly the CONASA and the SAENC. In a very real sense, the employers' reorganization of their own bargaining structure reflected and highlighted the power of the union to force a new scale of bargaining on the industry. Nevertheless, despite these gains, the union had still not achieved its ultimate goal of sitting down with employers from Maine to Texas at one time to negotiate a single industrywide contract. ILA officials now set their sights on achieving this next goal.

The Job Security Program as National Contract

The geographical uniformity of the 1971 contract and the ILA's satisfaction that the master agreement now included the Rules on Containers seemed to have laid the basis for the more peaceful resolution of future contract negotiations. Certainly the 1974 talks marked a radical break with past bargaining experiences. For the first time since the master contract's inception, the terms of the new agreement were settled without a strike. The only stumbling block of any significance was concern by New Orleans and Mobile employers that the agreement was rather expensive, perhaps an early portent of the growing schism between the Gulf ports and the International union's negotiators in New York. However, if the 1974 talks were remarkable for their peaceful resolution, the 1977 talks were reminiscent of the more usual negotiation and strike pattern. The catalyst for this reversion to a confrontational style of bargaining was the political conflict generated by the NLRB's 1975 nullification of the master contract's container-handling rules in New York, Norfolk, and Baltimore on the grounds that they were illegal work acquisition measures (see Chapter 4; also Herod, 1998c). Together with the GAIs operating throughout the industry, the Rules formed the core of the union's strategy to preserve waterfront work and minimize income dislocation in the face of rapid technological change. Their invalidation in these three ports threatened the ILA's entire job preservation strategy.[20]

During talks for the 1977 contract the ILA's primary concern was that the funds that supported each port's GAI, pension, and welfare schemes might be subject to shortfalls as a result both of the Rules' nullification and of the industry's continued geographic restructuring caused by containerization. The lack of any industrywide mechanism to make up shortfalls in individual ports' schemes (particularly the GAIs) meant that in those areas where the effects of the Rules' rescission would be hardest felt, funds might dry up at the very moment dockers needed them most. Furthermore, ILA leaders feared that the economies of scale to be gained by concentrating cargo-handling operations at the larger containerized ports such as New York might encourage steamship operators simply to close down their facilities at smaller, less automated ports. If operators concentrated cargo handling in only a few large ports, many smaller ones would run out of funds to support their GAIs. Only with a national scheme to ensure the viability of individual ports' funds, union officials argued, could dockers' benefits be guaranteed.

Throughout the negotiations for the 1977 contract ILA International President Thomas "Teddy" Gleason insisted that no progress could be made on the master contract until the issue of a national job security program had been settled once and for all. However, the ILA's determination to impose on the industry what was, in effect, a national income guarantee mechanism deepened the already significant geographical divisions among the employers. Although they were prepared to offer guarantees for pension and welfare funds, all North Atlantic outport employers opposed including the GAI in the master contract for fear they would be forced to subsidize the increasingly expensive New York scheme. Likewise, operators beyond the North Atlantic range expressed unwillingness to countenance a national GAI. The ILA, on the other hand, refused to accept anything less than a national GAI agreement. Only the NYSA indicated it was prepared to contemplate the union's request. This in itself was an important intraemployer disjuncture. The Association had historically argued against national agreements because it feared that strike settlements in New York, traditionally the first port to begin and end contract negotiations, would be held up as the union tried to force other employers both to accept New York's master terms and to negotiate their own local contract provisions. However, the ILA's success during the 1971 talks in compelling all employer associations to negotiate and sign the master and local contracts at the same time—a union goal designed to emphasize its national bargaining structure—had removed this fear and encouraged the NYSA instead to consider an industrywide agreement that would equalize the GAIs' terms and thereby similarly burden other employer associations with the same high costs of maintaining dockers' incomes. Ironically, agreeing to the ILA's demands for a national GAI might actually be a means for the NYSA to shift some of the costs of containerization in New York to other ports and thereby improve its own competitive position.

The growing schism between the NYSA and other CONASA members over the GAI issue had crucial implications for the ILA's efforts to develop a national contract. Now that the Rules on Containers had, for the moment, been overturned in New York and more generally cast into doubt elsewhere, New York employers concluded they could expect little waterfront peace as long as the union was without some degree of protection from job loss. Led by its container carrier members, in mid-October 1977 the NYSA finally withdrew from the CONASA after marathon negotiating sessions failed to settle the issue of an industrywide job security plan. This had momentous implications for the union's ongoing struggle to forge a national contract. The NYSA's secession provided the opportunity for renewed discussions between the ILA and New York employers on the job security issue. Now that the NYSA no longer had to accommodate the interests of other CONASA employers, its members were free to put forward a new proposal on job security that the NYSA was prepared to implement in all East Coast ports in which its members did business. In late October the Association announced that New York containerized carriers had agreed to underwrite union benefit funds in all those ports from Maine to Texas in which their ships called. Together with their subsidiaries outside New York, the 95 or so NYSA members who consented to the new fund guarantee program constituted the vast majority of all East Coast employers. However, because the carriers were themselves not legally authorized to bargain collectively for ILA contracts—in the vast majority of instances the NLRB only recognized the separate port associations as legitimate bargaining agents—they could not include the new program as a contractual item. Instead they would adopt it as a supplementary agreement to be known as the Job Security Program (JSP).

The JSP seemed to offer a way to negotiate the geographical tensions between the union's national ambitions and the employers' local concerns. It achieved this by establishing a common shortfall pool that was industrywide in scope yet that allowed employers' associations to continue negotiating and funding ports' benefits programs at the local level. It would, the union hoped, provide a measure of protection against the effects of containerization while also allowing employers and dockers the flexibility to respond to changing local conditions. A key concern for union officials, however, was how to link the JSP negotiated by the containerized carriers with the master contract negotiated by the employers' associations. They would achieve this, they hoped, by including as an eighth point in the master contract a codicil in which the union claimed the right to refuse to work any ships owned by containerized carriers who did not subscribe to the master contract and the JSP. Officials offered employers' associations a choice: either accept the eighth point and the JSP or risk a lengthy strike over the terms of the master contract. Although North and South Atlantic employers quickly signed the new master contract (including the codicil) and the JSP agreement, Gulf operators were decidedly reticent to do the same. Three major Gulf carriers refused to adopt the JSP, while the West Gulf Maritime Association, the New Orleans Steamship Association, and the

Mobile Steamship Association all questioned its legality, even though all three employers' groups ultimately did accept the North Atlantic master contract and the JSP.

Dockers all along the East Coast overwhelmingly endorsed the new contract. But when the dust had settled it soon became apparent that only the New York container carriers seemed happy with the contract. Gulf employers seethed. Arguably, the JSP marked the crowning achievement of the union's struggle to expand the scale of bargaining in the industry and develop a uniform national contract. In just over two decades the ILA had completely transformed the geography of collective bargaining from a parochial system of port-by-port wage and benefits agreements to a system of regional contracts and, subsequently, to a nationally coordinated and enforced job security scheme. Yet, forced to accept the JSP, together with what they saw as the master contract's ever more extravagant monetary packages, Gulf employers in particular increasingly began to look to cheaper non-union dock labor to work the piers, a practice that right-to-work laws in every coastal state from Virginia to Texas facilitated.[21] Ironically, the 1977 contract seemed simultaneously to have achieved uniformity while sowing the seeds of the destruction of that very uniformity, as would become evident during the 1980s.

Schism in the South

The union's satisfaction that dockers were now protected from the effects of job displacement regardless of where they worked laid the foundation for smooth contract talks in 1980 in which, for the first time, Atlantic and Gulf Coast employers' associations joined together to hammer out the master contract. In itself, this was a historic development, for it marked a move away from the traditional pattern in which North Atlantic employers alone had negotiated the master agreement.[22] Likewise, in 1983 South Atlantic and Gulf employers were actively involved in shaping the master contract terms rather than simply accepting a master pact worked out by North Atlantic operators. Furthermore, seeking to build on their recent achievements, ILA negotiators attempted to amend the master contract to provide uniform vacation and holiday pay. Although they were ultimately unsuccessful, their efforts are nevertheless indicative of the union's commitment to equalizing contract conditions throughout the industry so that competition between ports would be on the basis of worker efficiency rather than differentials in ILA contract provisions.[23]

However, whereas these developments marked an apogee in the ILA's efforts to standardize its contract in ports from Maine to Texas, during the mid-1980s the increasingly expensive agreements negotiated largely on the basis of conditions in the North Atlantic provoked the emergence of a schism in the Gulf and, later, South Atlantic that threatened the International union's ability to negotiate and enforce master contracts for the entire industry. Concerns by southern employers about the cost of contracts were increasingly played out through a renewed struggle to redefine the scale of bargaining, this time with the ILA's national agreement on the defensive. While the International office in New York continued during the mid-1980s to insist that employers concede ever more generous contracts, dockers in several Gulf and South Atlantic ports were forced to consider breaking with International officials in New York and offering concessions as a means of retaining work in the face of competition from nonunion labor.

Perhaps the first indication that Gulf dockers were feeling pressured to respond to local conditions rather than conform to national union goals came in 1985, when several Gulf locals agreed to forgo a $1-per-hour pay raise for loading bulk grain and granted a number of other concessions pertaining to the local contract. Although ILA International President Gleason subsequently nullified these locals' actions by arguing that wages were a master contract item and therefore not subject to modification by individual locals, Gulf dockers' actions were indicative of the political and economic problems they faced in trying to follow the terms of any national contract in the face of employers' increasing use of nonunion labor. This potential rift in the union was intensified by the growing calls of several employers—and even some union leaders—in South Atlantic and Gulf ports to negotiate separately from the North Atlantic region. For employers this was a means to avoid the ever more costly provisions of the master contract (which they felt largely reflected conditions and concerns in the North Atlantic ports), whereas for southern ILA locals it was a means to stem the loss of ILA work to nonunion operations. This raised the question of whether southern dockers and their leaders would continue to accept the master contracts negotiated by the New York-based leadership of the International union or whether they would prefer to implement their own independent agreements.

The growing tension between Gulf dockers' local worries and the desires of North Atlantic ILA leaders were soon manifested geographically. In late April 1986, and in defiance of ILA President Gleason, 11 West Gulf locals agreed to a longshore contract that slashed wages by $3 per hour on noncontainerized cargo, abolished their GAIs, and made concessions on a number of other issues. This represented an abandonment by these locals of the master contract's wage rates for bulk and break-bulk cargo, although West Gulf ILA negotiators did leave the matter of wages for container work to be settled by the traditional master contract talks, thereby raising the issue of whether southern locals might increasingly adopt some elements of the master contract while rejecting others. Soon thereafter, South Atlantic dockers also agreed to reduce wages for bulk and break-bulk cargo, although by

a smaller amount than had dockers in the West Gulf, where nonunion operations were a greater threat (*Journal of Commerce*, June 9, 1986). Such decisions represented both a significant break with the pattern of contract settlements that had been put in place during the preceding three decades and a
significant rift within the union—at the same time West Gulf locals were
abolishing their GAIs and South Atlantic dockers were voting to cut back
their GAI program, Gleason was attempting to negotiate an even more generous nationwide GAI as part of the master agreement.[24] Such actions in the
Gulf and South Atlantic signaled a split in the union that, some felt, might
lead ultimately to the total disintegration of the master contract as northern
and southern dockers began to respond in their own ways to the different
economic and political situations within which they found themselves.

Evidently, whereas at one time northern and southern dockers both
saw their interests best served by a national contract, changed conditions
and the growing use of nonunion labor in the South meant that by the mid-
1980s this was no longer the case. Equally, the fragmentation of the system
in the South provided a precedent for northern employers to demand their
own matching concessions from the union. Thus, the 1986 master agreement froze wages throughout the industry for the first time since 1949 and
abolished the JSP, which southern dockers had felt negatively impacted their
competitive position relative to northern ports.[25] However, the issue of competition from nonunion operators also split employer groups geographically.
Although it was forced subsequently to abandon the effort, the New Orleans
Steamship Association, for example, announced early in the negotiations for
the 1986 contract that it would withdraw from the master agreement and negotiate everything locally, while in the North Atlantic divisions between employer groups led the NYSA, the Boston Shipping Association, and the Carriers Container Council to negotiate with the ILA separately from the
CONASA (which represented employers in Hampton Roads, Baltimore,
Philadelphia, and Providence).[26] Although by 1989 the employers had
patched up some of their internal differences—master contract talks
involved all management groups, though Gulf employers attended merely
as observers—tensions still ran high, particularly on the issues of containerization, wage rates, and the dominance of the NYSA in negotiating the
master contract. At the same time, divisions within the union saw the ILA in
1989 abandon its traditional policy of "one port down, all ports down," as
officials allowed ports that had reached agreements on local contracts to
work even if others had not—a sign, perhaps, of the union's growing inability to enforce national agreements and standards in the industry.

The 1980s, then, represented a period of fragmentation and localism in
contract bargaining in the industry as many employers—but also many union officials and dockers—sought to break free of what they perceived to be
the confines of the national agreement. Thus, for instance, the resolution of

bargaining for the 1986 agreement left dockers in the North Atlantic, in the West Gulf, in the East Gulf, in South Florida, and in the Carolinas, Georgia, and North Florida working under very different contracts with regard to wages for noncontainerized work and certain benefits (*Journal of Commerce*, October 9, 1986). Such developments clearly illustrate how groups may seek at different times to develop different geographical solutions to the problems they face. Indeed, the decision of Gulf and South Atlantic dockers to abandon aspects of the master agreement may, in fact, be seen in the contorted story of contract bargaining in the industry as one more attempt by some workers to develop spatial and "scalar" fixes that they perceive to be advantageous to them at particular times and places. Whereas Gulf and South Atlantic dockers had seen their own interests best served by seeking to impose a national agreement during the period between 1953 and 1986, changing local conditions led them to decide that "going local" to develop a new scalar fix that was sensitive to local conditions in the South might be more appropriate to their continued ability to find work in the industry. Equally, it shows that divisions between workers located in different regions may lead them to adopt different scalar fixes. For North Atlantic dockers, the decision of Gulf and South Atlantic dockers to offer wage givebacks represented a worrisome situation, for it provided the steamship operators with an opportunity to ship through cheaper southern ports and to transport cargo up to the Northeast by truck or train. While North Atlantic dockers saw the national wage agreement as a way of eliminating the potential for the transfer of work to cheaper ports in the South, for Gulf and South Atlantic dockers offering locally negotiated contracts that included wage cuts and other concessions was a way of reducing the loss of work to nonunion operations—a challenge dockers in the more heavily unionized North Atlantic did not face to nearly the same degree—and of generating a competitive advantage that might ensure the continuance of work on their own waterfronts.

Despite the fact that West Gulf dockers continued to make local concessions related to noncontainerized cargo during the 1980s and 1990s, and South Atlantic dockers ultimately agreed to scrap their GAIs in 1990 in exchange for higher wages on certain types of cargo (*Journal of Commerce*, December 4, 1990), to date the ILA has nevertheless managed to maintain the geographical integrity of its national master agreement with regard to wages for container handling and a number of other conditions.[27] Equally, the tensions that emerged between employers in the mid-1980s soon led to efforts to develop greater unity. Indeed, the 1996 master contract negotiations (for a five-year contract) saw unprecedented congruity among the employers, a congruity which ultimately led to the 1997 formation of a new multiemployer coastwide entity—the U.S. Maritime Alliance, Ltd. (USMX)— which represents carriers, stevedores, marine-terminal operators, and local port employer associations from Maine to Texas and which bargained for employers for the 2001–2004 contract. Likewise, by the mid-1990s the different factions of the ILA appeared to have worked out some of their differences such that in the 1996 contract negotiations union and management agreed to implement a new coastwide managed healthcare system to replace the myriad local health plans existing up and down the coast, while both the 1996 and 2001 contracts included a number of coastwide wage increases. As the particularities of the economic and political contexts within which they must operate have changed, then, employers and dockers have found that organizing at different geographical scales has brought both challenges and solutions to their efforts to make their way in the world.

WORKING SCALE: LABOR'S REMAKING
OF THE GEOGRAPHY OF THE LONGSHORING INDUSTRY

Evidently the production of scale is both a highly political and a highly laborious process. The fact that such conflicts over the scale at which contracts are to be enforced have been so virulent and so crucial to the dynamics of the East Coast longshoring industry indicates quite clearly that the ability to make scale in particular ways and not in others confers with it great economic and political power, power that both labor and employers have sought to harness. For the ILA, constructing a national scale of contract bargaining was a key element in the union's efforts to confront the spatial integration of the industry during the postwar period and its tremendous transformation augured by the adoption of various labor-saving technologies, particularly containerization. The union's pursuit of an explicitly geographical strategy to address the issues of job security and income maintenance for dockers not only highlights a particular sensitivity to the spatial restructuring of the industry and the geographical diversity of conditions found throughout East Coast ports, but it also suggests that rank-and-filers and union officials were acutely aware that space and scale can be shaped in particular ways to serve political and economic ends. In this sense, their actions may be said very much to represent a workers' "applied geography."

However, dockers' struggles to make the scale of bargaining in ways that they perceived to be advantageous were not simply concerned with expanding or, in the case of some southern dockers during the 1980s, with contracting the scale at which contracts were negotiated per se. They were also concerned with crucial decisions about what provisions would be included in the different contracts. Thus, whereas some basic contractual items (such as contribution rates for dockers' welfare and pension plans) were included in the master agreement, other items continued to be negotiated as local issues (e.g., the actual benefits available to dockers under these programs). Such a strategy enabled dockers to take certain contract terms (i.e., those in the master contract) out of competition while also retaining flexibility with regard to others so as to be able to respond to local, perhaps unique, conditions. Equally, some employers fought to keep particular issues (e.g., GAIs) in local contracts so as not to be tied into more expensive industrywide schemes that might threaten the competitive advantages they enjoyed precisely because they had not agreed to the higher cost programs that ports such as New York had implemented.

The twists and turns of constructing geographic scales in the context of an unevenly developed economic landscape also sometimes saw employers and union solidly opposed to each other, whereas at other times they developed unexpected cross-class alliances and intraclass disjunctures. Hence, the ILA's decision in 1971 to negotiate both the master and local contracts at the same time meant that the NYSA no longer needed to worry about finalizing contracts in New York and then having to wait for all other ports to settle their local terms before New York dockers could return to work. Consequently, after 1971 the NYSA, which had long opposed an industrywide contract for this very reason, increasingly came to support the union's position on some issues related to national bargaining and supported the ILA's efforts to develop an industrywide GAI because such a provision would burden other ports with the same costs faced by New York employers, thereby removing a competitive advantage that these ports enjoyed. Similarly, whereas in some contexts it was vital for the ILA to try to negotiate all contracts simultaneously as a means of illustrating the union's national organization and bringing pressure to bear on all employers equally, at other times union spatial strategy was to allow some ports to return to work while continuing to strike others. With such a strategy, the union hoped to reward cooperative employers and to force recalcitrant ones to agree to the contract (as happened during negotiations in 1965, for instance).

The narrative presented above also shows the importance of the context within which actors create and renegotiate geographical scale. The geographic variations in conditions throughout the industry—in terms of types of goods shipped (consumer durables or agricultural products), size of port, work traditions, gang sizes, and the like—were significant shapers of dockers' and employers' attitudes, goals, and strategies, and illustrate the difficulties of trying to equalize conditions throughout an already unevenly developed industry. For example, when the Philadelphia Marine Trades Association finally agreed to accept the 1968 master contract's $1.60-per-hour monetary package, local dockers complained that the package would not provide the same level of vacations and holidays as dockers enjoyed in New York because Philadelphia had a smaller volume of work and, hence, of work hours. Philadelphia local officials argued that the volume of work in their port necessitated a higher monetary settlement if they were to be able to match New York's benefits package. International leaders, on the other hand, feared that such a demand, if successful, would simply lead New York dockers to ask for an even larger monetary package to match the new Philadelphia rate in a never-ending upward spiral and so pressured Philadelphia officials to accept the same package as New York (*Journal of Commerce*, January 28, 1969). This situation clearly illustrates what is often a central paradox in the process of building new geographic scales of social life, namely, that efforts to equalize some conditions (e.g., wages) among different locations may actually rely upon continued or even exaggerated differences in others (e.g., vacation and holiday packages) because of the uneven nature of preexisting patterns of economic development.

By telling the story of how the ILA developed a specifically spatial strategy to address the issue of technological innovation in the longshoring industry, I hope to have done three things. First, I hope to have shown how workers and union officials pressing for the equalization and differentiation of contract conditions have produced new geographical scales of bargaining in the industry. Local, regional, national, and even international scales of contract bargaining (or anything else) do not simply exist, waiting to be used; rather, they are actively created through the spatial praxis of social agents. Neil Smith (1993) has suggested that such a creative process involves actors "jumping scales" as they seek to expand geographically their struggle. While sympathetic to his intents, I would argue that this is, perhaps, an inappropriate metaphor, for it implies the existence of premade scales between which actors can jump—a position seemingly in contradiction to his argument about scale being socially produced. The ILA did not simply "jump" from one scale of bargaining to another in some sort of inevitable progression, but actively and laboriously created a regional and national system of labor relations through its struggles, compromises, and defeats. In the process, the union also forced the employers to develop regional and national organizations to match its own structure and to adopt a number of common practices and policies that helped integrate the industry geographically by extending to other ports many of the contract provisions first developed in New York.

Second, the concrete connections between actors operating simultaneously at different scales (local and national union officers, for example) meant that success at one scale often shaped the possibilities for success at another. Sometimes such officials worked in tandem (for instance, to pursue a national contract), whereas at other times they worked in opposition to one another (for instance, when local officials and dockers in the Gulf defied the union's national leadership located in New York). How these political relationships played out differently at different times and in different places clearly had implications for the ways in which the ILA and its constituent parts constructed the "scales of engagement" that linked dockers up and down the East Coast and enabled the union to implement a new national scale of bargaining in the longshoring industry. Furthermore, examining the ILA's decision to negotiate at the national scale over certain conditions and at the local scale over others, together with the union's practice of pursuing a strategy of enforcing national strikes ("one out, all out") on some occasions and at other times allowing dockers to return to work in those ports that had settled their contracts while continuing to strike recalcitrant employers, provides insights into how union policymakers at different levels within the organization—supported by rank-and-file dockers whose willingness to strike in support of such policies was necessary to achieve their goals—were able to shift the focus of their political praxis between different scales, depending upon their particular goals at various times. Such flexibility not only allowed them to pursue different goals at different times but in the process continually restructured the relationship between such scales of organization.

Third, the ILA's activities are crucial for understanding how the economic geography of the longshoring industry has developed during the postwar period. Paradoxically, the union's securing of job and income protections undoubtedly facilitated the spread of containerization because, once having secured such provisions, the ILA was less inclined to fight adoption of the new technology itself and focused instead upon negotiating its impacts.[28] Additionally, by increasing wages, especially in the South, and securing minimum gang sizes and other job preservation agreements, the union had a significant impact on shaping the economic geographies of waterfronts from Maine to Texas, while the multiplier effects of such higher wages and the preservation of work in the face of containerization helped bolster the economies of the local communities in which dockers lived and worked. Again, these issues force a rethinking of the way in which the industry's economic geography has evolved and the role traditionally attributed to the employers (i.e., "capital") in driving this evolution.

In summary, then, what I have tried to do in this chapter is to take the argument laid out in Chapter 2 one step further by illustrating how workers may shape the production of the geography of capitalism not just at the local or even regional scales—scales at which some may, grudgingly, concede that workers' activities can play a role in shaping the geography of capitalism—but also at the national scale. Succeeding chapters in this volume go one step further again to reflect upon how workers and their organizations may shape the production of capitalism's global spaces—a scale which theorizers of the geography of capitalism usually reserve for the activities of capital.

CHAPTER 6

Labor as an Agent of Globalization and as a Global Agent

Globalization is arguably the central political and intellectual issue of our age. There appears to be very little in either the political or the intellectual arena that has not been shaped in some way by the debates surrounding globalization as either a material or a discursive force. Whether it is in terms of the apparent loss of national sovereignty, the growth of global cultural homogeneity and hybridity, or exhortations about how workers and capital must now compete globally, governments, businesses, and workers have been impacted by globalization both as process but also as idea. Typically, the process of globalization itself has been seen to result from the activities of transnational corporations and the international flows of capital more generally. Indeed, to date, and with the exception of studies of international labor migration flows, there has been relatively little examination of how workers and their organizations have played roles in shaping globally the economic geography of capitalism. While workers and unions have been theorized as capable of exerting sometimes quite powerful influences on local and regional space-economies and, to a lesser degree, on national ones, there has often been little faith that they can do so globally. In fact, it is fair to say, I think, that the power of workers to shape the geography of capitalism has often been implicitly assumed to vary inversely with the scale of analysis adopted, that is to say that, as one moves up the spatial scale from local to regional to national to global, workers' capacities are seen to diminish—witness, for instance, the admonition so popular with many on the left to "think globally" but to "act locally." In such a view, the global economy's creation has been seen as the outcome of capital's need to incorporate ever greater amounts of living labor, raw materials, and consumers/ markets within the circulation process, such that "going global" is presented as a means for individual capitals to secure greater economic and political flexibility than if they remain "local," while it may also allow capital to displace spatially to other parts of the world economic crises (see Harvey, 1985c). The result is that the global scale has invariably been represented and understood as the unchallenged—and unchallengeable—domain of capital (see Gibson-Graham, 1996).

Yet, international labor organizations and international links among workers have existed in some form since at least the middle of the nineteenth century when, in response to the dramatic economic and political transformations associated with the geographical spread of the industrial revolution across Europe and North America, many socialists, anarchists, trade unionists, and others began to become increasingly concerned that labor protections in one country would be undercut by the lack of equivalent provisions elsewhere. Such concerns were particularly brought home to workers by the growing synchronization of economic booms and crises that became evident during the 1840s and 1850s (Hobsbawm, 1975) and the increasing use, especially by British employers, of foreign strikebreakers during periods of industrial unrest. Lattek (1988), for instance, has suggested that the first efforts to develop a formal structure to encourage internationalism among laborers can be traced at least to 1844 and the founding by British Owenites and Chartists, together with refugees from France, Germany, and Poland, of the London-based "Democratic Friends of all Nations," while Southall (1989) has shown how the migration overseas by British union members and artisans in the latter half of the nineteenth century and their subsequent founding of branches of British trade unions in Australia, South Africa, New Zealand, Canada, the United States, and elsewhere helped transmit ideas about unionism across vast distances and provided some of the earliest structures around which international networks of union activists could develop. During the 1850s and 1860s some half-dozen international democratic and workers' organizations came into being, including the International Association (founded in 1855), the Congrès Démocratique International (1862), the Association Fédérative Universelle (1863), the Ligue de la Paix et de la Liberté (1867), and the Alliance de la Démocratie Socialiste (1868) (Devreese, 1988). And, of course, the 1864 founding of the International Workingmen's Association (the "First International") brought together trade unionists and socialists from several countries to discuss, among other things, strategies to facilitate international labor cooperation.

Prior to 1870, in what van der Linden (1988) has described as the "prenational phase" of labor internationalism, efforts to develop greater ties among workers in different countries were largely carried out by a relatively small number of political activists who, though deeply committed to the principles of international labor solidarity, were often unaffiliated with the trade unions in their respective countries. Through their endeavors they attempted to establish an international organization of workers before nationally structured trade union federations—which might fuel the growth of nationalist tendencies—had had the opportunity to consolidate themselves. However, the First International's decline and subsequent collapse in the 1870s, combined with the setting up in Western Europe and North America of permanent national trade union centers during the last three decades of the nineteenth century (the first being the British Trades Union Congress, established in 1868), marked something of a "transitional phase" in the history of international labor affairs.[1] Although rival institutions such as the Second International (founded in 1889) still sought to speak as the international voice of workers, after 1890 it was the various nationally constituted trade union movements that increasingly came to dominate the field of international labor affairs. Indeed, the three decades prior to World War I were marked by a veritable flurry of international trade union activities which included the establishment of some 30 international trade secretariats (ITSs), designed to foster cooperation between unions in particular industries (Busch, 1983; Price, 1945; Segal, 1953), and the 1901 formation of the International Secretariat of Trade Union Centers, to which national trade union federations could affiliate as a means to promote greater labor ties across national boundaries, to collect data of use to unions, and to deliver some modest financial aid during strikes (Sturmthal, 1950).[2] Such was the extent of these efforts to develop international labor links that van Holthoon and van der Linden (1988: vii) have described the hundred years prior to the outbreak of World War II as the "classical age" of working-class internationalism.

During the twentieth century trade unions have developed international labor cooperation and contact primarily through four sets of institutions. First, workers have acted globally through the international trade secretariats. Currently there are 11 such secretariats, of which the International Metalworkers' Federation (IMF), the International Federation of Chemical, Energy, Mine and General Workers' Unions (ICEM), and the International Union of Food, Agricultural, Hotel, Restaurant, Catering, Tobacco and Allied Workers' Associations (IUF) have been most active in seeking to meet the challenge raised by the globalization of economic relations.[3] Second, the national trade union centers in particular countries have affiliated with a number of international labor federations that serve to address broader economic and political issues affecting workers than do the secretariats, whose activities are more focused on particular industries. During the post-World War II period the two most important of these have been, arguably, the Western-oriented International Confederation of Free Trade Unions (ICFTU) and its Communist rival, the World Federation of Trade Unions (WFTU), although other similar global institutions have also played important roles—for example, the religiously oriented Confédération Internationale des Syndicats Chrétiens (International Confederation of Christian Trade Unions-CISC).[4] These labor federations have also either set up their own regional organizations covering various parts of the globe or worked closely with such organizations in various parts of the world. Third, trade unions have pursued labor and human rights issues through a number of intergovernmental bodies. Chief among these are the International Labour Organisation (ILO), the Organisation for Economic Cooperation and Development (OECD), and the Centre on Transnational Corporations, all of which are specialized agencies of the United Nations (for more on the ILO see Alcock, 1971; Ghebali, 1989; and Johnston, 1970).[5] Fourth, various unions and groups of shopfloor activists have developed direct links with their counterparts abroad, frequently attempting to bypass the ITSs and other international organizations that they see as either too bureaucratic or too beholden to particular interests (I shall return to the establishment of some of these links in Chapter 8).

My purpose in recounting this activity in some detail here is to note that it raises important questions about the process and implications of economic and political globalization, and workers' roles therein. Clearly, not only has capital become more global in its operations during the past century and a half but so, too, have workers. Whereas writing concerning the process of globalization usually portrays the emergence of the global economy as capital's creation, the logical outcome of the expansionary nature of capital, and the new economic and political reality to which labor must respond, in this chapter I argue that such a view is problematic for (at least) three interconnected reasons.

First, there is little sense in this literature that workers are themselves capable of proactively shaping economic landscapes through their direct intervention *at the global scale* in the geography of capitalism. The literature has tended to ignore the role that workers themselves have played in the actual genesis and subsequent integration of the global economy, and thus in shaping the geography of capitalism at the global scale. Capital has not been the only actor operating at the global scale. Indeed, given the history outlined above, it might even be suggested that the formal transnationalization of labor in many ways predates that of capital, at least with regard to the arrival on the world stage of the transnational corporation and the global assembly line, two entities that are often seen as emblematic of globalization.

Second, and relatedly, there has been a tendency to portray workers as structurally defenseless in the face of a hypermobile, rapidly restructuring, and globally organized capital. They are seen as the bearers of global economic restructuring, not as active participants in the process. However, while the transnationalization of capital has certainly presented workers with new problems as they now frequently must negotiate with corporations whose operations are located in many different parts of the globe, I would maintain that labor's structural position is not always and necessarily that of the passive victim of globalization. Although their relative immobility in the face of more mobile global capital may make some workers that much more amenable to corporate arguments about the necessity for contract concessions and the like, theoretically (and politically!) we cannot concede the global scale of economic and political organization to capital and present labor as structurally confined to the local, regional, or national scale. Reserving the global scale of action for capital in such a way is problematic, for not only does it ignore the long history of trade unions' international activities but also it denies workers their agency by assuming that they are incapable of creating structures that enable them to operate globally. In the process it naturalizes, rather than explains, the creation of the global scale of economic activity because it fails to recognize that geographic scales, like geographic spaces, are actively created social structures. Corporations have not simply colonized some preexisting global scale but have actively created the global scale of economic relations through their actions. Yet, they have not been the sole agents bringing about the globalization of economic relations. Workers, too, have the capacity to act supranationally and, indeed, have a lengthy history of so doing. To assume that only capital can act globally is not only to neglect labor's agency but, ultimately, it is to conceive of the world undialectically, for it leaves capital as the sole actor on the global stage.

Third, examining labor's role in processes of globalization opens up the black box of "the global economy" to alternate visions. Indeed, when talking about "globalization" we should be cognizant that there are multiple varieties of this process and that globalization does not come in a single flavor. Marx's admonition that the workers of the world should unite in international proletarianism reminds us that there are forms of globalization that could serve as alternatives to the neoliberal version that appears to have millions of workers across the planet running for cover. Recognizing how workers have shaped various aspects of our contemporary global economy—from actually showing support, in some cases, for certain brands of neoliberal globalization to developing international organizations as a way of building global links of worker solidarity—allows us both to understand that there are many different strands to the processes of globalization that we are currently witnessing, and, I hope, to imagine forms of globalization that are more humane than are those desired by the imagineers of neoliberalism.

In this chapter, then, my goal is to lay out a general critique of how labor's role in globalization and shaping the geography of capitalism at the global scale has been ignored. The chapter itself is in three main sections. The first examines some of the international activities of the U.S. labor movement (particularly with regard to Latin America and the Caribbean), both in support of its own goals and in connection with U.S. foreign policy. I use this historical narrative to highlight the U.S. labor movement's complicity in bringing about a globalization of economic relations and how the movement was instrumental in structuring geopolitical discourse and practice after World War II. The narrative is intended to serve as a counterpoint to accounts that see globalization only in terms of the transnational activities of capital. In essence, it allows a much more active role for workers to be written into analyses of the globalization of economic and political relations. The second section discusses how some international trade union bodies have attempted to develop transnational strategies and structures to confront globally organized capital. This section, too, suggests that workers should not be conceptualized simply as passive flotsam driven by the powerful currents of global capital restructuring but that, through their actions, workers actively define and circumscribe the very possibilities for capital's global activities. In sum, both sections serve to illustrate the active roles played by organized labor in shaping the historical geography of the global economy. The third section ponders some of the conceptual issues that these activities raise for the way in which globalization is represented as a process, particularly with regard to the genesis of the global economy and labor's structural position within it.

U.S. LABOR'S ROLE IN THE GLOBALIZATION OF CAPITALISM

The U.S. labor movement has been involved in international affairs since the last half of the nineteenth century.[6] Primarily the movement's foreign affairs have been conducted by the leaders of the national trade union federations (the American Federation of Labor [AFL], the Congress of Industrial Organizations [CIO] and, since these federations' 1955 merger, the AFL-CIO) rather than individual national or local union affiliates.[7] The main exception to this has been a number of unions' activities in the various trade secretariats. A handful of unions, such as the International Ladies' Garment Workers' Union, the United Auto Workers, and the United Steelworkers of America, have also maintained staff on their own payrolls who are responsible for international affairs.[8] Certainly this list does not form an exclusive accounting of the international activities of U.S. workers and their organizations. Not all unions have historically been affiliated with these national federations, and not all labor activists and workers operating in the international arena are members of AFL-CIO unions. Furthermore, AFL-CIO foreign policy has frequently been challenged by dissident elements within the labor movement and so should not be viewed simply as the product of the leadership's will (e.g., see Battista, 1991). Nevertheless, the making of the labor movement's "official" foreign policy has traditionally been the preserve of the president and executive council of the AFL-CIO, who control the Federation's Department of International Affairs.[9]

As we shall see below, the U.S. labor movement's international activities during the twentieth century have been shaped to a large degree by the economic and political position of the United States vis-à-vis other countries. The international activities of U.S. capital and labor have been conditioned, ultimately, by economic self-interest—an expanded market for U.S. goods abroad would generate profits for U.S. business and jobs for U.S. workers— and a belief that it is their historical mission to bring U.S.-style liberal democracy and modern economic systems (i.e., capitalism) to those parts of the world (especially Latin America) left economically and politically "backward" by European colonialism.[10] This belief itself is deeply engrained in U.S. ideologies concerning American exceptionalism and Manifest Destiny to which both U.S. capitalists and U.S. workers have frequently subscribed. It is nowhere illustrated better, perhaps, than in the 1898 statement by Samuel Gompers (who led the AFL for most of its first half-century of existence) that "The nation which dominates the markets of the world will surely control its destiny. To make of the United States a vast workshop [for the world] is *our manifest destiny*, and *our duty*" (quoted in Scott, 1978a: 92, emphasis added). Thus, whereas the labor movement's goals have sometimes been at odds with those of the government and U.S. business, frequently they have been mutually reinforcing. Consequently, labor officials have played crucial roles in enforcing, and indeed shaping, U.S. foreign policy.[11] Equally, though U.S. labor has often virulently opposed the domestic actions of U.S. corporations, the AFL-CIO's Executive Council has frequently been prepared to work hand-in-hand with those very same corporations in the international arena. Representatives from the AFL and from the CIO were involved, for instance, in implementing the Marshall Plan that so dramatically reshaped Europe's economic geography and in reorganizing the trade union movements in postwar Germany, Italy, and Japan (Heaps, 1955; Maier, 1987; Windmuller, 1954). AFL-CIO officials have been active in supporting pro-U.S. trade unions in Europe, Latin America, the Caribbean, Asia, and Africa (we shall return to some of these activities in Chapters 7 and 9). This support has had a dual purpose. Certainly, Federation officials have sought to help weak local unions to become strong enough to provide services to, and to bargain on behalf of, their members. But their intervention has also often been designed to oppose more militant labor activists and the perceived threat of Communist control over local unions (Cox, 1971; Spalding, 1988). Presently, with the apparent end of the Cold War, the AFL-CIO is working with various Central and East European governments and labor officials in an effort to mold industrial relations and labor law in these countries in the image of U.S. trade unionism (for more details see Herod, 1998d, and Chapter 9).

Ideological Underpinnings of U.S. Labor's Foreign Policy

The often close relationship between AFL-CIO policy and that of U.S. capital and the federal government in the international arena stems from the underlying philosophy that has guided the labor organization's affairs both domestically and abroad, namely, the acceptance of capitalism as, essentially, a munificent economic system for (U.S.) workers. Emerging, in part, out of the material conditions in which U.S. workers found themselves during the late nineteenth and early twentieth centuries, especially the growing dominance of the U.S. in the world economic system, this philosophy became enshrined in AFL policy and practice largely through the efforts of Gompers and others to promote what he called "trade unionism pure and simple" (also sometimes referred to as "bread and butter trade unionism") in the face of the more militant and politically-oriented brands of trade unionism advocated by various European and U.S. socialists. For Gompers, workers and capitalists were idealized as partners in progress, and so the purpose of trade unions was to ensure, through organization of the paid workplace, that workers gained the highest income and best working conditions possible within the given (and accepted) distribution of power in capitalist societies (Gompers, 1925). Although he sincerely believed in the principle of international links between national trade union movements and was often critical of European and U.S. government officials for pursuing colonial policies of outright territorial acquisition, Gompers was himself not necessarily opposed to U.S. economic expansionism abroad.[12] Rather, he believed that the international supremacy of U.S. capitalism, with its capacity to generate large quantities of relatively cheap commodities for both domestic and foreign consumption, would bring ever higher standards of living to both U.S. and foreign workers (Simms, 1992). Furthermore, the "civilizing" mission of the United States would bring democracy and modernization to those workers of the world still toiling under the control of the various European imperial powers and/or local dictatorships.

The consequence of such a worldview during the twentieth century has been a frequent correspondence in the international arena between the goals of U.S. organized labor and those of the U.S. government and big business, namely, the defense of U.S. capital's interests abroad, opposition to militant "socialistic" organizations that threaten the hegemony of this capital, and the promotion internationally of a brand of trade unionism that takes capitalism as a given, rather than contestable, political and economic system. Although U.S. labor officials and representatives of business and the government may disagree on tactics, there is little fundamental disagreement on issues of basic philosophy: faith in the capitalist system and the need to make the world safe for U.S. investment.[13] Arguably, three elements have dominated U.S. labor's foreign policy during the past hundred years.

First, the Federation has exhibited a certain ambivalence toward affiliation with international labor organizations, many of which it sees as too politically radical—particularly those that profess an anticapitalist agenda. Such ambivalence was evident at an early time. In 1910 the AFL somewhat reluctantly affiliated with the International Secretariat of Trade Union Centers, which, under pressure from Gompers, soon thereafter changed its name to the International Federation of Trade Unions (IFTU).[14] This marked the first time the AFL formally participated in an international labor organization. However, by the end of World War I Gompers felt himself increasingly isolated from the more radical goals of several other member federations (especially concerning international labor policy toward the Versailles Peace Treaty), and, at his urging, in 1921 the AFL withdrew from the IFTU (see AFL, 1921, for more reasons behind this decision). For most of the next two decades the Federation remained largely outside the ambit of the international labor movement.[15] Virtually the only organic link between U.S. organized labor and the international labor scene during this period remained the ITSs, with which a number of individual AFL affiliates were still connected. Although in the post-World War II period the AFL-CIO has participated more fully in various international labor organizations, it has never really committed itself to the principle of ceding any of its own sovereignty to pursue collectively stated goals of the international labor movement and has tended to pursue an on-again/off-again relationship with such organizations.

Second, the Federation has often operated in close cooperation with the federal government in the international arena. The AFL has always been committed ideologically to the success of U.S. capital abroad since, until fairly recently at least, the success of U.S.-based firms in opening up foreign markets was usually a guarantee of employment for Federation members. This commitment is highlighted in a compact made between Gompers and President Woodrow Wilson's administration during World War I. In exchange for government recognition of the Federation's right to organize, Gompers worked to ensure that industrial production (particularly in munitions factories) would proceed unhindered by union-sponsored disruptions (Larson, 1975). Although during the 1920s Republican administrations broke this compact domestically and backed a corporate open-shop drive against the AFL, in the international sphere the Federation continued to receive federal support for its policy of exporting Gompers-style "trade unionism pure and simple" (Simms, 1992). Isolated from the IFTU, and sharing many of the same international political and economic objectives as the U.S. government and big business, the AFL increasingly pursued the goal of implementing labor's own Monroe Doctrine in Latin America and the Caribbean. To achieve this goal the AFL sought to destroy radical anti-U.S. union and political organizations in the region, to fund labor organizations more favorably oriented toward U.S. geopolitical interests in the hemisphere, and, generally, to ensure the supremacy of U.S. capital. While during the 1920s and 1930s the AFL was much more isolationist and cautious in its international dealings than had been the case prior to World War I (Roberts, 1995), as World War II approached both the AFL and the upstart CIO—which had in fact been more international in its point of view than had the AFL during the mid-1930s—increasingly favored a global outlook on economic and political issues.[16] This outlook ultimately evolved into an interventionist labor stance which has sometimes seen the AFL-CIO work closely with the Central Intelligence Agency (CIA) to suppress movements (themselves often led by trade unionists) opposed to the spread of U.S. political and economic influence, and to support the development of a form of trade unionism that more closely corresponds with the Federation's own view of unions' "proper" roles in capitalist economies (for more on this, see Barry & Preusch, 1990; Morris, 1967; Radosh, 1969; and Scott, 1978a).

Third, since Gompers's time the AFL, and subsequently the AFL-CIO, has pursued a virulently anti-Soviet policy that has fundamentally shaped international labor politics, especially in the post-World War II period. The outbreak of World War II weakened the IFTU significantly, and in late 1945 a new international organization, the World Federation of Trade Unions (WFTU), was established to replace it. Whereas the newly formed CIO—initially less hostile to the Soviets and Communism in general—played a leading role in creating the WFTU, the AFL rejected membership because, unlike the IFTU, the new organization included Soviet trade union representation. Although its refusal to join the WFTU left the AFL as the only major North American or European labor federation to remain unaffiliated, this did not mean it was inactive in international labor affairs. In light of the perceived Stalinist domination of the WFTU, the AFL's international policy after 1945 was increasingly structured by three objectives (Windmuller, 1954): (1) to oppose the WFTU wherever and whenever possible; (2) to maintain contact with the ITSs (which the AFL saw as less tainted by Soviet influence) and to build from within them an organization to rival the WFTU; and (3) to provide direct assistance to non-Communist unions and workers' organizations around the world, even to the extent sometimes of supporting socialists against Communists (Scott, 1978a). The State Department also largely supported these objectives.

U.S. Labor and the Making of the Geopolitical and Geoeconomic Order of the Cold War

In pursuing its anti-Soviet foreign policy, the AFL (and subsequently the CIO, too) played a significant role in fueling the Cold War and so in shaping the global geopolitical and geoeconomic order of the post-World War II period. By 1948 the Cold War was gaining momentum and the national federations making up the WFTU had begun to ally themselves with either Washington or Moscow. A year later most of the Western trade union federations withdrew from the WFTU, which they accused of having become simply an instrument of Soviet foreign policy, to form the International Confederation of Free Trade Unions (ICFTU) (Sturmthal, 1950; MacShane, 1992).[17] The ICFTU's formation marked a clear political and ideological victory for the AFL. Not only had the Federation achieved its international objective of splitting the WFTU into "free" trade unions and those sympathetic to the Soviet Union, but also domestically AFL leaders now asserted the moral superiority of their stance over the CIO, which, they gloated, had been duped into believing it could work with Communist-dominated unions. Combined with the AFL's larger membership, this moral victory allowed the Federation to claim the position of chief architect of the U.S. labor movement's international affairs in the immediate postwar period. The CIO, somewhat more ambivalent than the AFL toward the ICFTU's goals and preoccupied with expelling several of its own left-wing affiliates (Zieger, 1986), was prepared largely to concede this international leadership to the Federation (Windmuller, 1954). This was significant not only for U.S. labor's international activities, for it left the generally more conservative AFL at the helm of foreign policy, but also for its domestic agenda. With the AFL once more the dominant voice of U.S. labor abroad, the Federation's leadership, long critical of the CIO, felt that the time was now right for a rapprochement with its old nemesis that would enable the rival organizations to join forces in the looming struggle against the spread of Soviet Communism. Indeed, rapprochement in the international arena played an important role in laying the groundwork for the two labor organizations' eventual merger in 1955.

Both the AFL and the CIO were eager participants in helping to secure U.S. postwar geoeconomic and geopolitical interests around the globe as a means to ensure access for U.S. products to foreign markets. Perhaps one of the earliest indications of their commitment to this Cold War agenda, and hence of their role in facilitating the globalization of U.S. capital and culture, was the support they provided to the European Recovery Program (the "Marshall Plan"). The centerpiece of the newly articulated Truman Doctrine, the Marshall Plan was designed to prevent the further spread of Communist influence and to reintegrate Western Europe into a liberal capitalist international trade system dominated by the United States. Both the AFL and the

CIO were called upon to play vital roles in undermining Communist-leaning and/or anti-U.S. labor groups, particularly in France, West Germany, Italy, Austria, and Greece (see Radosh, 1969: 304–347, for more details).[18] By providing material aid and organizational support, all wrapped up in a major propaganda campaign extolling the virtues of U.S.-style economic unionism, the AFL and the CIO hoped to encourage the development of a pro-U.S. stance among European workers and labor leaders, an aim which often necessitated working closely and surreptitiously with the State Department and the newly formed CIA.[19] The AFL also encouraged its affiliate unions, which traditionally had not been that active in the ITSs, to step up their activities to ensure that the secretariats remained non-Communist and supportive of the Plan (Godson, 1976). Through these activities the AFL and the CIO were deeply involved in enabling the growth of U.S. political and economic influence in Western Europe and so of ensuring the emergence of a postwar economic system that marched to the tune of Bretton Woods and U.S. capital.

In its vision for remaking the postwar world under the aegis of the Pax Americana, the AFL was eager to see the ICFTU become the foremost international labor organization in the fight against Communism (e.g., see the tone adopted in AFL, 1952: 114–117). As part of this, the Federation urged the ICFTU to develop programs to preach the benefits of U.S.-style economic unionism geared toward workplace collective bargaining rather than broader political and social struggles, struggles that smacked of socialist agitation. To provide an ideological bulwark against domination of the ICFTU by more socialist-oriented European union centers, the AFL also encouraged closer ties with the predominantly Catholic Confédération Internationale des Syndicats Chrétiens (CISC) while at the same time it sought to develop strong regional organizations that would prevent the concentration of power in Europe, a situation too reminiscent for the AFL of the prewar IFTU. Yet, almost immediately, these goals proved to be a source of disharmony in the new organization. European trade union officials appeared less committed to the fight against Communism than the AFL leadership desired and were even prepared at times to work with Communist-controlled labor organizations (such as the French Confédération Générale du Travail) in pursuit of common economic or political goals. Furthermore, whereas U.S. labor leaders saw Soviet Communism as the greatest threat to "free" trade unionism, European union leaders seemed more preoccupied with the threats posed by right-wing regimes in Spain, Portugal, Argentina, Venezuela, and elsewhere (Radosh, 1969). They were also more inclined to see the ICFTU as a vehicle for political agitation, whereas the AFL wanted to use the Confederation as a vehicle to propagate its ideas about Gompersian "bread and butter" trade unionism. Finally, European union leaders were concerned about U.S. control of the ICFTU's agenda and the possibility that the regional organizations would take power away from the central Confederation, thereby weakening its ability to implement coordinated international labor policy.

Throughout the 1950s and the 1960s the AFL-CIO maintained a strained relationship with the ICFTU and U.S. labor leaders continued to clash with their counterparts abroad (particularly those in the unions beginning to form in postcolonial Africa, Asia, and Latin America) over the ICFTU's position on several matters, including its stance toward the Soviet Union and contact with Communist-dominated unions, the balance of power within the ICFTU, "economic" versus "political" unionism, the role to be played by AFL-CIO controlled regional labor organizations such as the American Institute for Free Labor Development (see below and Chapter 7), and whether such organizations should be subject to ICFTU dictate. After nearly 20 years of tense relations, in 1969 the AFL-CIO formally disaffiliated from the ICFTU.[20] Although it reaffiliated in the early 1980s, during the past three decades the AFL-CIO has largely pursued its foreign policy through its own Department of International Affairs and the four regional labor institutions that it established in Europe (the Free Trade Union Institute), Latin America (the American Institute for Free Labor Development), Africa (the African-American Labor Center), and Asia (the Asian-American Free Labor Institute).[21] These regional organizations funded a number of projects designed to strengthen pro-U.S. forces abroad, including supplying technical assistance, grants, and training to local unions and supporting political campaigns (see AFL-CIO, 1987a and 1987b, for official statements about the Federation's foreign policy and organs working in the international arena). Although all the AFL-CIO's regional institutes were active in this regard (e.g., see Foner, 1989; West, 1991), given the value of U.S. investments in Latin America, it is perhaps not surprising that none was more so than the AIFLD.[22]

Making the World Safe for (U.S.) Capitalism: Latin America and the Caribbean

The AFL has a long history of intervention in Latin America and the Caribbean. Under the guise of putting an end to old-style European colonialism in the region (and in the process making way for new-style U.S. economic hegemony), Gompers and others on the Federation's Executive Council actively supported anti-Spanish labor activities in Cuba, Puerto Rico, and elsewhere during the 1880s and 1890s. When war with Spain finally came in 1898, William Randolph Hearst may have supplied the headlines, but it was the AFL that provided moral and technical support to U.S. military operations in the Caribbean. However, it was the outbreak of World War I that really served as the catalyst to the Federation's deep intervention in the region in tandem with U.S. capital and the State Department. As the war disrupted trade between Europe and Latin America, President Wilson saw an opportunity to expand control over the region's economies. The Federation took a leading role in this initiative. Working with members of the Confederación Regional Obrera Mexicana (CROM— Mexican Regional Labor Confederation), Gompers played a key part in helping to establish in 1918 a new regional labor organization for the region—the Pan-American Federation of Labor (PAFL)—which, he hoped, would secure U.S. interests in the hemisphere in the face of more radical anticapitalist labor movements in countries such as Argentina, Chile, and Uruguay (Snow, 1964).[23]

Publicly Gompers averred that the PAFL was designed to encourage solidarity among all Latin American workers—and, indeed, in his mind this was probably true. But it soon became clear that the PAFL had been established largely with the goals of U.S. labor in mind. There is evidence to suggest, for instance, that Gompers's preliminary efforts to promote such a regional labor federation under AFL tutelage were designed to undermine attempts by the Industrial Workers of the World to create a similar, though more radical, organization. The PAFL would also be a way to control German influence in Mexico during the war.[24] Furthermore, by organizing workers (particularly in Mexico) the AFL hoped to discourage U.S. capital from locating in the region to take advantage of cheap labor. Finally, the PAFL would serve as a mechanism, Gompers hoped, to stabilize Mexico in the wake of the 1910 revolution and his fears that the revolution might radicalize along the lines of the recent Bolshevik actions in Russia (Andrews, 1991). For their part, the corporatist-minded leaders of the CROM saw the PAFL as a way of easing growing tensions between the U.S. and Mexico (particularly in light of General John "Black Jack" Pershing's 1916 invasion) and of trying to improve conditions for Mexican migrant workers in the United States. A number of Latin American trade unions—principally those in Argentina, Chile, and Uruguay—refused to join the PAFL, however, arguing that it was little more than an arm of the U.S. government (which had provided substantial financial support for its establishment) and an instrument of the *monroismo obrero* ("worker Monroism") practiced by the AFL.[25]

Although the PAFL continued in existence on paper until 1940, Gompers's death in 1924 and the loss of most of its U.S. financing following the crash of 1929 were essentially death knells for the organization. In its place a far more militant Latin American-led regional organization, the Confederación de Trabajadores de América Latina (CTAL—Latin American Workers' Confederation), emerged in 1938. The CTAL's agenda was much more radical than was that of the PAFL. CTAL officials advocated massive land-redistribution programs, nationalization of foreign-owned facilities, a more clearly class-based, politically oriented labor politics that called for proletarian internationalism, and opposition to U.S. imperialist designs on Latin America.[26] Arguing that Latin America needed to develop free from U.S. interference, the CTAL leadership specifically excluded the AFL from the new organization, although it did, however, develop closer links with the CIO.[27] While the CTAL was initially quite ideologically diverse, within a short time Communist elements came to dominate the organization, something which the AFL found particularly distressing. As a result, the AFL stepped up efforts to oppose the Confederación and even attempted to revive the moribund PAFL, which had not had a congress of its members in over a decade, an attempt that ultimately failed to prevent the PAFL from formally going out of existence.

Meanwhile, the CIO, whose leaders were, arguably, seen as being less tainted with the brush of the AFL's *monroismo obrero*, developed quite strong links with many of the trade unions and labor federations in Latin America, establishing significant relationships with the CTAL and the (at the time) Marxist-led Confederación de Trabajadores de México (CTM— Confederation of Mexican Workers), the main rival of the more conservative CROM (Roberts, 1995). Despite their differences during the 1930s with regard to relations with unions in Latin America, the outbreak of World War II saw both increased U.S. labor activity and increased cooperation between the AFL and the CIO in Latin America as both labor organizations worked to secure essential raw materials. Close links were forged between the AFL and the CIO, on the one hand, and the U.S. Office of Inter-American Affairs (OIAA) headed by Nelson Rockefeller—whose family owned vast holdings in Latin America—on the other. The U.S. government channeled funds through the OIAA to aid efforts by the AFL and, subsequently, the CIO— which was increasingly coming to be dominated by its right wing—to undermine the CTAL. As part of these efforts, the AFL took the leading role in founding in 1948 a new regional body—the Confederación Interamericana de Trabajadores (CIT—Inter-American Confederation of Workers)—to rival the CTAL.

The CIT, however, was relatively short-lived. When the World Federation of Trade Unions split in 1949, the CIT made provisions to dissolve itself and to reconstitute as the Organización Regional Interamericana de Trabajadores (ORIT—Inter-American Regional Organization of Workers).[28] Through the ORIT, AFL leaders hoped once again to use a regional labor organization as cover to encourage the formation and growth of Latin American trade unions that were more favorably inclined toward U.S. economic influence and/or disinclined toward Communist politics in the region. While the ORIT primarily carried out its task through a wide range of organizational and training activities, it did also engage in more clandestine operations. ORIT members, for instance, participated in the 1954 Guatemalan coup that toppled the government of Jacobo Arbenz after he nationalized much of the country's land (including some owned by his family) and redistributed it to the local peasantry—whereas giving his own land away to peasants was one thing, giving away land owned by the United Fruit Company was, clearly, quite another!

Although ORIT enjoyed several successes and was able to affiliate the majority of Latin America's largest unions into a non-Communist confederation, Fidel Castro's seizure of power in Cuba in 1959 greatly alarmed the architects of AFL-CIO foreign policy. Quickly, they came to believe that the ORIT was too independent a body to effectively prevent the further spread of Communism throughout the Western hemisphere, especially because some its Latin American officials actually expressed support for Castro (Scott, 1978a). Rather, what the AFL-CIO desired was a body that was totally accountable to its own Executive Council. It was with such a desire that in 1961 the AFL-CIO launched the American Institute for Free Labor Development (AIFLD). Operating in conjunction with President Kennedy's Alliance for Progress initiative—which was itself designed to undermine radical anti-U.S. social movements in Latin America and the Caribbean—the AIFLD was set up to serve as the AFL-CIO's principal foreign policy arm in the region. Not only would the Institute gather intelligence for use by the State Department, but it would also train Latin American unionists in the practices of U.S.-style "economic" unionism and oppose more militant social/political unionism that was frequently anti-U.S. and anticapitalist in orientation (Barry & Preusch, 1990; Spalding, 1988).[29]

Founded "primarily in response to the threat of Castroite infiltration and eventual control of major labor movements within Latin America" (U.S. Senate Committee on Foreign Relations, 1968: 9), and initially conceived of as a tripartite corporatist entity, the AIFLD included elements from organized labor, business (e.g., officials of W.G. Grace, United Fruit Company, Pan American Airlines, and other large U.S. corporations with interests in Latin America), and the U.S. government.[30] Its principal publicly stated goal was twofold: (1) "provide necessary training to members of democratic trade unions in order to develop their organizations as responsible, free, and democratic unions and to install in the leaders the qualifications required to develop and maintain organizations dedicated to social and economic development"; and (2) to "administer and conduct a program of labor leadership training and/or labor leadership seminars" (U.S. Senate Committee on Foreign Relations, 1968: 57). In some years over 90% of the funding for the Institute's central operations and 100% of the funding for its special programs in Latin America came from various U.S. government agencies (e.g., the United States Agency for International Development [USAID], the United States Information Agency [USIA], the Department of Labor, and the National Endowment for Democracy), which, together with the Inter-American Development Bank, have provided funding levels greater than the AFL-CIO's entire domestic budget.[31] Much of this funding went toward the AIFLD's training facility in Front Royal, Virginia (to which came some 4,300 Latin American and Caribbean trade unionists between 1962 and 1990), to training local trade union leaders "in country" (some 500,000 during the same period, according to AIFLD estimates) who would then train members of their own unions, and to funding a number of labor colleges throughout the hemisphere (such as the Trade Union Education Institute of Jamaica, the Critchlow Labour College in Georgetown, Guyana, and the Caribbean Congress of Labour's education program). In 1966 AIFLD contracted with the Inter-American Center at Loyola University in New Orleans to teach a nine-month course in labor economics.[32] This was to give those unionists (particularly from what were seen as key economic sectors such as postal, telecommunications, dock, and transport workers) who had been identified as having great leadership potential the opportunity for more in-depth study so that they might play a role in national economic planning upon return to their home countries (Doherty, 1966).

Whereas AIFLD proclaimed that its goal was to help organize Latin American workers so that they might improve their lives through unionization and collective bargaining, several critics have accused the AIFLD of seeing its role primarily as one "of trying to make the investment climate more attractive and more inviting to [U.S. corporations]" (AIFLD official, quoted in Spalding, 1988: 264), a policy intended to encourage U.S.-style "development" and thus political stability in the region but one that has undoubtedly encouraged the movement of capital out of the U.S. and into Latin America and the Caribbean. In addition to the labor education programs mentioned above (which usually focused on issues related to collective bargaining, labor movement structures, how to spot and counter "Communist propaganda," union finances, political theory from a liberal, procapitalist perspective, labor legislation, etc.), the Institute also carried out its goals through aiding in setting up credit unions, workers' banks, producer/consumer cooperatives, and building housing for members of pro-U.S. local unions (Romualdi, 1967; for more about such housing programs, see Chapter 7). Latin American and Caribbean trade unionists who have been trained by AIFLD representatives either in country or at the Institute's school in Virginia have also been involved in several coups against leftist governments— including those in the Dominican Republic (1963), Brazil (1964), Chile (1973)—and other political activities aimed at destabilizing regimes which are seen to be anti-U.S.[33]

Although the 1980s saw growing rank-and-file opposition in the U.S. labor movement to official AFL-CIO foreign policy in Latin America, it is only recently, with the 1995 election of a new leadership, that the AFL-CIO's Executive Council appears to be moving away from the policies that drove Federation activities during the Cold War. Hence, for instance, while during the Cold War AFL-CIO policy with regard to Mexico was to cultivate relations with anti-Communist allies in the labor wing of Mexico's ruling party (the Partido Revolucinario Institucional [PRI]), in a visit to Mexico in early 1998 the AFL-CIO's new president, John Sweeney, scheduled several meetings with independent unions that are pushing for a more radical form of trade unionism that challenges the neoliberal agenda pursued by recent Mexican governments and supported by the main progovernment unions (*New York Times*, January 23, 1998). Whereas, of course, part of the AFL-CIO's objective was to help stem the hemorrhage of jobs south of the border, Sweeney's actions indicate that a new politics is perhaps beginning to emerge among the Federation's foreign policy strategists, a politics more concerned with the geoeconomic questions of trade and jobs than with the geopolitical questions of the fight against Communism (I shall return to the theme of such cross-border labor linkages in Chapter 8).

In this section of the chapter, then, I have attempted to do three things. First, I have examined the AFL-CIO leadership's basic philosophy regarding the position of workers in a capitalist economic system—a philosophy that owes much to the economic dominance enjoyed by the United States during much of the twentieth century. Second, I have shown how this philosophy has led the AFL-CIO frequently to work hand-in-hand with U.S. capital and the federal government in the international arena, especially in the fight against Communism and anti-U.S. political activity. Third, I have tried to raise the issue of labor's role in the globalization of economic and political relations. The AFL-CIO—and before that the AFL and the CIO—were not simply taken along for the ride by U.S. corporations as they sought to expand operations in Latin America and the Caribbean. Rather, the Federation's own philosophy has led it to adopt an interventionist role in the pursuit of its own political and economic agenda, namely, implementation of a workers' Monroe Doctrine. The AFL-CIO and U.S. capital were, I would argue, frequent partners in the expansion of U.S. economic and political influence in Latin America and the Caribbean. Not only were U.S. corporations responsible for processes of economic and political globalization, particularly with regard to Latin America and the Caribbean, but so, too, was the U.S. labor movement. Such international activities by the AFL-CIO, its surrogate organizations, and other labor bodies demand that a much more active role be accorded workers in explanations of the genesis of the global economy.

LABOR AND THE CHALLENGE OF THE TRANSNATIONALS

As we have seen, during the post–World War II period the U.S. labor movement's fear of the spread of Communism has dramatically shaped its foreign policy. However, this has not been the only concern driving U.S. labor's international activities. As many corporations have shifted their production overseas to previously unheard-of degrees, and as growing imports threaten jobs domestically, in recent years U.S. unions have increasingly concerned themselves with networking internationally and have turned to the international trade secretariats (Uehlein, 1989) and other entities to do so. While some AFL-CIO officials, perhaps wary of the waning of their own political influence, have expressed fears that growing involvement with the ITSs might "reinforce [not] a common policy for the AFL-CIO but one that is fragmented" (quoted in Gershman, 1975: 72), many have come to recognize the potential role the secretariats can play in confronting globally organized corporations. Paradoxically, a further stimulus in the United States for strengthening the secretariats has come from dissident elements within the labor movement who see the possibility of using them to challenge official AFL-CIO foreign policy. After the United Auto Workers broke with the AFL-CIO in the late 1960s over the latter's support for U.S. involvement in the war in Vietnam, for instance, the UAW increasingly used the International Metalworkers' Federation (IMF) as the conduit for its foreign aid contributions (Foner, 1989; Windmuller, 1970). Additionally, the AFL-CIO's disaffiliation from the ICFTU and the U.S. withdrawal from the ILO in 1977 encouraged many individual unions to step up their activities in the secretariats as one of the few means left open to maintain organic links with workers in other countries.[34]

Arguably, the secretariats' organization on the basis of industrial structure rather than territoriality makes them best suited of all the international labor bodies to confront transnational corporations' global production structures and strategies.[35] The most active secretariats have pursued four main policies to counter the effects of the growth of transnational production. First, they have had some success in developing a degree of international coordination in the realm of contract bargaining. Although the bureaucratic impulse among various secretariats means they are sometimes slow to act and may be detached from the rank-and-file union member (Moody, 1997), they have nevertheless successfully conducted a number of international activities to pressure recalcitrant corporations when an impasse has been reached by an ITS affiliate in its collective bargaining, including organizing boycotts, cessation of overtime work in sympathy, providing financial support, and implementing negative publicity campaigns against the company (Cox, 1971; Bendiner, 1987; Moody, 1988). In 1969, for instance, the International Federation of Chemical, Energy and General Workers' Unions (ICEF) trade secretariat played a key role in coordinating contract negotiations between the French glass manufacturing giant Companie de Saint Gobain and local unions in France, Italy, West Germany, and the United States, in the process ensuring that the company bargained on the basis of its global profits rather than national conditions in each of these four countries (Cox,

1971).[36] This success undermined the company's efforts to play off these local unions against one another during collective bargaining talks. In the mid-1980s West German unionists lobbied executives at the chemical transnational BASF to settle a dispute with workers locked out of one of the company's plants in Louisiana, a campaign that was partially coordinated through the ICEF (see Bendiner, 1987: 62–102, for more on this and other such examples; also, Barnet & Muller, 1974: 312–319).[37] More recently, a group of U.S. steelworkers locked out of their aluminum smelting plant in Ravenswood, West Virginia, successfully used the IMF, the ICEF, and the International Union of Food, Agricultural, Hotel, Restaurant, Catering, Tobacco and Allied Workers' Associations (IUF) to articulate a global corporate campaign that forced the company to allow the union back into the plant (this case is examined in Chapter 8). Similarly, when the U.S. telecommunications company Sprint fired Latino/a workers organizing at its facility in San Francisco in 1994, the Postal Telegraph and Telephone International trade secretariat (now part of the Union Network International—UNI) organized an international campaign to force Sprint to rehire the workers and allow them to organize. As part of this campaign, telecommunications unions in Germany, France, Nicaragua, and Brazil lobbied their own governments to reject efforts by Sprint to buy parts of their national telephone systems unless the company negotiated with its fired workers (MacShane, 1996)—itself an example of how organizing at the scale of the nation-state can still be used successfully to challenge a global corporation such as Sprint.

Second, although ITSs recognize that economic conditions around the globe realistically preclude establishing uniform wage rates (particularly given fluctuating exchange rates), they have argued that such differential conditions should not prevent workers from enjoying similar health and safety conditions at work, nor rights to union representation and collective bargaining.[38] In this light, the IMF has pioneered the concept of world industry councils as permanent structures to help coordinate international bargaining by providing information and resources to aid in strikes and other activities for promoting international solidarity. Beginning in the mid-1960s World Auto Councils (WACs) were established under IMF auspices at each of the large producers to pressure them to adopt a number of policies concerning union rights and working conditions throughout their plants worldwide. Initially four such councils were established to coordinate negotiations in the auto industry with Ford, General Motors, Chrysler, and Volkswagen, but subsequently they have been extended to other auto producers and to other industries (particularly those represented by the IUF and ICEF), including mechanical engineering, electronics, petroleum refining, tire manufacturing, food production, tobacco processing, and metalworking (see Olle & Schoeller, 1977).[39] In 1971 the WACs initiated a campaign to harmonize several nonwage issues (such as length of break-times) in an effort to discourage manufacturers from playing off workers in different parts of the globe against one another in a continuous process of downward concessionary whipsawing.

However, this strategy is not without its problems. Cultural differences among Japanese, European, and North American auto workers, particularly concerning attitudes toward work and labor–management cooperation, have made it difficult to pursue such harmonization to the fullest extent. Also, whereas auto production until fairly recently was characterized by relatively integrated production systems concentrated in North America and Europe, the spread of production into low-wage countries such as Mexico, South Korea, and Brazil, combined with the auto producers' assault on national bargaining in North America (Herod, 1991a) and elsewhere, have made it increasingly difficult to secure more harmonized contracts. Furthermore, different groups of workers have different pressing needs. Latin American and Southeast Asian delegates to WAC meetings have consistently stressed the need to ensure basic trade union rights in the face of authoritarian regimes, whereas in North America and Europe job security, wages, and shorter working hours have often been the priority (Bendiner, 1987). Nevertheless, the WACs do see harmonization as the most viable strategy for confronting the growing transnational challenge.

Third, several secretariats have encouraged mergers between unions in particular industries as a means of presenting a more unified position against management (especially in countries such as Britain and Japan, where several different unions might be present in a single manufacturing facility), although this has been a slow process because of jurisdictional and ideological rivalries. In 1972, for instance, the IMF undertook a campaign to promote the merger of several Japanese auto unions (Curtin, 1973). The secretariats themselves have also engaged in such mergers, with several amalgamating in the 1990s to form larger secretariats whose membership represents workers in a variety of industrial sectors in which their employers have interests. Despite some success, engaging in such mergers has proved problematic, though, because the varying ideological and political affiliations of the unions representing workers in particular industries often limit cooperation, let alone the possibility of mergers. Even in cases where politics and ideology are not issues, particular unions' unwillingness to cede their autonomy to other unions can also be a stumbling block.

Fourth, the secretariats have invested time and money in developing data bases to provide affiliates with information on corporate operating and bargaining practices in different parts of the globe, ownership of subsidiaries, contract details in different countries, and location of operations. The industry councils have generally been responsible for gathering this information, while the secretariats have served as the coordinating bodies for its dissemination to the appropriate affiliates. By the early 1990s the IMF Economic and Social Department, for instance, had established files on more than 500 multinational companies (IMF, 1991). Such information has proven particularly valuable to IMF affiliates seeking to implement campaigns (increasingly global in extent) against corporations. Indeed, it is somewhat ironic, perhaps, that international communication among workers has been greatly facilitated by unions' use of the very same new technologies by which transnational corporations coordinate their global operations. Whereas many of those who write about the global economy have argued that the telecommunications revolution has now made possible the micro-management of distant operations by corporate decision makers, few seem to appreciate the fact that electronic mail, chat rooms, and bulletin board systems also allow workers in different parts of the globe to exchange information about strikes, contracts, and working conditions almost instantaneously (e.g., see *Labor Notes*, 1994; Froehling, 1997; Lee, 1997). Access to this type of information has proven vital for unions seeking to trace patterns of corporate ownership and control as a starting point for political campaigns and for coordinating international activities. The rise of the Internet and the ease with which information can be retrieved and disseminated have greatly facilitated such activities by unions and have led at least one commentator to suggest that international trade union activities will be increasingly conducted by those (often, rank-and-file) "networkers" who have the linguistic and communication skills to link together workers in different parts of the globe rather than by the professional "agents" who work for various international trade union organizations (Waterman, 1993). In 1994, for example, the ICEM and the United Steelworkers of America (USWA) initiated a two-year-long cybercampaign—presented by the ICEM as the first corporate campaign to be conducted on the Internet—against the Bridgestone/Firestone tire company in its dispute over the illegal firing of 2,300 workers at five of the company's U.S. subsidiaries. The Steelworkers and the ICEM effectively used the Internet to globalize the dispute and to enlist the help of "cyber pickets" across the planet who were urged to bombard the company's public relations offices with messages of condemnation, which ultimately contributed to the company's rehiring of almost all of the fired workers (see Herod, 1998e, for more details).

In order to concentrate their efforts on these four issues the secretariats have delegated to the ICFTU many of their traditional activities concerning propagating trade unionism and representing workers' interests before intergovernmental agencies (Casserini, 1993).[40] By working with agencies such as the International Labour Organisation, the Organisation for Economic Co-operation and Development, and the United Nations's Centre on Transnational Corporations, the ICFTU has tried to regulate corporations' international activities by exerting influence on the shaping of international economic and social policies. For example, the ICFTU developed a Code of

Conduct for multinationals investing in South Africa during the apartheid period, and trade union representatives have pushed hard for inclusion of a social clause guaranteeing basic labor and human rights in the General Agreement on Tariffs and Trade (GATT) and, subsequently, in negotiations over the creation of the successor to GATT, the World Trade Organization. Such efforts to link labor standards to trade have, however, been decried by some (such as Malaysia's Prime Minister Mahathir Mohammed) as "imperialistic" since they threaten to limit developing nations' "sole advantage," that is, their "comparative advantage" in cheap labor (MacShane, 1996: 16)—a claim that illustrates the difficulties for unions in trying to develop codes that protect workers when one group of workers' loss (those in the global north) is frequently another group's gain (those workers in the global south). Nevertheless, the unions have successfully managed to incorporate a number of guidelines and codes into several international agreements and resolutions (e.g., formulation in 1976 of the OECD's "Guidelines for Multinational Enterprises"), although the lack of an effective enforcement mechanism has limited the practical effects of such regulations (the ILO's Tripartite Declaration of Principles concerning Multinational Enterprises and Social Policy, for example, is entirely voluntary). In other cases, some corporations have agreed voluntarily to adopt such regulations, largely out of concern that failure to do so might damage their corporate image and hence ability to market their products (Weinberg, 1978). When the German metalworkers' union IG Metall developed a 14-point Code of Conduct regarding trade union rights, for example, many West German multinationals followed voluntarily, as did several Swedish corporations who were themselves under pressure from their own unions (MacShane, 1996).

Whereas some labor dissidents in the United States have turned to the international trade secretariats as an alternative to official AFL-CIO foreign policy, still others have tried to develop international connections directly between plants, a task made much easier through modern computer technologies such as e-mail and one that allows them to bypass altogether the secretariats, which they see as too often beholden to the bureaucracies of their various national affiliates. The past two decades have seen a number of such organizations spring up. One of the most important has been the Transnationals Information Exchange. Composed of some 40 research and activist labor groups, the TIE tracks transnationals and disseminates information concerning their activities (for more on these activities, see Peijnenburg, 1984). In 1981, 24 top AFL-CIO officials and hundreds of rank-and-filers formed the National Labor Committee in Support of Democracy and Human Rights in El Salvador to oppose official Federation policy toward Central America. Because many of the officials on the Committee headed large AFL-CIO affiliates, they have been able to use the size of their membership to initiate some debate concerning the foreign policy goals of the Federation's Executive Council and Department of International Affairs, particularly concerning support for U.S. military intervention in Central America. During the mid-1980s several labor-led, grassroots antiapartheid groups also emerged in efforts to build links with the growing number of black trade unions in South Africa, and apartheid was an issue in a number of strikes in the United States against transnationals that had operations in that country (Moody, 1988). A number of grassroots labor groups in the United States (such as the North American Worker-to-Worker Network and the Coalition for Justice in the Maquiladoras) have also developed contacts with workers in Central America and several other less developed countries. Although such groups are usually lacking in funding and are often relatively small, they do nevertheless represent an increasingly important challenge to official AFL-CIO foreign policy and an alternative means by which rank-and-file workers can communicate with, and support, one another globally (for other examples of grassroots international labor cooperation, see Hecker & Hallock, 1991).

To summarize, then, there are a number of ways in which workers and their organizations in various countries have attempted to develop transnational structures to coordinate their activities globally. As this brief account suggests, they have been quite active in seeking to build structures and organizations to challenge *at the global scale* the operations of transnational corporations. Although bodies such as the trade secretariats have a long lineage, during the past three decades they have become particularly important as capital has become transnational to a much greater degree. Equally, numerous new worker networks are being formed in the wake of the telecommunications revolution. Through the ITSs and other mechanisms such as the Internet, workers across the globe have been able to network with their confederates and to organize successful transnational labor campaigns. These efforts raise important conceptual issues about how globalization is represented in the literature, particularly with regard to the frequently presumed impotence of workers in the global economy.

IMPLICATIONS OF INTERNATIONAL TRADE UNION ACTIVITIES FOR DEBATES CONCERNING WORKERS' ROLES IN GLOBALIZATION

The international activities on the part of U.S. and other workers outlined above raise three critical issues that have broader bearing on the process of globalization and how it is represented in the literature, namely: (1) how the economic and political context within which various social actors find themselves affects the ways they behave; (2) what workers' roles are in shaping the (international) division of labor; and (3) what workers' roles are in shaping the geography of capitalist uneven development in the global economy.

The Contingent Nature of Capital–Labor Coalitions

U.S. unions have generally subscribed to a view that their historical mission on the global stage is to help bring "modernization" and liberal democracy to precapitalist and/or authoritarian societies, with both gifts to be delivered under the umbrella of U.S. economic hegemony. However, this belief has had different implications at different times. U.S. labor had much to gain during the first half of the twentieth century by paving the way for U.S. economic expansionism in Latin America and the Caribbean at a time when production was organized nationally and the U.S. was the dominant global economic power. As long as these conditions prevailed, U.S. workers could enjoy relatively secure, well-paid jobs producing commodities that were ultimately destined for export and consumption south of the border. Yet, in working frequently hand-in-glove with corporations to destroy militant anti-U.S. trade unions and precapitalist economic systems, ironically the U.S. labor movement itself played a significant part in changing these very conditions and, ultimately, in undermining this accord. By helping to make Latin America and the Caribbean safer for U.S. investment, labor's foreign policy actually encouraged and facilitated the spread of offshore production by many U.S. corporations. In turn, this has meant that, instead of U.S. workers producing commodities for consumption in Latin America and the Caribbean, workers in countries like Mexico and Guatemala are increasingly producing commodities for consumption by U.S. workers. Combined with the decline of U.S. economic power generally, this transformed economic situation has encouraged many U.S. workers to turn to the trade secretariats and other grassroots international organizations during the past few years as a means of trying to save their own jobs by bringing pressure to bear on the global operations of U.S. transnational corporations. Likewise, the AFL-CIO's Department of International Affairs has also become much more involved in developing solidarity links with workers in the region in response to changing economic conditions and the rise of neoliberalism as manifested, for example, in the North American Free Trade Agreement.

This change in relative emphasis in U.S. workers' international activities suggests a number of things.[41] First, there is clearly a need to be historically and geographically specific when seeking to understand the political and economic decisions made by workers and their organizations. Rather than imputing a universal behavioral trait to particular social groups because of their structural relationship to other groups—labor is automatically assumed to be opposed to capital because it is, in terms of the extraction of surplus value, exploited by the latter—behavior can, in fact, be dramatically influenced by contingent factors (e.g., a nation's relative economic strength at particular historical junctures). Whereas U.S. labor and capital have often fought bitter battles domestically during the twentieth century, U.S. global economic dominance meant they both had a vested interest in working closely internationally to secure common objectives. This suggests that not only may social relationships between the same actors take different forms in different places (domestically versus abroad, for instance) but that as the political and economic environment changes so, too, may the relationship between those actors. The transnationalization of U.S. capital has redefined the boundaries within which U.S. workers once lived. In turn, this is leading many to seek to develop new relationships with workers abroad.

Second, the existence of such a capital–labor accord in the international arena contradicts much of the literature on globalization that sees unions attempting to "go global" in an effort to match the transnationalization of capital and so to maintain some degree of economic and political leverage vis-à-vis corporations. Not only did U.S. labor largely go global *in collaboration with* U.S. capital, but much of the impetus to do so came as a result of labor's efforts to stop the spread of Communism internationally and to secure U.S. geopolitical interests rather than from the transnationalization of capital. Indeed, U.S. labor's international activities during the first three-quarters of the twentieth century probably had as much, if not more, to do with foreign and domestic political concerns (the fight against Communism, rivalries between the AFL and the Industrial Workers of the World concerning creation of a Pan-American regional labor organization, and between the AFL and the CIO concerning membership of the IFTU and the WFTU) as it did with the emergence of the global assembly line. This suggests a need to integrate more closely economic explanations of labor's attempts to go global (e.g., the need to confront transnational capital) with political concerns (e.g., the geopolitical intrigue of the Cold War).

Third, the literature on Fordism and the apparent emergence of a post-Fordist mode of regulation has made much of the importance of the domestic capital–labor accord as a prop for the emergence of a mass production/mass consumption economic and political system in the U.S. (and elsewhere) (e.g., see Aglietta, 1979; Amin, 1994). The existence of such an accord in the international arena suggests that the argument may be extended geographically in potentially fruitful ways. In particular, it raises questions concerning to what extent U.S. Fordism was based on the nation's dominant position within the global economy and the ability of U.S. capital and labor to get foreign workers to consume U.S.-produced goods, and to what extent it was based on factors internal to the United States. Equally, the collapse of the domestic capital–labor accord in the United States and the transition to what some have called a post-Fordist system appears to have an international aspect to it, as the interests of U.S. capital and labor are increasingly diverging abroad and as many U.S. workers have begun to develop new cooperative relationships with foreign workers.

Labor and the International Division of Labor

The division of labor is one of the most fundamental social structures that defines any economic system. The transnationalization of capital, particularly during the past three decades or so, has augured a dramatic transformation in the organization of capitalist production and has brought with it a new international division of labor (Fröbel et al., 1980). Many writers on this topic conceive of the forces that have led to this unprecedented transnationalization of capital and new international division of labor as deriving from the internal logic of capital itself. It is the "tendencies inherent in the nature of capitalist development," so the argument goes, that have led to the emergence of "multinational corporations [as] the dominant institutions of advanced capitalism" (Peet, 1987: 42). Competitive pressures (including, often, the desire to avoid well-organized labor in the core capitalist countries) have forced many companies to go global by enlarging the geographical scope of their operations in search of new markets and sources of raw materials, cheaper labor, less regulated environments in which to do business, and generally lower production costs. In this interpretation, transnational corporations are seen as the agents of contemporary globalization while workers are seen as (usually) passive bystanders in this process.

Such a focus on the activities of capital as the primary agent bringing about the globalization of economic relations has a long lineage. Classical trade theory, for instance, was largely concerned with examining the international flow of capital in the form of traded commodities and portfolio investments. Classical capital theory sought to explain international flows of capital in terms of interest-rate differentials. Marx (1967 [1867]) examined the roles played by colonies as providers of working capital for European industrialization, suggesting (though not in so many words) that the economic crises inherent to capitalism to some extent could be geographically displaced to other parts of the planet through a process of capital's becoming increasingly global in its operations. Equally, Lenin's (1939 [1900]) analysis of imperialism focused on the role played by cartels and multinational financial capital as an integrating and globalizing force. Similar preoccupations with capital as a globalizing agent are to be found in the contemporary geographical literature written by both radical and more mainstream analysts.[42]

Three points are important here. First, the historical narrative outlined in the first part of the chapter illustrates that workers, too, have played an active role in shaping the new international division of labor. They have done this both indirectly and directly. Indirectly, workers in the industrialized economies have helped propel capital toward transnationalized operations through their organizing activities. Indeed, one might even argue that labor's success in organizing in such economies has forced capital to look to the global south for new investment opportunities. Directly, workers have actively shaped flows of capital globally, either by encouraging/discouraging capitalist penetration into particular regions of the world or, alternatively, by organizing to prevent capital from leaving particular localities. This suggests that analyses that portray the spatial and social division of labor as evolving in response solely to the way *capital* needs to organize the production process in particular ways are problematic, for they marginalize workers' roles in this process—such as the historical willingness of workers in the global north to maintain their status as part of the global labor aristocracy by supporting imperialism and neocolonialism. The international activities detailed above suggest the need to conceptualize workers as active participants in the creation of the (international) division of labor and the shaping of the global economy through, for instance, directly influencing the geography of foreign direct investment (as we shall examine more closely in Chapter 8).

Second, the contemporary actions of some North American and European workers relating to the trade secretariats raise interesting questions concerning the intersection of class relations and spatial relations within the international division of labor. On the one hand, workers in the industrialized economies have increasingly been trying to use the secretariats to encourage the organization of workers in the less developed countries so that such workers may protect themselves against the predations of transnational capital. This practice can be seen as a *class*-based response to globalization, with the secretariats attempting to develop global links between workers as a means to limit corporations' ability to play different parts of the world against each other in a downward concessionary spiral. On the other hand, however, it is clear that the impetus for many workers in the industrialized nations to develop such global campaigns often comes from their desire to protect their own particular niches in the international division of labor. In this sense, developing global links through the secretariats and other organizations can be read as a *geographical* response to capital flight and restructuring by which North American and European workers are simply attempting to preserve their own industrial spaces in the face of the globalization of production. This raises significant questions (to which we shall return in Chapter 8—see pp. 215–217) concerning political strategy and the extent to which geography may cross-cut class-based social structures (see also Herod, 1991c, and Herod, 1994, especially pp. 91–93). It particularly suggests the need to pay closer attention to how class practices are geographically constituted and how space is infused with class relations. Any attempt to develop a class politics clearly must be sensitive to issues of geographic variation between places. Equally, geographic variation between places has profound implications for the process of class formation and the articulation of class politics.

Third, the spread of transnational manufacturing into the less developed countries (LDCs) has been marked by the growing employment of female workers in the "host" nations. Indeed, some commentators have estimated that in the export-processing zones of the "Third World" up to 90% of all workers are women. This raises a number of theoretical and practical issues for unions from the core industrial countries who seek to develop international solidarity, chief among which is the fact that most such unions have historically been male-dominated and have frequently paid scant attention to issues that affect female workers—a situation that has led Haworth & Ramsay (1984: 75) to suggest that to date "international solidarity [has been] a male preserve." Since women workers invariably have significant domestic responsibilities in addition to their paid work activities, effective organization for international solidarity, Haworth and Ramsay suggest, will require strategies that are sensitive to the different social realms and contexts within which men and women live their lives in different parts of the world. Given that gender relations vary spatially in form, such strategies inevitably will need to be geographically specific. How unions and workers seeking to develop international links deal with issues of gender, then, will undoubtedly affect the efficacy of their efforts and, consequently, the emerging structure of the global division of labor.

Labor and Globally Uneven Development

The third issue raised by the examination of workers' international activities concerns the link between these activities and the globally uneven development of capitalism. Typically, the uneven geographical development of capitalism is seen in one of two ways (Smith, 1990 [1984]). The first and most widespread view in the geographic literature is to see uneven development as simply inevitable, given the impossibility of *even* development. However, this conceptualization is problematic and has little explanatory power because it presents uneven development as an ahistorical, universal process in which everywhere develops unevenly all the time. The second approach portrays uneven development as inherent to the dynamic of capitalism. In this view (Smith, 1990 [1984]: xiii) "uneven development is the systematic geographical expression of the contradictions inherent in the very constitution and structure of capital. . . . [It is] structural rather than statistical" and is seen to emerge from the internal dynamic of capital itself.

Certainly, there are definite theoretical advantages to be gained by conceptualizing uneven development as an expression of deeper structural forces at work under capitalism rather than as simply an ahistorical inevitability. For instance, by doing so it becomes possible to make powerful theoretical links between the dynamics of capitalist accumulation and the production of economic landscapes in ways that the more traditional concept of uneven development does not allow. However, the international activities of workers outlined above suggest that explanations that see uneven development solely as an expression of contradictions within capital alone do need to be revised. U.S. workers and their unions, I would argue, have played an important role at certain historical junctures in maintaining Latin America and the Caribbean in a state of dependent development vis-à-vis the United States, both by helping to undermine political organizations that have challenged this dependency and by ensuring markets remained open to U.S. goods and services. In this sense, they were active players in the underdevelopment of the region. Furthermore, they played this role in their own right in pursuit of their own economic and political agenda (principally job creation in the United States and fighting Communism) rather than as simply adjuncts to U.S. capital. Equally, by seeking to restrict the flow of capital out of the industrialized nations and to maintain their own niche in the new international division of labor, North American (and, for that matter, European) workers are at the moment actively engaged in trying to maintain present levels of economic development in their own economies. At the same time, workers who are organizing themselves in the less developed countries are also shaping patterns of production and consumption and thus, ultimately, of global (uneven) development.

WORKERS AND GLOBALIZATION: CONCLUSION AS PROLEGOMENON

In this chapter I have tried to do three things. First, I have suggested that workers have been actively involved in the processes that have brought about the globalization of economic and political relations. Although I have focused much on the activities of U.S. labor in this regard, I do not doubt that similar arguments could be made about the relationship between, for example, the various European labor movements and their own national governments' policies toward their (former) imperial possessions (e.g., see MacShane, 1992, and Thomson & Larson, 1978, on the British Trade Union Congress's relationship with unions in India and the West Indies; Weiler, 1988, for his account of how the TUC and the British Colonial Office worked hand in hand during the Cold War to ensure that "responsible" and not "radical" trade union leaders emerged in Britain's colonial possessions; Jagan, 1972 [1966] on TUC support for British counterinsurgency policies in Malaya; Bédarida, 1974, on French unions' perspectives on colonialism;

Mergner, 1988, on German labor's attitudes toward colonialism in Africa; Tichelman, 1988, on Dutch unions' support for colonialism in Indonesia; see also Wedin, 1984, and Hatch & Flores, 1977, on the Israeli labor federation Histadrut's links to the nation's Foreign Ministry and efforts both to stimulate Israeli exports throughout the globe and to limit the spread of pro-Soviet Communism in Latin America). The creation of organizations such as the international trade secretariats and the various global labor federations, together with U.S. labor's activities relating to the implementation of its own version of the Monroe Doctrine in Latin America and the Caribbean, are indicative of workers' roles in shaping international geoeconomics and geopolitics during the nineteenth and twentieth centuries. Rather than resulting solely from the internal logic of the necessary expansion of capital, the emergence of a global scale of capitalism has clearly been shaped, at least in part, by the activities of workers and their organizations.

Second, whereas much of the literature avers that workers are now seeking to go global simply in response to the globalization of capital, the narrative presented here shows that this argument is problematic. Not only have workers been active in creating international labor organizations and operating internationally for over a century, but the impetus to do so frequently has had more to do with political concerns (e.g., anti-Communism) than with countering the activities of transnational corporations. Therefore, while the contemporary globalization of capital is obviously an important process to which workers are having to respond through international solidarity campaigns, it should not be seen as the only (or even the most important) factor that has driven labor's international activities. To do so suggests a highly economistic reading of workers' activities in the international arena (i.e., that it has been just economic concerns that have shaped their international praxis) but also results in a historical foreshortening of such international activities in the narrative. In other words, if it is the globalization of capital in the past three decades that is seen as the stimulus to labor's international activities, then not only will labor's international activities be interpreted as being conducted solely in response to those of capital but, moreover, such activities will be seen to have emerged only in the post-1960s era when the globalization of capital has become such a topic of intellectual and political interest. In such an approach, earlier labor internationalism will be ignored. In contrast, by examining some of the political issues that have shaped international labor activities during the past 150 years or so, we can develop a much more historically and geographically nuanced account of these activities.

Third, I have attempted to show that, rather than being *necessarily* structural victims of globalization, workers have often successfully challenged *at the global scale* the actions of transnational corporations. Thus, I would suggest, workers' immobility in the face of capital flight should not been conceptualized as *necessarily* leading to political impotence. While developing global solidarity is probably more difficult than is developing national or regional practices of solidarity, it is nevertheless still possible. By building global networks to exchange information and provide material aid, immobile workers in different parts of the globe have, in fact, been successful in challenging more mobile capital's attempts to play workforces against each other. The greater availability of information and ease of communications resulting from innovations in computer technology have facilitated increased worker contacts around the globe. Although the greater variety of levels of development, legal structures, and cultural attitudes undoubtedly make it more difficult for workers and unions to confront corporations that are organized at the global scale than it does to confront those that are national or regional firms, the evidence suggests such difficulties are not insurmountable. The ability of workers to further develop such contacts and links will have immense implications for international labor politics in the twenty-first century.

In sum, then, the issues raised in this chapter lead to a broad theoretical questioning, namely: what are workers' roles in making *globally* the geography of capitalism and how have they been able to construct global scales of engagement in pursuit of goals that have sometimes been defined in class terms (i.e., uniting workers in different parts of the world) and sometimes in spatial terms (i.e., in using labor internationalism to further the interests of workers in some places at the expense of those located elsewhere)? While economists, economic geographers, and other social scientists have traditionally relied on understanding the dynamics of capital to explain geographies of global economic development (e.g., theory of the firm, circuits of capital circulation, patterns of foreign direct investment, etc.,) and have tended to assume a priori that capital may act globally but that workers may not, the historical record does not sustain such a supposition. Moreover, understanding workers' roles in processes of globalization, and recapturing this history and geography, can provide important theoretical insights in the struggle to build a political project that challenges neoliberal versions of globalization by exploring alternative visions of globalization that might be realized through workers' praxis. Whereas to date the term "globalization" has usually been taken to be synonymous with neoliberalism, a globalization orchestrated by entities such as the World Trade Organization bent on creating a superconductive, friction-free capitalism (Gates et al., 1995) in which places are played against one another in a race to the bottom, this is only one version of globalization. Instead, globalization could also be taken to mean greater transnational links among workers. Thus, perhaps, the question to be asked is not whether or not workers and others should mobilize against globalization but, rather, what kind of globalization should we have? "Globalization" should not be conceded to the neoliberals in the way in which it is when workers' roles in shaping patterns of global development and international relations are ignored. By recognizing the part of workers and their organizations in shaping contemporary processes of globalization, we can see how they may in fact play important roles in developing different geographies of globalization, geographies shaped not by the tenets and practices of neoliberalism but by those of proletarian internationalism. Contesting notions and processes of globalization in this way, then, is no mere academic exercise, but can serve as the foundation for a global politics that challenges neoliberalism. In such a contestation, the stakes could not be any higher. Indeed, there is, quite literally, a world to be won.

CHAPTER 7

Engineering Spaces of Anti-Communism

Connecting Cold War Global Strategy to Local Everyday Life

> If the working man has his own house, I have no fear of revolution.
>
> —Attributed to ANTHONY ASHLEY COOPER,
> 7th Earl of Shaftesbury, 1860s

> Revolución es construir [Revolution is to build]
>
> —Slogan used at CONATRAL
> union housing project opening ceremony,
> San Pedro de Macorís, Dominican Republic

> No man who owns his house and lot can be a Communist. He has too much to do.
>
> —BILL LEVITT,
> developer of Levittown, New York[1]

In the preceding chapter I took something of a macro approach to examining how, through its international activities, the U.S. labor movement has shaped global geographies of development during the twentieth century. In this chapter, in contrast, I want to focus on the impacts such activities have had at the very local scale of individual locales and particular neighborhoods. Whereas a great deal of analysis of Cold War labor geopolitics has focused on the global scale in terms of rivalries between transnational organizations (e.g., the ICFTU vs. the WFTU) or national-level concerns (as manifested in conflicts within the AFL and the CIO between Communist and

161

non-Communist elements), in this chapter I show how Cold War international geopolitics and the politics of economic development were also played out through very local struggles over the spatial engineering of the built environment in various countries of Latin America and the Caribbean. Specifically, the chapter examines a number of housing and infrastructure improvement programs implemented by the American Institute for Free Labor Development (AIFLD)—the AFL-CIO's regional organization for the hemisphere (see Chapter 6)—and local partner unions and community groups beginning in the early 1960s.[2] In a classic case of using spatial engineering to pursue a social and political agenda, these programs were geared toward members of labor unions that were either anti-Communist or were threatened with Communist infiltration and takeover. Some partner unions were chosen on the basis of their strategic role in the economy (e.g., communications workers, transport workers, port workers, local government workers), while locations for other projects were chosen because they were considered by AIFLD to be in "key geographic sections" of various countries (AIFLD, 1964b: iv).[3] Quite simply, the programs' goal was to stimulate local economic development in those neighborhoods and locales in which projects were located, thereby to transform the local geographies of union members' and their families' everyday lives, and thus to limit the appeal of Communist ideology in the barrios, slums, and favelas of the hemisphere.

In examining these projects, I pursue two related issues. First, I show how workers and their organizations—both within, but also outside, Latin America and the Caribbean—sought to secure their social and political interests through spatial praxis that shaped urban geographies and patterns of development in many countries of the region. Indeed, AFL-CIO President George Meany boasted that during the late 1960s AIFLD was the "largest builder of worker-sponsored, low-cost housing" in Latin America and the Caribbean, providing "American labor money . . . mortgage money from American labor . . . consist[ing] of money from welfare and pension funds" at rates much lower than could be obtained from local banks and other financial institutions (*AIFLD Report*, August 1967a). Given that part of the impetus to provide housing for workers was tied up with Cold War goals of containing the spread of Communism after Fidel Castro seized power in Cuba in 1959, I show here how the grand strategies of superpower labor geopolitics were played out through a deliberate spatial strategy to remake the built environments in numerous places. Struggles over housing and infrastructure were both shaped by international geopolitical goals but, in turn, were also designed to structure the global geography of the conflict between East and West from the local up. In other words, while the global conflict between superpower ideologies shaped local politics and patterns of development, these strategically located housing and infrastructure improvement projects served to produce a local scale of community development and anti-Communism that simultaneously molded the global geography of this ideological conflict by providing bulwarks in the local landscape for U.S. and local labor's strategic geopolitical goals. For both the AIFLD and the local unions involved, such spatial praxis to reshape the built environment was a central part of their broader political and strategic objectives.

Second, whereas much of my argument so far has involved analysis of the construction of material scales in the landscape—the special zone in New York City's garment district that has been readily visible to the eye through the patterns of conversion and nonconversion of lofts, the physical location of work along waterfronts, the legally enforceable national scale of bargaining affecting wage rates and working practices as spelled out in ILA contracts—in this chapter I argue that part of the AIFLD's and its allies' goals revolved around creating ideological scales of community identity and anti-Communism. The creation of a local scale of social development and opposition to Communism that could be seen in the material landscape in terms of the physical construction of buildings and neighborhoods was closely bound up with the hard ideological work of creating a scale and a sense of local identity and commitment to anti-Communism that was focused upon these economic development projects. Thus, the bricks and mortar of the housing projects would serve as a local daily reminder that, in the words of one AIFLD (1964i: iv) publication, "free trade unions can produce results, while the Communists produce only slogans." In this way, ideology would quite literally be anchored to the local built environment.

The chapter itself is organized into three sections. In the first section I outline some of the AIFLD's social programs, specifically those involving housing, and how these fit into the Institute's geopolitical objectives both of stimulating economic development in its own right and of providing decent housing for workers as a means of limiting the spread of Communism. The second section provides a more detailed case study of two such housing developments, one in San Pedro Sula, Honduras, and the other in Georgetown, Guyana. Whereas the first section, then, may in some ways be read as an example of what Sayer (1984) has called "extensive" research, the second provides a more "intensive" investigation of the specific politics surrounding the projects in Honduras and Guyana, together with the impacts that these projects have had on the built environment and workers' everyday lives in their respective locations. The final section brings the specifics of these case studies back to the larger conceptual issues of how workers and their institutions make space and scale as a central part of their social praxis.

SPATIAL ENGINEERING FOR SOCIAL ENGINEERING

We are not afraid of words like "revolution." We are not afraid of them because
we have come to realize that through our continuously evolving private
enterprise system—able to provide more and better things, better spiritually and
materially, to an ever greater number of people—we in the United States are in
fact the revolutionaries. We are proud of this. We don't take any back seat to
those who, in following antiquated, hundred-year-old philosophies invented by
old men with beards, claim that they are revolutionaries, because we know they
are not. We are helping to fight and win the revolution in Latin America—
because this revolution does exist. . . .

                    —WILLIAM C. DOHERTY, JR., AIFLD Executive Director, Speech to Council
                    for Latin America (*AIFLD Report*, April 1966c: 4)

Attempting to mold the built environment in certain ways as a means to
shape individuals' social behavior and to attain certain social goals—what
Fishman (1977: 7) has called putting "social thought in three dimensions"—
has a history stretching, arguably, back to the construction of the first human settlement. Certainly the nineteenth and twentieth centuries witnessed
numerous efforts—both for radical and for reactionary purposes—to use urban planning and architectural design for the purpose of furthering social
and political agendas.[4] Such thoughts were clearly evident, for example, in a
1904 U.S. Bureau of Labor report that argued that the "provision for improved and sanitary . . . living conditions" for workers was key to the "establishing of cordial relations between employers and employees [which] invariably results in a greater industrial efficiency on the part of the workman"
(quoted in Petersen, 1987: 86). Similarly, Mitchell (1996) shows how, during
the 1920s, the California Commission of Immigration and Housing exhorted
agricultural producers to build clean and sanitary labor camps as a way of
undercutting the purchase "outside labor agitators" might have on disgrun-
tled migrant farm workers (see also Mitchell, 1998). Likewise, Hayden (1984:
33) suggests that during the last decades of the nineteenth century, when
strikes and other workers' protests led some to fear for the future of the
American city, many social reformers saw the provision of better housing for
slum dwellers—preferably to be owned rather than rented—as essential for
the maintenance of urban social order and openly argued that better dwell-
ings would "help foster 'a conservative point of view in the working man.' "
Such views were not confined to the United States. Upon the unveiling of
the British government's 1919 proposal to encourage local authorities to
build subsidized housing for workers, the Parliamentary Secretary to the Local Government Board is reported to have commented that "the money we
are going to spend on housing is an insurance against Bolshevism and revolution" (Porter, 1994: 309).

In taking to heart Lefebvre's (1991 [1974]: 59) admonition that "new social relationships call for a new space, and vice versa," many unions and working-class groups have seen the building of housing as central to their political and social aspirations for creating a more emancipatory urban landscape. In Germany, right after World War I, unions became involved in creating low-income housing using their own funds and those of the newly established *Arbeiterbank* (Workers' Bank), accounting for some 10% of housing built in Berlin in some years (Homann & Scarpa, 1983). During the same time period, socialist elements in the U.S. labor movement began to agitate for the establishment of worker cooperative housing, which they saw as a step toward the construction of a "workers' city" in which the working class would ultimately be empowered (Vural, 1994). Such developments would not only serve workers' material needs by providing lower cost and better-quality housing to union members than could be found via the market, but they also represented a political vision of securing for workers a space for relaxation and leisure. Hence, Plunz (1986: 49) maintains that the construction in 1927 in the Bronx, New York, of the first union-run housing project in the United States by the Amalgamated Clothing Workers of America (ACWA) "took on an extended social meaning as a symbol of the successful departure of the working classes from the densely packed tenement districts of Manhattan."[5] Indeed, as part of an ongoing political project, the ACWA built even more housing in New York's Lower East Side during the 1930s.[6] Likewise, in Latin America there exists a rich history of unions building cooperative housing developments for both material and ideological purposes. In addition to providing shelter for workers, such projects have long been seen, as one official of the Brazilian National Housing Bank has put it, as "an instrument of social justice, a tool for the emancipation and independence of the working classes" (*AIFLD Report*, April 1968d: 5).

Drawing on these and other experiences, and fearing that poor living conditions might make Latin America and the Caribbean susceptible to Communist agitation, in 1962 the AIFLD established a Social Projects Department as its contribution to President John F. Kennedy's Alliance for Progress. Although the AIFLD had been involved since its founding in a number of educational and training activities with workers in the region, it was the Institute's "social projects" that, arguably, would have the greatest material impact. U.S. labor officials had several reasons for supporting the Alliance for Progress. As one former high-level official has put it (Anon, 1998; see also the statement about the AFL-CIO's "great humanitarian responsibility" by President George Meany in U.S. Senate Committee on Foreign Relations, 1969: 6):

Why did AIFLD get involved in Social Projects as well as education? One theory was that in the "developing countries" unions could be hard put to get standard benefits for their members. Therefore fringes such as housing and some of our smaller "impact projects" could show the workers tangible benefits of belonging to their unions, thus strengthening democratic unions, and of course assisting them to resist the siren call of the Reds. We got government support because it was worried about Castro and because Kennedy supported and got support from the AFL-CIO. We did it to help fight the Cold War—no question—but also because of the AFL-CIO's deep-seated belief in the dignity of free workers and the enlightened self-interest philosophy that the well-being of the American worker depended on the well-being of all workers.

From a more pragmatic point of view, AFL-CIO President George Meany argued that the U.S. labor movement should be involved in the Alliance for Progress so that the money made available by the federal government for development projects in the region would not all go to business interests but that "some of it should be channeled through trade unions, that trade unions should sponsor projects which are under the Alliance" (*AIFLD Report*, August 1967a).[7] Through such involvement, Meany hoped, employment might be generated for union members in the United States, particularly in the construction trades.

Using monies and pension funds from AFL-CIO affiliate unions, during the 1960s and 1970s the AIFLD provided grants and loans—backed by the Inter-American Foundation, the Inter-American Development Bank (IDB), the United States Agency for International Development, and various local governments in the region—to establish community centers and rural credit cooperatives, to engage in small construction projects involving bridges, roads, and schools, and to help in the building of numerous housing projects throughout the region. Many such "self-help" projects were very small-scale ventures in terms of their investment on the part of the Institute but could have very dramatic impacts on the lives of large groups of people in the communities chosen to receive such aid. For example, the AFL-CIO provided a loan that was used to buy 15 looms for a textile workers' cooperative in Goiana, Brazil, made up of people who had lost their jobs when the local factory closed (*AIFLD Report*, July 1968b). Similarly, the AIFLD provided grants to purchase six sewing machines for a sewing cooperative in Las Charcas de Garabito, Dominican Republic (*AIFLD Report*, July 1968b), while providing the transport workers' union in Uruguay with a mimeograph machine, typewriter, and filing cabinets (*AIFLD Report*, April–May 1967b) and making a loan to the letter carriers' union in Lima, Peru, to help in the purchase of a storage tank, pump, and generator for its kerosene cooperative (*AIFLD Report*, May 1968c).[8] Publicly, AIFLD officials stated that, by bringing economic benefits to workers, by increasing the resources available to non-Communist unions, by increasing the capacity of such unions to service their members, and by increasing rank-and-file participation in union affairs through "self-help" activities, the AIFLD's social projects would "contribute to the strength of a free labor movement" (American Technical Assistance Corporation, 1970a: 39) and bring about a peaceful social revolution in the region.[9]

Although such small-scale projects could have very significant political payoffs for little financial investment, the Social Projects Division's biggest projects involved community building, both through constructing housing and through engaging in other community-development activities. Houses were not only important from the material perspective of providing shelter, but they were also important symbolically, for they were a physical manifestation of unions' and workers' power and ability to shape the urban landscape to their advantage on a relatively large scale. For William Doherty, Jr., the AIFLD's Executive Director at the time, housing projects represented "a living affirmation of the Alliance [for Progress] in action achieving its goal of social revolution without violence" (*AIFLD Report*, July 1968a: 2). The provision of housing was designed, according to an AIFLD official, "not just to give [workers] better places in which to live, but to give them a new base for stronger participation in national life [and so to] create new strength for democratic unionism" (*AIFLD Report*, May 1966c: 4). AIFLD considered housing a productive investment that would energize national development "both directly through provision of jobs and stimulation of materials industries, and indirectly through improving public health, worker motivation, and social stability" (AIFLD, 1964f: vi).

In seeking to make the schemes viable, AIFLD attempted to ensure that no more than 25% of a worker's salary went toward housing, and workers often did much of the infrastructure building themselves as their "self-help" contribution to various projects. In addition to cheapening the cost of construction, two social objectives appear to have been achieved through such "self-help" work. First, encouraging the incorporation of this type of "sweat equity" gave workers both a sense of involvement in the projects but also undoubtedly instilled in them a notion of ownership in the project, a sense that hard work and self-reliance produced results, and that they were not simply the recipients of a munificent bounty. Second, as Hayden (1976) has argued in a different context, such building together frequently helps create a greater sense of community, a consideration that was perhaps important given the ideological and development goals of the projects (see also Vural, 1994).

In the remainder of this section of the chapter, I provide an overview of some of the projects in which AIFLD and its local allies were involved.[10] By examining a number of projects in countries throughout the region, I want to give a sense both of the variety of projects undertaken but also of their geographical spread and impact upon a diverse set of locations. The section that follows this one provides a more in-depth analysis of the politics surrounding two specific projects, one in Guyana and the other in Honduras.[11]

Mexico

The first major housing project involving AIFLD was built in Mexico City. Using USAID guarantees, a $10-million loan from the pension funds of AFL-CIO affiliates (principally the ILGWU, the International Brotherhood of Electrical Workers, and the Brotherhood of Locomotive Firemen and Enginemen), and funds saved by members of the local Graphic Arts Workers' Union, the AIFLD helped build some 3,100 apartments as part of the 96-building "John F. Kennedy Housing Project" at Colonía Balbuena.[12] Stressing the worker-to-worker nature of the project, AFL-CIO President Meany was keen to emphasize that, although the project enjoyed a USAID guarantee on the investment, "these moneys were provided by American workers to the Mexican union" (U.S. Senate Committee on Foreign Relations, 1969: 8). Dedicated by Robert Kennedy in November 1964 in a ceremony attended by Mexican President Alfonso López Mateos and carried live nationally on Mexican television (*AIFLD Report*, December 1964b), the project not only represented an important step in helping to deal with a housing shortage in the capital region—by the end of 1971 the project was housing some 22,000 people who were paying between $34 and $72 a month to live there (*AIFLD Report*, December 1971b), and by 1997 was estimated to house close to 40,000 (Doherty, 1997)—but it was also important symbolically, a fact noted by President Johnson in his message for the ceremony: "This housing project," Johnson stated, "is more than just mortar and wood, more than foundations and walls; it is a symbol of what free men working together can achieve" (*AIFLD Report*, December 1964b: 6). Indeed, with the subsequent construction of an elementary school adding to the growing community, U.S. Ambassador Fulton Freeman was able to argue some years later that

> These thousands of apartments have truly become homes, each with its distinctive character that reflects the individual taste of its occupants. The entire area has taken on a human warmth—with the shouting of children on the playground, the crying of babies and the laughter of the families, as well as the savory kitchen aromas. But there has been another important transformation: *The Kennedy Housing Project has developed a community spirit, a framework of conviviality, cooperation and civic responsibility.* And it is this community spirit in action that has brought about the construction of this fine school. (*AIFLD Report*, June 1967b: 3, emphasis added)

With the construction of this project, the AIFLD argued that it had put in place "a program which will show the Mexican workers that free trade unions can produce results, while the Communists produce only slogans" (AIFLD, 1964i: iv). Its success encouraged AIFLD and other Mexican unions subsequently to plan several additional developments, including those for airline workers and telephone workers in the capital region, for state employees in Sonora and Guadalajara, and for sugar workers in Morelos—although owing to political and economic problems these were never completed.

Argentina

Although the Mexico City project was the first large-scale one involving AIFLD, the Institute subsequently helped build housing in several other countries. In Argentina AIFLD worked principally with the unions representing the light and power workers (Luz y Fuerza) and municipal workers (the Unión de Obreros y Empleados Municipales), largely because they were strong unions, and with FOECYT (Federación de Obreros y Empleados de Correos y Telégrafos—the Federation of Postal and Communications Workers), principally because AIFLD Executive Director Doherty knew key local and national leaders from his days with the Postal, Telegraph, and Telephone International (PTTI) trade secretariat (Holway, 1998a). AIFLD also worked with the railway workers' Unión Ferroviaria (*AIFLD Report*, July 1968a).[13] Six of the projects were in the Greater Buenos Aires region, with others in Bahía Blanca and Pergamino in Buenos Aires province.[14] AIFLD both sponsored the projects and provided the local unions with technical assistance (Holway, 1972; *AIFLD Report*, July 1968a). The units in these developments were all either 10-to 16-story apartment buildings or three- and four-story walk-up apartments, with some garden-type apartments (AIFLD, 1964a; *AIFLD Report*, June 1968). The projects incorporated a small facility service fee, the proceeds of which were expended on various local activities designed to foster a sense of community (*AIFLD Report*, February 1972a). Each owner was expected to make a down payment equal to 10% of the sales price and to take out a 25-year mortgage. In all, some 1,667 apartments were built using a USAID-guaranteed and AIFLD-managed loan of $13.5 million made to the National Mortgage Bank. The homeowners were charged an interest rate of 9% at a time when the lowest housing rate available from local sources was about 13%.

Upon completion of this first stage of "Operación AFL-CIO," a second stage was planned with the Unión Ferroviaria and other unions in Buenos Aires, and with the bank workers' union in Córdoba, though, as in Mexico, these were never completed due to political and economic problems (*AIFLD Report*, February 1972a). Although after the first project AIFLD did not help build any more houses, Executive Director Doherty did help a number of unions secure some allocation of units in an IDB project that was being conducted at the time, bringing the total units in which the Institute claimed some involvement to 6,227 (AIFLD, 1972; Holway, 1998b). As part of its community-building activities, AIFLD also made a grant to the leather workers' federation to buy equipment for a children's playground in Exaltación de la Cruz and provided grants and loans to the Rural and Stevedore Workers' Union to build and equip a medical and community center in Chilibrost, Córdoba (*AIFLD Report*, July 1968b). In addition, AIFLD provided money to the Leather Workers' Union of Morón to build a classroom and library at the union's headquarters that could be used for worker education (*AIFLD Report*, April–May 1967c).

Columbia

In Colombia the AIFLD worked with the two main labor federations, the Confederación de Trabajadores de Colombia (CTC—Confederation of Colombian Workers) and the Catholic-dominated Unión de Trabajadores de Colombia (UTC—Unity of Colombian Workers). In July 1963 the CTC and UTC had formed, with help from AIFLD, the Alianza Democrática Sindical para Vivienda (Democratic Trade Union Alliance for Housing) to provide low-cost housing for workers in 15 cities throughout the country (AIFLD, 1964g).[15] One of the largest of these was a 1,400-unit project (which included 800 single-family houses) in the outskirts of Bogotá named for John F. Kennedy (the "Ciudad Kennedy"). The funding for this came from a consortium that included the IDB, local unions, the Colombian government, local employers, and the AIFLD. AIFLD's role was primarily one of supplying the CTC and UTC with technical assistance concerning the physical planning of the project and helping to establish cooperatives and other community services in the postcompletion period (*AIFLD Report*, February 1967a). The Bogotá housing project consisted principally of three- or four-story walk-up apartment units, since individual housing would have proven too costly to construct (*AIFLD Report*, May 1966a). The first section (288 apartments) was completed in December 1966, and in January 1967 these were assigned to their owners via lottery in a ceremony broadcast on television, thereby assuring maximum publicity for the projects and AIFLD's activities (*AIFLD Report*, January 1967b). The total price for each unit was approximately 53,000 pesos (then about $3,300) (*AIFLD Report*, February 1967a). By late 1971 the UTC and CTC, with AIFLD help, had built over 2,100 homes in various Colombian cities (*AIFLD Report*, December 1971b). In addition to helping in the construction of housing, AIFLD also made loans for structural improvements to union buildings, giving $5,000 to the UTC to install an elevator in the union's new headquarters in Bogotá and making a $3,000 loan to the steelworkers' union in Paz del Rio, Boyacá, to purchase equipment for its vacation center (*AIFLD Report*, July 1968b).

Housing represented an important component of the AIFLD's and local unions' efforts to improve the material well-being of the CTC and UTC unions' membership, but it was not the only one. AIFLD and local unions also established and/or supported numerous consumer cooperatives. In Colombia, the UTC had set up supermarkets in many of the major towns and cities, which had enabled workers to buy food and other goods at prices much lower than they could get elsewhere (*AIFLD Report*, May 1972). Thanks in part to a $50,000 loan from AIFLD, the UTC cooperative system was able to expand its existing network of stores in Bucaramanga, Medellín, San Gil, Neiva, Cúcuta, Barranquilla, and Bogotá to other cities in Colombia, including Cartagna, Pereira, Soggamoso, and Barrancabermeja. This had a snowball effect on local markets. Not only did such stores offer union members lower prices, but in many cases they also forced others to reduce prices to remain competitive, thereby reducing the cost of living for many more workers, even those not belonging to UTC unions (*AIFLD Report*, December 1971a). Additionally, AIFLD helped a number of unions improve their service-provision capacities, such as when it made a loan to the Union of Workers of Atlántico (UTRAL-UTC) to help stock its pharmacy (*AIFLD Report*, January 1967b).

Brazil

The case of Brazil, as one AIFLD document suggests, "present[ed] the Alliance for Progress with a tremendous task" (AIFLD, 1964c: iv). Under the governments of Janio Quadros and later João Goulart, Communist parties allied with the U.S.S.R. and China had been quite active in the labor movement as part of the Comando Geral dos Trabalhadores (General Workers' Command), and social conditions in the country had fostered tremendous political unrest (for more on the labor situation under Quadros and Goulart, see Alexander, 1965; and Greenfield, 1987). Following the removal of Goulart from power by the military in April 1964, AIFLD was particularly concerned to develop projects that would improve conditions in both urban and rural areas, especially in northeast Brazil. The Institute engaged in a number of small-scale projects, including providing technical assistance and a brick-making machine to build a school in a small community in rural Municipio do Bezerros, Pernambuco (*AIFLD Report*, June 1967e), giving a grant to help build an auditorium for a school attended by trade unionists' children in Rio de Janeiro, and making a loan to the National Federation of Telephone Workers to help them acquire a building to serve as a union headquarters (*AIFLD Report*, January 1967b). With regard to housing, AIFLD sponsored the "Samuel Gompers Memorial Housing Project," a 448-unit costing $2,272,000 and located in the municipality of São Bernardo do Campo, which forms part of the industrial heart of São Paulo (*AIFLD Report*, May 1966b). Members of the local unions involved made down payments and then purchased the housing units from the labor cooperative that had been set up to manage the project. Other housing was planned for several other cities, including Rio de Janeiro, Recife, Porto Alegre, and Salvador (AIFLD, 1964c), though the inability of local unions to handle financial matters concerning readjustments of mortgages, especially when tied to U.S. dollar loans, meant these projects ultimately died (Holway, 1998b).

In addition to involving itself in urban housing, AIFLD was also very much interested in improving conditions for rural workers. Brazil was the first country in which AIFLD helped with the construction of rural service centers geared toward combining worker education with practical experience on individual field projects. AIFLD's rural development program consisted of a three-stage model: (1) a leadership training program for potential rural union leaders; (2) application of the training by these leaders through "community action projects" in their home towns (often involving small-scale infrastructure improvements); and (3) establishment of rural worker centers (*AIFLD Report*, January 1966a; *AIFLD Report*, April 1966a). In 1966, in partnership with the local union federation, a Catholic priest, and USAID and other agencies, AIFLD opened three union-community centers (the first of 10 planned centers) in the state of Pernambuco in northeastern Brazil, one at Carpina in the northern coastal sugar zone (an area dominated by *latifundia* and wage-earning workers), one at Ribeirão in the southern coastal sugar zone (also dominated by *latifundia* and wage-earning workers), and one at Garanhuns, a small-farmer region in the interior of the state (an area dominated by *minifundia* and sharecroppers and tenant farmers).[16]

Funding for these centers came from a USAID grant made to the AFL-CIO in response to a request from Brazilian trade union leaders (*AIFLD Report*, August 1965). AIFLD's role was one of planning the program and acting as the catalyst for its implementation through negotiating the financing. The centers were designed each to service about 15,000 families, providing literacy training, teacher preparation, training in arts and crafts such as carpentry and sewing, trade union training and instruction in the operation of cooperatives, agricultural extension in the areas of horticulture and livestock husbandry, first-aid and other health services, and legal assistance to unions and their members (*AIFLD Report*, April 1968b). They would also serve as a base for establishing consumer and producer cooperatives, rural credit unions, and land leasing and purchasing cooperatives. For AIFLD, the goal of such activities was "to help these peasant workers fight poverty, illness, inequality, ignorance, and injustice," while the activities were "carried out through existing workers' organizations so that these [would] be strengthened, encouraged, and enabled to act as a constructive force towards modernization and a better general welfare" (*AIFLD Report*, August 1967c: 3). According to AIFLD Executive Director Doherty, such centers represented "tangible evidence to the workers of the region of the concern of the North American labor movement for their plight" (Doherty, 1966: 43).

Dominican Republic

The political instability in the Dominican Republic after the 1961 assassination of the dictator Rafael Trujillo threatened to open the door for many Communist-dominated and other revolutionary political and labor groups to seize power. The U.S. government became particularly concerned when, in 1962, Juan Bosch was elected president on the Dominican Revolutionary Party's ticket. Fearing "another Cuba," the U.S. State Department, the CIA, USAID, the AFL-CIO, and various U.S. multinationals such as Gulf + Western soon began to play active overt and covert roles in the Dominican labor movement, principally with the goal of encouraging more moderate unions and countering the influence of unions with revolutionary and/or Castroite tendencies (Murphy, 1987). For its part, AIFLD backed CONATRAL (the Confederación Nacional de Trabajadores Libres/National Confederation of Free Workers) against the Communist and revolutionary unions.[17] As elsewhere, the provision of housing and other small impact projects not only was intended to improve the material lives of those workers associated with CONATRAL unions but also was designed to counter the influence of workers' groups allied with leftist political organizations, who saw such projects as part of U.S. imperialist strategy toward the island and engaged in propaganda campaigns against them (Pita, 1997).[18]

The principal housing project with which AIFLD was involved was located in the southern coastal town of San Pedro de Macorís, which itself had symbolic value because it was the only town to host a significant strike against the Trujillo regime. Located in the town's Porvenir neighborhood, the 110-unit "John F. Kennedy Housing Project" was completed in December 1966 and was soon housing some 700 sugar workers (*AIFLD Report*, January 1967b).[19] The houses themselves sold for between $3,800 and $4,100, with a 10% down payment, monthly payments of about $25, and a 25-year mortgage (*AIFLD Report*, February 1967b). The project, using USAID and AFL-CIO funds administered and coordinated by AIFLD, included schools, parks, playgrounds, and commercial areas and was provided with clean drinking water, sewage systems, and electricity. In addition to coordinating the physical construction of the project, AIFLD appointed a community services officer to set up a continuing community-development program. This led to the establishment of a vaccination program for children to guard against several common diseases (*AIFLD Report*, June 1967c; *AIFLD Report*, May 1968d). As part of the community development process, AIFLD gave money to buy equipment for the playground and built a plaza to serve as a sports field (*AIFLD Report*, May 1968c; *AIFLD Report*, May 1968d). The project was considered a great success by AIFLD, since it both "strengthened the housing cooperative of the [CONATRAL] and enabled it to form a community services co-op" (*AIFLD Report*, February 1967b: 5).

In addition to the project in Porvenir, AIFLD was also involved in planning and doing the groundwork for housing projects (with a total of some 17,400 units) in several other Dominican cities, including Caei, Angelina, Colón, Rio Haina, and Sabana Grande de Boya (all for sugar workers), La Romana (for local government workers), Santiago (for electrical workers), and Santo Domingo (for telephone and electrical workers), although the weakness of the unions involved meant that these projects ultimately never materialized (AIFLD, 1964d; Holway, 1998b). The success of the San Pedro de Macorís project did, however, later allow it to provide a $20,000 loan to the local telecommunications and postal workers' union to establish a savings and loan institution that allowed workers to borrow money at rates much lower than were available from local moneylenders. Additionally, a $25,000 loan from the housing project allowed the sugar workers' union to establish a supermarket with lower prices than those available in other markets locally (*AIFLD Report,* May 1972), while a rural school at Villa Riva and the leprosy institute in Bajos de Haina received furniture as part of an AIFLD small grants program (*AIFLD Report,* June 1967c).

Ecuador

With the exceptions of Cuba, Haiti, and Paraguay, in which AIFLD did not try to work due to hostility from the governments of those countries, arguably no country posed greater problems for the AIFLD's social projects than did Ecuador. The principal reason for this was that, unlike in other countries in which AIFLD worked, in Ecuador the country's major national labor federation—the Confederación de Trabajadores del Ecuador (CTE)/Confederation of Workers of Ecuador, which represented some 65% of all organized workers in the early 1960s—was affiliated with the Communist-backed World Federation of Trade Unions. Communist elements had a strong hold in both the labor movement and among intellectuals, despite repression from the military government. In 1962 AIFLD had helped establish a counter to the CTE, namely, the Confederación Ecuatoriana de Organizaciones Sindicales Libres (Ecuadorian Confederation of Free Labor Organizations), and sought to attract unions to its ranks. As part of this goal, AIFLD saw its principal role as one of "strengthen[ing] democratic unions by helping them to provide social services to their members" in the form of such things as savings institutions, credit unions, consumer and producer cooperatives, medical clinics, and other community development projects that did not require the political support of the country's central labor federation (AIFLD, 1964h: 5).

As part of this strategy to "translate 'international worker solidarity' from a slogan into concrete achievements which will accelerate the growth of trade union strength and responsibility" (AIFLD, 1964h: 5), AIFLD initially planned a 500-unit housing project in Quito and Guayaquil to be financed by a 20-year AFL-CIO loan. Believing that "housing and other social projects for the democratic sector of the labor movement [would] demonstrate the advantages of belonging to unions dedicated solely to the responsible advancement of the worker rather than those with policies swayed by partisan influence" (AIFLD, 1964h: 4), upon completion of the first group of housing units the Institute hoped to build more. Despite such high hopes, political and financial problems meant that by 1972 AIFLD had only managed to construct 14 units (AIFLD, 1972). The Institute did, though, fare better in implementing some of its other social projects, including giving a grant to unionists in Los Rios to help establish a sewing and embroidery training center (*AIFLD Report*, May 1968c).

Uruguay

Due to its relatively advanced economic state, one of the lowest population growth rates in Latin America, and the fact that rural-to-urban migration was not as significant as it was in neighboring countries, the housing situation in Uruguay was (relatively) much better in the early 1960s than it was in many other countries in the region. Although USAID and the IDB had made loans as part of the Alliance for Progress to the Uruguayan government to build housing, as a body organized labor in the country had not participated in such programs. Furthermore, a shortage of credit institutions also made it difficult for workers as individuals to arrange financing for home purchases (AIFLD, 1964j). As a result, AIFLD cooperated with local unions in a project to build 1,020 units in Montevideo's Malvín Norte district, 410 of which were under construction by late 1971 (*AIFLD Report*, April 1971; *AIFLD Report*, December 1971b). In addition to providing housing for union members, the project was also designed to stimulate employment in the construction industry, in which Communists had some influence among workers and which was experiencing an unemployment rate estimated to be as high as 50% (AIFLD, 1964j:7).

AIFLD also engaged in a series of other impact projects whose influence was seen as meaning "the difference between democratic and Communist control of the labor movement in strategic parts of Uruguay" (AIFLD, 1964j:7). One of the most significant projects involved helping the rural sugar workers' union of Artigas (SURCA) to establish a clinic in Bella Unión near Uruguay's border with Argentina and Brazil (*AIFLD Report*, August 1971). SURCA had been involved in a long struggle with a rival sugar union, the Unión de Trabajadores Azucareros de Artigas (UTAA), which was affiliated with the Communist-influenced Convención Nacional de Trabajadores (National Convention of Workers). Upon completion of the clinic and its stocking with medicines, the SURCA made the facilities available to the community as a whole rather than for its own members solely, thereby securing more local support. Indeed, the clinic's success enabled the SURCA to conduct an organizing drive that attracted so many members of the UTAA that the latter union was no longer considered a serious threat by AIFLD. Several other projects included making a loan to the musicians' union to help remodel its headquarters in Montevideo to house a medical center (*AIFLD Report*, February 1971b), and providing funds to the Graphics Arts Union to establish a credit union for its members, with AIFLD giving technical assistance in setting this up (*AIFLD Report*, June 1971).

In this section of the chapter, then, I have given a broad overview of a number of housing and community development programs throughout Latin America and the Caribbean to show the breadth of the activities in which AIFLD and local unions engaged. Specifically, I have tried to show how the building of housing and other infrastructural developments was seen as key to fighting Communism, virtually on a neighborhood-by-neighborhood basis. In what follows, I present two case studies in greater depth to provide a more nuanced account of the contexts surrounding the building by AIFLD and its local allies of housing designed to transform the local spatial relations within which workers lived their lives and thereby to fight Communism. I have chosen to examine these two projects—one located in San Pedro Sula, Honduras, the other in Georgetown, Guyana—principally because in the early 1960s both countries represented strategically important situations for the U.S. government and the AFL-CIO. In the case of Honduras, Communist sympathies among banana workers in particular had appeared to strengthen during the 1950s, a development that posed threats to various U.S. interests, especially those of the United Fruit and Standard Fruit companies. Guyana also represented a strategically important country. The United States had leases to territory as a result of the "bases-for-destroyers" arrangement it had made with the British government during World War II, while for both Britain and the United States Guyana represented an English-speaking outpost in the South American continent that possessed important resources such as bauxite, sugar, and rice. From the perspective of the AFL-CIO, in both countries the lack of decent housing not only made workers' everyday lives harder but threatened to tip their labor movements toward the Communist camp.

HONDURAS AND GUYANA:
SOCIAL REVOLUTION OR SOCIALIST REVOLUTION?

San Pedro Sula, Honduras

AIFLD's goal in Central America was to "strengthe[n] the free, democratic and independent labor unions" through the implementation of social programs that would ameliorate the "intolerable conditions that perpetuate the misery of the workers" of the region and "provide social justice that [would] restore human dignity for [the worker] and his family" (AIFLD, 1964e: iv).[20] The nucleus of this program consisted of two housing projects, a 128-unit project in San José, Costa Rica and a much larger project in Honduras that, the Institute hoped, would greatly "strengthen the [participating union] members' loyalty to their unions" (AIFLD, 1964e: v). This was of particular concern in the case of Honduras, long a country of great political instability, poverty, and, arguably, the archetypal "banana republic." Since gaining independence in the early nineteenth century Honduras has experienced some 400 or so domestic actions, more than 120 changes of government, and more than a dozen constitutions, and has consistently had the lowest per capita income in Central America and one of the lowest in all of Latin America. Furthermore, the country's economy has largely been shaped by the interests of external capital such as the infamous United Fruit and Standard Fruit companies. Struggles between Communist and non-Communist labor factions date back to the 1920s but they erupted into violence with a massive strike by banana workers in the northern coastal region in 1954 (for an account of the strike—the "Gran Huelga"—by one of those involved, see Robleda Castro, 1995; for more on Honduran labor history, see Pearson, 1987).

AIFLD's first involvement with constructing housing in Honduras came in 1962 when Oscar Gale Varela, the president of the Union of Workers of the Tela Railroad Company (SITRATERCO/Sindicato de Trabajadores de la Tela Railroad Company)—the Tela Railroad Company was the legal name of the United Fruit Company in Honduras at the time—approached the Institute for technical assistance to establish 187 units of housing in two projects (*colonías*), one located in the town of La Lima (headquarters of the United Fruit Company) and the other in the coastal city of Tela (Romualdi, 1967; AIFLD, 1965: 30).[21] AIFLD helped SITRATERCO negotiate a $400,000 USAID loan, while SITRATERCO also obtained $110,000 worth of land and $45,000 of infrastructure investment from the United Fruit Company through collective bargaining. Although a military coup in October 1963 led USAID to suspend temporarily the loan, construction subsequently continued and the first residents took possession of the houses in late 1963. In March 1965, two Peace Corps volunteers were assigned to the SITRATERCO project to do community development work (AIFLD, 1965: 30). As part of its community building program, SITRATERCO also ran child daycare centers in these *colonías*, having obtained a small grant from AIFLD to purchase playground equipment for the housing project in La Lima (*AIFLD Report*, April 1968c) and a $2,000 grant from the AFL-CIO to purchase kitchen and laundry equipment (*AIFLD Report*, February 1972b). AIFLD also provided a $50,000 loan to build a warehouse, a consumers' cooperative, and a community center at La Lima (*AIFLD Report*, May 1972; American Technical Assistance Corporation, 1970c), while the Institute helped fund a vacation colony at Puerto Cortés on the north coast containing a restaurant, a large classroom building, and 48 cottages that SITRATERCO rented out to both members and nonmembers (*AIFLD Report*, February 1972b).

The SITRATERCO housing project was the first one sponsored by AIFLD in Honduras (*AIFLD Report*, June 1967a) and by the early 1990s had grown to include 1,375 houses (Galdámez, 1997b). Its success soon led the Institute to involve itself in a number of other such projects. In 1967 Gale Varela and Céleo González y González, president of the Union Federation of National Workers of Honduras (FESITRANH/Federación Sindical de Trajabadores Nacionales de Honduras), of which SITRATERCO was an affiliate, arranged a loan from the IDB to construct 1,001 houses as part of a much larger cooperative project (the "Colonía FESITRANH") in the suburbs of San Pedro Sula in northern Honduras.[22] The model house, constructed so that the final design for the *colonía*'s units might "more accurately reflect the workers' desires" (American Technical Assistance Corporation, 1970c: 37), was begun in 1967 using an interest-free loan from AFL-CIO pension funds (*AIFLD Report*, August 1967b). Funding for the remaining 1,000 units came from an AIFLD-sponsored IDB loan of $2,270,000 (at an interest rate of 2.25% per annum, to be paid back over 25 years), $224,000 from the city of San Pedro Sula (which partially paid for some municipal services), $339,000 from the local Chamber of Commerce (which was mostly used to pay for new infrastructure such as paving of streets), and $10,000 from AIFLD's Regional Revolving Loan Fund. The national government of Honduras guaranteed the loan to FESITRANH and also gave one hundred *manzanas* of land for the project (*AIFLD Report*, June 1967a; Galdámez 1997a).[23]

AIFLD's role in this project largely was to coordinate the whole undertaking, to ensure loans were made, and to provide technical assistance with regard to the design of the housing floor plans, the *colonía*'s sewage system and processing plant, and the provision of clean drinking water. When war and floods threatened to delay the project early on, AIFLD also made a $34,000 loan to FESITRANH until the IDB funds became available. Such was the propaganda value of the project for the union and the government that the May 7, 1972, loan-signing ceremony was broadcast nationally over a radio hookup. At the ceremony T. Graydon Upton, Executive Vice President of the IDB, stated the project was "part of an intensive effort on the part of the workers of Honduras to mobilize their savings towards the construction of dwellings, and is, therefore, an expression of their desire to help themselves and reach the goals of economic and social development of the Alliance for Progress" (*AIFLD Report*, June 1967a: 3). As one independent observer remarked (American Technical Assistance Corporation, 1970c: 37):

> The major AIFLD contribution has perhaps been the provision to FESITRANH of help that gave [the] Federation the nerve and knowhow required to turn an inexperienced labor group into a capable and accept-

able borrower. Informed observers . . . assert that the AIFLD role was critical in bringing the labor group to [IDB's] attention, establishing its
credibility, and pressing its claims for resources.

The FESITRANH project contained two types of houses. The larger type
(750 of which were built) had 60m^2 of living space, consisting of a living
room, a dining room, a kitchen, three bedrooms, a bathroom, and a storage
closet, and were priced at Lps (Lempiras) 6,600 ($3,300).[24] The smaller size
(of which 250 were built) had similar facilities, although only two bedrooms, had 54m^2 of living space, and were priced at Lps 6,200 ($3,100).
Houses were bought by members of the union cooperative on a 20-year
mortgage at 6.5% per annum interest, usually with a 5–10% down payment,
which members could save through their cooperative and the Workers'
Bank in San Pedro Sula (*AIFLD Report*, June 1967a). The money generated
by charging a higher interest rate to the house purchasers than had been
charged by FESITRANH's creditors was used for organizational expenses
and as a capital fund to continue building (Galdámez, 1997b). The project
also contained a commercial center with 30 buildings, a civic center, a secre-
tarial college, a sports ground, a health center, union office buildings that included space for a labor education center, a kindergarten and public school,
a baseball field, and a building out of which broadcast the 5,000-watt Christian radio station "Fraternidad" ("Brotherhood") (Galdámez, 1997a). To enable workers to get to work, FESITRANH ran its own bus service between
downtown San Pedro Sula and the *colonía*. In 1972, a $50,000 loan from
AIFLD allowed FESITRANH (which itself contributed $10,000) to purchase
and remodel a building near its housing project for use as a union center,
thereby allowing the Federation to take full responsibility for conducting basic and intermediate labor education seminars and generating income
through renting the center out to other organizations, such as the Honduran
Ministry of Labor's Agricultural Extension Service DESARRURAL (*AIFLD Report*, October 1972). By 1997 some 6,000 people were living in the *colonía*
(Galdámez, 1997b).

Following completion of this part of the project, a second stage—the
Colonía Céleo González—was begun in 1990. Funding for this project came
from the national government's *Fundo Sociale de Vivienda* (Social Fund for
Housing) and from contributions made jointly by workers and their employers under the *Régimen Aportaciónes Privadas* (Private Contributions Rule) of
their collective bargaining agreement, under which both parties contribute
equally to a housing fund. Reflecting changed conditions in Honduras, interest rates were much higher in this project than in the first, ranging between
16% and 24% (Galdámez, 1997b). This stage was planned to include some
2,400 houses, sanitation and electric generation plants, paved streets, a telephone system, recreation areas, and a water purification center. In an effort

to produce housing more quickly and at lower cost, FESITRANH chose to use prefabricated construction materials. However, financial problems meant that, whereas FESITRANH successfully purchased all 2,400 lots of ground, the construction proceeded slowly and by April 1997 (when I visited the *colonía*) only 430 houses had been built, though the Federation had every intention of completing the remaining units and facilities (Galdámez, 1997a). Indeed, such was its confidence that it has begun to build a third stage consisting of a further 690 houses, the first of which became available for purchase in 1998 (Galdámez, 1997b). In this development two types of houses are being built, a 40m² 2-bedroom model to sell for approximately Lps 70,000 ($5,400), and a 56m² 3-bedroom model for Lps 94,000 ($7,400). (In 1997 cooperative members had an average monthly income of about $120–$170 [Galdámez, 1997b]. Note the huge devaluation in the value of Lempiras during the 25-year span [see note 24].)

Such housing projects have been copied by numerous other labor unions in Honduras. Banana workers appear to have been particularly aggressive in pursuing union-run housing options. In 1971 SITRATERCO, the Standard Fruit Company, and the AIFLD presented the IDB with a proposal to build a further 1,740 units for members of the Standard Fruit Company workers' union, SUTRASFCO (Sindicato de Trabajadores de la Standard Fruit Company), in La Ceíba in northern Honduras (*AIFLD Report*, December 1971b). This was undoubtedly driven, at least partly, by a desire to use housing to undermine growing Communist sympathies within SUTRASFCO (Pearson, 1987). In addition, AIFLD prepared the plan and acted as engineering consultant for construction of a union building for SUTRASFCO (*AIFLD Report*, April–May 1967c). By 1997 the United Fruit (Chiquita brand) banana workers had established the La Lima *colonía* and the Standard Fruit (Dole brand) workers' *colonía* was operating in La Ceíba, while port workers, hotel workers, cardboard manufacturing workers, and La Ceíba hospital workers also all had their own *colonías*.

Although FESITRANH is the largest labor federation in Honduras and its member unions have been responsible for building the greatest number of houses, other federations have also engaged in such building projects. Perhaps seeking to emulate the success of FESITRANH's projects, both those built with and without AIFLD help, the Federación Unitaria de Trabajadores de Honduras (FUTH/Unitary Federation of Honduran Workers), a federation with strong links historically to the pro-Soviet Communist Party of Honduras (PCH), has begun a housing project for university workers and other public sector workers, while the Federación Central de Sindicatos Libres de Honduras (FECESITLIH/Central Federation of Free Unions of Honduras), which has also had a history of pro-Soviet and Maoist sympathies among many of its members, has begun a small *colonía* in Tegucigalpa (Galdámez, 1997b).

In addition to its input in providing housing, AIFLD also engaged in

Member of consumer cooperative shopping for groceries, Catia, Venezuela (1968) (AFL-CIO).

Coracrevi Homes, a cooperative of the Venezuelan CTV, Caracas, Venezuela (1968) (AFL-CIO).

Aerial view of original FESITRANH housing project, San Pedro Sula, Honduras
(1972) (AFL-CIO).

Alley in new FESITRANH housing cooperative, San Pedro Sula, Honduras (1997)
(photo by author).

Lots for new housing laid out, new FESITRANH housing cooperative, San
Pedro Sula, Honduras (1997) (photo by author).

Example of smaller house, new FESITRANH housing cooperative, San
Pedro Sula, Honduras (1997) (photo by author).

Example of larger house, new FESITRANH housing cooperative, San
Pedro Sula, Honduras (1997) (photo by author).

Interior of larger house, new FESITRANH housing cooperative, San Pedro Sula,
Honduras (1997) (photo by author).

Workers in Georgetown, Guyana, prepare to receive keys to the first 35 units in Tucville (1968) (AFL-CIO).

House in Tucville, whose owners have done much renovation and upgrading (1997) (photo by author).

House in Tucville showing stilt building typical of the area (1997) (photo by author).

Interior of house in Tucville (1997) (photo by author).

several other small-impacts projects designed to bring about closer links between Honduran unions and the AFL-CIO. Students from the AIFLD's national course for Honduran *campesinos*, for instance, built primary schools
in several small communities such as Pinolapa and Yojoa, installed water
purification systems in La Sarrosa and La Guacamaya, completed a community center in El Tablón, and built a fence around a primary school in El
Milagro (*AIFLD Report*, July 1968c). Other examples include the AIFLD giving a grant to the Lard and Soap Workers' Union (SITRAFA) to help equip a
clinic and meeting hall for the union in La Ceíba—with the hall being used
for adult education classes in the evening and as a kindergarten during the
day (*AIFLD Report*, April 1968a)—and providing a grant to the anti-
Communist *campesino* union ANACH-FESITRANH Medical Brigades to buy
medical supplies for their rural health program (*AIFLD Report*, May 1968c).[25]

In sum, by constructing these *colonías*, the local unions involved, together with AIFLD, have managed to create virtually self-contained munici-
palities that provide everything from schools and commercial buildings to
bus services (in the case of FESITRANH this takes workers from the *colonía*
into the center of San Pedro Sula) and radio stations. Through such activities
the unions have not only created better living conditions for those workers
able to buy into the projects, but they have attempted to create a sense of
community and collective identity that connects the spheres of the paid
workplace and of the community in workers' everyday lives.

Georgetown, Guyana[26]

During much of the post-World War II period Guyana has been a strategically important country for both Britain and the United States. However, the
political situation in Guyana in the early 1960s was decidedly unsettled, in
terms of both Cold War ideological rivalries and of ethnic tensions among
the three main population groups (East Indians, Afro-Guyanese, and Euro-
Guyanese). In 1950, before independence from Britain, the largely East In-
dian-backed People's Progressive Party (PPP) had been established with the
ideological aim of building "a just socialist society in which the industries of
the country shall be socially and democratically owned for the common
good" (Premdas, 1993: 115). Led by Dr. Cheddi Jagan (an East Indian) and
Linden Forbes Burnham (an Afro-Guyanese), the party had as its primary
goals opposing British rule and bringing about greater cooperation among
the various ethnic groups in the country. The rise of the PPP also allowed
the British Guiana Industrial Workers' Union (BGIWU), which Jagan had
helped found in the 1940s and which was made up largely of East Indian
sugar workers, to gain strength and, ultimately, a larger membership than
the conservative, employer-favored Man Power Citizens' Association that
had been the main sugar workers' union since the 1930s (Cumiford, 1987).

In 1953 the PPP won an overwhelming electoral majority and introduced legislation to reform the country's labor law. However, fearing a radicalization of the local working class, the British government soon intervened, suspended the constitution, and removed the PPP from power. In 1957 Jagan and Burnham split over a number of issues and Burnham formed a new, predominantly Afro-Guyanese, party, the People's National Congress (PNC). With the PNC's formation the country's two largest ethnic groups were split politically, with the numerically superior East Indians largely supporting Jagan's PPP, the Afro-Guyanese population largely supporting Forbes Burnham's PNC, and the small number of Euro-Guyanese and business interests largely supporting a third, smaller party, the United Force.

Although both Jagan and Forbes Burnham (who was also president of the British Guiana Labour Union, which mostly represented waterfront, sawmill, and construction workers) publicly espoused socialist politics, Jagan was seen to be much more anti-U.S. and pro-Soviet and pro-Cuban than was Forbes Burnham (for a collection of speeches that outline some of his policies, see Jagan, 1979). In 1961 the PPP won a national election, and Jagan served as Prime Minister until 1964, when he was defeated by a coalition of the PNC and the United Front, a situation that led Forbes Burnham to become the country's leader.[27] Jagan's Communist leanings, and the fear that he might regain power and "turn British Guiana into another Cuba," worried both Washington, DC, and London. For its part, the AFL-CIO began to fund those segments of the Guyanese labor movement favorable to the PNC (particularly the ICFTU-affiliated, Afro-Guyanese dominated Trades Union Council [TUC]), in opposition to those favoring Jagan (for more on the Guyanese labor movement, see Cumiford, 1987). As elsewhere, housing soon came to occupy a central role in efforts to sway the local working class's political loyalties.

Several unions had long recognized the importance of providing decent housing for their members. For instance, the Commercial and Clerical Workers' Union had operated a small housing scheme in Berbice County (one of Guyana's three counties) in the early 1960s (Todd, 1997). When Jagan had become premier in 1961 the anti-Jagan TUC began actively to pursue housing as a means of improving its membership's living standards and of providing a political bulwark against what it perceived to be Communist tendencies among the pro-Jagan forces. As TUC General Secretary Joseph H. "Polly" Pollydore (1997) has put it: "The [housing] project came about because at the time we had a serious ideological problem. In Guyana we had a regime [Jagan] that was very close to the Soviets and so we came together with AIFLD to do something about housing." However, the political rivalries between, on the one hand, the PPP and the Guyana Agricultural Workers' Union (successor to the BGIWU) and, on the other, the PNC and the TUC, meant that the plan was effectively vetoed by Jagan. Five years later, however, with the TUC-supported Forbes Burnham now in power, the government purchased 102 acres of land in La Penitence and Ruimveldt, two suburbs of the capital Georgetown, from a number of sugar plantations and then sold the land to the TUC Cooperative Housing Society, which had been set up with AIFLD help to initiate the housing project (Todd, 1997; American Technical Assistance Corporation, 1970b). In addition to helping to establish the cooperative, the AIFLD also secured from the AFL-CIO funding (guaranteed by USAID and the now favorably disposed national government) for the project. In November 1966, AFL-CIO President George Meany signed a $2-million loan agreement with the TUC. The 20-year loan was made at 5.75% interest, well below local market rates, and local participation brought the total financing for the project to about $2.2 million (*AIFLD Report*, January 1967a; *AIFLD Report*, June 1967d). Bidding on the contracts opened in January 1967, and a contract was finally signed at Transport House (TUC headquarters) in Georgetown on April 20, 1967. The project (called "Tucville") was formally opened at a 1968 May Day rally by Forbes Burnham, who handed over keys to the first 35 housing units to be completed (*AIFLD Report*, May 1968a).

The initial plan for Tucville called for the construction of 568 houses, together with primary and secondary schools for the residents' children, and the establishment of a bus service to enable residents to get to work in various parts of Georgetown and its environs. The central government put in the basic utility services such as electricity and sewer lines—Tucville was the first housing project in Guyana with its own modern sewage treatment facility—while AIFLD provided technical assistance in housing design, and the USAID made a $100,000 grant to construct the project's primary school to educate 800 pupils (*AIFLD Report*, June 1967d). Houses were built on stilts to avoid problems of flooding and were constructed out of wood with corrugated metal roofs. Three basic models were designed with two, three, or four bedrooms to accommodate different-sized families, with the cost of purchasing ranging from G$4,800 to G$8,200 (at the time about $2,400 to $4,100), depending on size. Purchasers (who had to be members of one of the TUC participating unions) put down a G$300 deposit, and mortgages were undertaken on a 20-year basis. A large proportion of those who bought houses in the project were waterfront and post office workers (Todd, 1997), unions that were largely made up of Afro-Guyanese and affiliated with the TUC and that were involved in key sectors of the economy (transportation and communications).

The Tucville project did run into problems, however. Early on, the U.S. contractor failed financially and AIFLD had to take over as the general contractor (Holway, 1997). Subsequently, the Housing Cooperative attempted to manage the project, but lack of skills led to mismanagement and the appointment of a new program administrator (American Technical Assistance

Corporation, 1970b: 53). Furthermore, after having constructed some 300 houses the TUC ran out of money, and the final group of houses had to be finished by the government (Holway, 1997). The financial failure of the contractor meant that the final costs of the houses were higher than anticipated, which made them too expensive for many workers and led to complaints that only wealthier workers could benefit from the project (American Technical Assistance Corporation, 1970b: 54; U.S. Senate Committee on Foreign Relations, 1968: 60). In turn, this led to some intraclass conflict between those workers who could afford to buy into the project and those who could not, with many of these latter subsequently setting up squatter settlements on the project's outskirts, which resulted in the backing up of drainage systems, causing problems of mosquito infestation and malaria (Todd, 1997). The plan also called for the establishment of a cooperative for mechanics, something that did not ultimately occur. Local AIFLD officials and USAID also had a number of disagreements over how the available funds should be spent on the project (American Technical Assistance Corporation, 1970b).

Nevertheless, despite such problems the project did undoubtedly benefit those workers fortunate enough to be able to buy into it—houses were considered to be worth well over their mortgage price (American Technical Assistance Corporation, 1970b: 54)—and also provided the TUC and AIFLD with a potent political exemplar to rebuff Communist propaganda, allowing the U.S. Ambassador to say on the occasion of the loan signing:

> I believe this project of the AFL-CIO, the AIFLD, and the Guyana TUC to decently house 568 families in Guyana, under the blessing and guarantee of the Governments of Guyana and the United States, will be lasting evidence that in democracy there is true world brotherhood; that in cooperation there is progress; and that in better conditions of living there will be increased stability and peace. (*AIFLD Report*, January 1967a)

In 1997 Tucville's population was some 2,500 people, and the majority of residents, many of whom see their homes as investments upon which they can draw during retirement, are the original owners of the houses (Pollydore, 1997; Todd, 1997). The construction of Tucville also helped stimulate other similar housing projects in Guyana, such as plans for a 16-unit workers' housing development for the Local Government Officers' Association union to be funded under an AIFLD loan (American Technical Assistance Corporation, 1970b: 55). More recently, in February 1997, TUC officials met with government ministers about the possibility of constructing an additional housing cooperative, and preliminary negotiations over land purchasing took place (Pollydore, 1997).

While AIFLD's largest monetary contribution in Guyana was in the field of housing, the Institute was also involved in a number of education programs in the country. Although it trained some top-level officials at its Front Royal, Virginia, facility in the United States, AIFLD's major educational role within Guyana was in helping to finance and construct a labor education college in Georgetown for mid-level officials (the Critchlow Labour College) and a vocational center for upgrading basic working skills (the Guyana Industrial Training Center). In cooperation with the TUC, AIFLD also provided a small-projects grant in 1967 to purchase some 200 school safety signs for use in the rural areas of Demerara County (*AIFLD Report*, August 1967d), a loan to build two bridges linking the La Penitence and Ruimveldt communities in Georgetown, a loan to help re-equip the Transport Workers' Union headquarters, a grant to install a drinking-water pipe system in a village in Berbice County, a grant to help buy a film projector for the Tucville cooperative, and a number of other small-impact projects throughout the country (*AIFLD Report*, January 1969). Its largest expenditure on small projects was for a printing press that helped the Man Power Citizen's Association sugar workers' union, the rival to Jagan's sugar workers' union, keep in touch with members. The overall result of such activities, as determined by the American Technical Assistance Corporation (1970b: 26), an independent entity contracted by USAID to evaluate the Institute's programs, was to "ear[n] in Guyana considerable good will and respect, both for the American labor movement generally and for the United States."

HOUSING, SPATIAL ENGINEERING, AND LOCAL GEOGRAPHIES OF THE COLD WAR

By the time AIFLD's housing program ended in 1978, when high global interest rates made it no longer a feasible proposition, the Institute had been involved in building 18,048 housing units in 14 nations valued (conservatively) at some $70 million and accommodating some 125,000 people (AIFLD, 1987) (see Table 7.1).[28] In addition, the Institute's actions had inspired a number of spin-off projects. As the American Technical Assistance Corporation (1970a: 41) maintained:

> AIFLD involvement in housing [has] not only contributed to construction of the specific projects associated with the Institute, but has prompted initiation of other housing projects for workers that are not directly linked to AIFLD activity. . . . While it is impossible to know what would have happened without AIFLD involvement, relationships between the social projects and governmental responses [elsewhere] are sufficiently close to support attribution of the responses to the program.

TABLE 7.1. Housing Units Completed or under Construction and Impact Projects as of December 1971 (AIFLD, 1972)

	Housing	Impact projects
Argentina	6,227[a]	21
Bolivia	0	16
Brazil	488[b]	50
Caribbean	0	4
Chile	0	26
Colombia	2,106	25
Costa Rica	128	15
Dominican Republic	110	42
Ecuador	14	29
El Salvador	0	13
Guatemala	0	9
Guyana	362	7
Honduras	1,185	31
Mexico	3,104	1
Nicaragua	0	5
Panama	0	4
Paraguay	0	1
Peru	3,229	24
Uruguay	425	12
Venezuela	920	5

[a]AIFLD was actually involved in constructing directly 1,667 units in Argentina. The higher figure presented here includes a number of units built by the IDB, some of which AIFLD Executive Director Doherty managed to secure for local unions and that he claimed as AIFLD project units (Anon, 1998).
[b]One of the former AIFLD officials with whom I spoke lists the number as 448 in the São Bernardo do Campo project (see p. 171). The additional 40 units claimed in the table above may be the result of a misprint or may actually exist in another project, the whereabouts of which I have been unable to determine.

The building of housing and the concomitant restructuring of the spatial relations of workers' everyday lives were intimately tied up with the politics both of the Cold War and of stimulating union and social development in the region. For the AIFLD, the provision of housing was a physical manifestation of its development objectives geared toward bringing about the U.S.-style modernization of social conditions in Latin America and the Caribbean, a position summarized by Angelo Verdu, Assistant to the AIFLD Administrator, at the dedication of the San Pedro de Macorís project in the Dominican Republic: "The [U.S.] labor movement," Verdu stated, "is philosophically and pragmatically on the side of the reformers and democratic revolutionaries who are trying to change the outmoded social structures which continue to plague the hemisphere. This is the fundamental objective of the Alliance [for Progress] and the immediate problem confronting the labor movement of the Americas. . . . [*The housing project*] *represents the Alliance in action*" (*AIFLD Report*, April–May 1967a: 5, emphasis added). Through the Institute's projects and the self-help programs it helped establish to foment its version of social revolution, local labor's role in the Alliance for Progress was "clearly defined: the direct, active participation of the workers and peoples of Latin America in their own social and economic development" (*AIFLD Report*, August 1967e).

AIFLD's activities, Executive Director William Doherty, Jr., has suggested, were designed to "get the unions strengthened to play a role against extremism from both the left and the right" and to inculcate in workers the ethos of private property and ownership as a bulwark against Communist inroads into various local labor movements. The result was that "the workers became owners or little capitalists because of their investment" (Doherty, 1997). Certainly, the housing projects were not designed to aid all those in need of housing, although they did house many tens of thousands of people. Most were initiated on the basis of their strategic geographic location or the economic and political role played by the unions involved with them. While the housing projects, together with the other smaller-scale social programs, may have had material benefits for only a relatively small proportion of those workers in need of decent housing, they did have very definite propaganda value that went well beyond those who were actually residents or beneficiaries of the projects, propaganda value that AIFLD officials sought to exploit in their conflict with Communist forces in the region. Thus, upon returning from a visit to the Dominican Republic in 1966, Doherty proclaimed in a speech over the ABC radio network (quoted in *AIFLD Report*, April 1966b: 7) that "I saw a situation where the communist propaganda weapon of blaming everything bad in life on 'Yankee Imperialism' had been taken away from them because there are an awful lot of people in San Pedro de Macorís who are now saying, 'If this is American Imperialism, then give me more of it, because we need houses.'"

Likewise, for the local unions involved the housing projects were seen as both a practical means of improving workers' material standards of living and as ways of strengthening local unions, either to fight against Communism or simply to strengthen the unions in and of themselves. But for local unions, union-built and -operated housing was not just about providing shelter. It was also about installing new traditions and "collective memories" (see Hobsbawm & Ranger, 1983; Alderman, 1998; Schwartz, 1987) in members that bound the workers and their unions together.[29] Housing and other social programs quickly took on symbolic meaning for workers in the region, representing concrete examples, built in wood and stone, of the unions' power to transform the daily existence of their members and to challenge prevailing social conditions. The creation of self-contained neighborhoods in San Pedro Sula, Georgetown, and elsewhere allowed unions to reinforce both at work and at home what Vural (1994) has called "unionism as a way of life." Through the process of building and naming projects after U.S. labor figures such as Samuel Gompers or local figures such as Céleo González, unions were able to show not only that they had the physical and financial power to overcome social problems faced by their members but that they had the spiritual power to do so as well. Whereas Jencks (1969: 13) has argued that "it is one of the basic assumptions of semiology that creation is dependent on tradition and memory," in these cases tradition and memory were shaped by the ways in which the built environment was created. The housing projects allowed unions to reinforce a shared sense of collective identity, community, and union loyalty that could take on spiritual qualities.[30] Thus, for Doherty (quoted in *AIFLD Report*, April 1966b: 7, emphasis added), the efforts of AIFLD and local unions were about "try[ing] to teach people how to live in a *modern* society, with better sanitation, better health, better community spirit. In effect, we are trying to help people live better *spiritually* as well as materially." Of course, whereas for some living better spiritually meant improving workers' sense of community and union loyalty, for others it undoubtedly meant living as non-Communists.

Although my purpose here has been to examine how housing was used as part of the geopolitics of the Cold War, it is not unreasonable to infer that the building of these housing projects also affected the ways in which gender relations played out in the region, though without intensive fieldwork and interviews it is difficult to determine exactly how. On the one hand, whether deliberately or unconsciously, the fact that housing designs were based upon fairly traditional gender roles within patriarchal families suggests that there was little overt attempt made to challenge historical gender divisions of labor. Thus, there were no plans (at least that I have been able to locate) that sought, for example, to transform domestic space in the ways in which nineteenth-century U.S. feminists and others had proposed, such as through the provision of communal kitchens and the like (see Hayden, 1981)—perhaps such communalism would have been too reminiscent of Communism! On the other hand, the provision of modern conveniences in homes undoubtedly affected many women's (and men's) everyday lives in very real ways. Having internal running water meant that many women, who frequently are responsible for hauling water from communal wells or rivers, no longer had to spend such a large part of their day on such basic tasks. Equally, for some the purchase of domestic appliances such as washing machines negated the necessity of having to spend time at commercial laundry services or washing clothes in local streams and rivers. The proximity of schools built as integral parts of several projects also meant that many women could spend less of their day taking children to and from classes, which perhaps freed up time to engage in full- or part-time paid work elsewhere (see Hanson & Pratt, 1995, who suggest that the time constraints frequently placed upon women by domestic chores limit their mobility in space).

With regard to the more general argument I am pursuing in this book, the situations outlined in this chapter show how workers and their organizations played important roles in making the built landscapes of many cities, towns, and smaller settlements as part of their proactive social and spatial praxis. Certainly, some of the funding to do so came from various government agencies, but much of it came from U.S. labor unions, local unions, and the local workers themselves who saved through various cooperatives and workers' banks to provide the money necessary to build the housing projects. Although they may have had different reasons for becoming involved in such projects (fighting Communism, providing shelter, local rivalries, self-aggrandizement, etc.), it was through the activities of thousands of trade union officials and ordinary workers who, either by managing to secure funding or through their own labor as part of the physical construction of various projects, that the urban geographies of many cities and towns in Latin America and the Caribbean were dramatically reshaped. In such projects social development and non-Communist social revolution were intimately tied up with transforming the spatial context within which workers lived their everyday lives.

AIFLD's social projects, of which housing was arguably the most important part, also illustrate the role played by unions in the processes of economic and political globalization, as outlined in the preceding chapter. For the Institute, its activities in the region were designed both to stimulate "modernization" and to fight Communism. Through AIFLD's international activities, closer links were developed between many U.S. unions and unions and workers in Latin America and the Caribbean. Likewise, AIFLD's activities shaped the flow of U.S. capital into the region, ensuring that federal and other funds were directed into certain locations and patterns of investment in ways that they would not have been had the Institute not been involved. But this was a two-way street. The processes of economic and political globalization and development were not just shaped by the activities of AIFLD in the region. Local unions within Latin America and the Caribbean also actively sought links with, and aid from, the AFL-CIO. Thus, they sent workers and officials to the United States to attend AIFLD-run training courses and appealed to AIFLD for financial, technical, political, and other sources of support to ensure that housing and other projects were implemented. Indeed, frequently it was the local unions in the region that initiated the contact with AIFLD and the AFL-CIO, rather than the other way around. This fact suggests the need for a further questioning of how the process of globalization has usually been conceptualized. Specifically, whereas much literature on globalization seems to see the process rather uncritically as one driven by the economic and political needs of "core" countries, the situations recounted here suggest that actors in "peripheral" countries may also play important roles in the process, in this case through Latin American and Caribbean unions' appealing for help from their U.S. confederates.

Finally, the examples shown here—with regard to both housing and other infrastructure improvements—highlight the interplay of the global and the local in understanding the geopolitics of the Cold War and how workers and unions actively shape landscapes. Certainly, the global dictates of the Cold War shaped the possibilities and processes of development locally throughout the region, at least with regard to the involvement of AIFLD. However, we should not forget also that what went on locally in these housing and other impact projects also directly shaped the global geography of the Cold War. As the struggle for the political hearts and minds of workers played out on a street-by-street, neighborhood-by-neighborhood basis, success or failure at this local scale shaped what occurred globally in this greatest of twentieth-century ideological conflicts. Put another way, the Cold War was fought out geographically not only globally but also locally, with what occurred at one scale simultaneously shaping possibilities for action at the other—if key groups of workers in key neighborhoods in key regions of key countries rejected or successfully fought off Communist insurgents within their ranks, this directly shaped the global geography of political ideology. As I have tried to stress elsewhere in this volume, such a truth should encourage us to question the supposed impenetrability of the global as represented in much of the discourse surrounding globalization and to recognize that the local can, in fact, impact the way in which the global is constructed and operates in quite powerful and direct ways. In itself this is, perhaps, a lesson for today's workers who are struggling to come to terms with globalization.

CHAPTER 8

Thinking Locally, Acting Globally?

The Practice of International Labor Solidarity and the Geography of the Global Economy

The idea of international solidarity, as good as it sounds, means nothing unless you can develop specific actions.

—PHILLIP JENNINGS,
General Secretary of the International Federation of Commercial, Clerical and Technical Employees

If we're going to be able to effectively challenge companies like Shell or Exxon or DuPont and other corporations which operate without regard to national boundaries, we have to redefine solidarity in global terms.

—RICHARD TRUMKA,
Secretary-Treasurer, AFL-CIO

The creation of this network [of unions representing workers at Goodyear Tire & Rubber Company] is an historic first step to build worker solidarity across borders in one multinational corporation.

—RICHARD DAVIS,
Vice President for Administration,
United Steelworkers of America

We are constantly reminded of the global economy and global trade. Well, we will add a new "global"—global unionism.

—GEORGE BECKER,
President of the United Steelworkers of America[1]

As we saw in Chapter 6, international labor solidarity has a history stretching back at least to the mid-nineteenth century. However, with the growing internationalization of manufacturing since the 1960s, talk of international labor solidarity has come to play an increasingly prominent role in many unions' conventions, publications, and political agendas. Indeed, in response to corporate strategies that seek to play workers in different parts of the globe against one another in a seemingly endless downward concessionary spiral, many workers have increasingly come to view "international labor solidarity" as the sine qua non for trade unions to retain economic and political power in the face of transnationally organized capital. At their 1986 convention, for instance, members of the United Steelworkers of America called for worldwide bargaining in the metals-fabricating industry. In the early 1990s several miners' unions—whose members in the United States, Canada, Britain, South Africa, Germany, and Australia accounted at the time for some 72% of the coal sold on the international market—began to explore how they might coordinate international strikes against transnational coal producers (*Labor Notes*, 1991). Electrical workers in the United States and Japan similarly began examining the possibility of joint industrial action directed against their common employers (*Labor Notes*, 1992). Likewise, workers in German auto producers' plants in Alabama have developed a working relationship with the German union IG Metall, which has pressured Mercedes-Benz not to oppose unionization at its U.S. plants, while the United Auto Workers has established a Council of Ford Workers to represent the interests of Ford workers in Canada, the United States, and Mexico (Howard, 1995). In 1998 the International Federation of Chemical, Energy, Mine, and General Workers' Unions (ICEM) Rubber Industry Section unveiled a "worldwide action plan" that included sharing information among member unions, developing an integrated global support system, global strategic planning, and reinforcing trade union networks within key rubber transnational corporations (ICEM, 1998). A year later representatives from ICEM member unions in Canada, Guatemala, Colombia, Venezuela, Brazil, Chile, the United Kingdom, France, Germany, Slovenia, Turkey, Morocco, South Africa, Malaysia, Japan, and the U.S. met to establish a networked database on the Goodyear Tire & Rubber Company's operations that will be used to assist one another in collective bargaining negotiations and to help organize the company's nonunion facilities (Gerdel, 1999). Myriad other examples could be found.[2]

Within this context, during the 1990s two international labor solidarity campaigns, more than any others, arguably came to epitomize workers' abilities to engage in global political actions. One of these was the 1995–1998 Liverpool dockworkers' strike. Caused by a dispute between the British Transport and General Workers' Union and the Merseyside Docks and Harbour Company relating to overtime work, the conflict saw Liverpool dockers who had been fired ask for, and receive, help from supporters throughout Britain, together with dockers' unions in the United States, Japan, Australia, and elsewhere who delayed the sailings of dozens of ships destined for the port. Dockers' unions from across the world also donated financial resources to the Liverpool unionists, with the New York-based International Longshoremen's Association, for example, sending some £3,000 at Christmas 1995 to aid their British confederates (*Daily Telegraph* [London], December 27, 1995; see Mann, 1962 [1923], and Schwantes, 1979, for more on earlier instances of dockers' expressions of international solidarity). Although the dockers ultimately failed to secure their objectives locally, their struggle highlights the important fact that they were able to build a grassroots organization that linked together workers from across the planet (see Castree, 2000, for more details on the dispute).

The second dispute that involved a long, globally organized campaign and that drew press coverage from around the world was a 20-month struggle between some 1,700 members of USWA Local 5668—who had been locked out of an aluminum-smelting facility in Ravenswood, West Virginia, in November 1990 in a contract dispute—and a multibillion-dollar metals trading empire run by international financier and fugitive from U.S. justice, Marc Rich. Indeed, some commentators have argued that the steelworkers' ultimate victory in this dispute is serving as the basis for a "revival of American labor" (Juravich & Bronfenbrenner, 1999) after two decades of, mostly, political defeats. It is this second campaign with which this chapter deals. Specifically, I examine how Local 5668 went about actually implementing a transnational solidarity campaign and what impacts this campaign had on the geography of the global investments of one particular corporation. Indeed, the fact that the plant itself was a subsidiary of a transnational metals trading corporation with headquarters in Switzerland played a crucial role in shaping the union's response to the lockout. In particular, the steelworkers' political strategy hinged on developing a well-articulated *global* network of support in opposition to the company's actions, a strategy that involved working closely with a number of international labor organizations, including several international trade secretariats and the ICFTU.

With regard to the general argument I am pursuing in this volume, then, in the preceding two chapters I have attempted to lay out a critique of much of the literature (particularly that of a neoliberal bent) on globalization on a number of counts: its failure to recognize that workers have been involved in the genesis of many of the processes that are reshaping the contemporary global economy; its failure to recognize that workers have created organizations and structures that are, in many cases, giving them real leverage upon contemporary processes of international economic integration (i.e., that workers are not always simply the passive victims of globalization that they are so often presented to be); and its frequent dismissal of how local praxis may challenge and shape global processes and globally organized entities. In the current chapter I want to provide another layer to this critique by examining how, through their success in developing a global solidarity campaign to defeat their employers' union-busting plan, these

1,700 steelworkers and their supporters not only gained readmittance to their plant but also played an active role in shaping the corporation's wider patterns of foreign direct investment (FDI). Furthermore, whereas in previous chapters I have argued that locally organized campaigns may sometimes be used successfully to challenge globally orchestrated and organized processes and institutions, in this chapter I show how workers organized globally to secure their local goals—in a twist on the mantra favored by so many on the left, this is a case of what we might call workers "thinking locally" but "acting globally." Indeed, it is my contention that the above-mentioned examples of the practice of international labor solidarity (for some other recent examples, see Moody, 1988, 1997; Moberg, 1989; Hecker & Hallock, 1991; Bendiner, 1987; Davis, 1995; Frundt, 1987, 1996; and Wills, 1998b) can be regarded as an effort by particular groups of workers to develop spatial fixes and to organize social relations between workers in different countries in such a way as to shape the manner in which the global space-economy is made. Building networks of solidarity is precisely about overcoming geographical and social barriers to cooperation between workers that, in the process, affects how the economic geography of capitalism evolves.

In terms of its organization, the first part of this chapter analyzes how local union members, the USWA International union, and the Industrial Union Department (IUD) at the AFL-CIO built an international campaign that ultimately forced the aluminum company to reinstate the locked-out Local 5668 union workers.[3] This campaign involved use of a number of techniques that were highly innovative at the time and that have subsequently been copied by other groups of workers. Following from this, I examine a number of conceptual issues relating to the practice of international labor solidarity and the geography of the global economy. The concluding section draws on the Ravenswood case study to reinforce and expand on the argument that analyses of the geography of the global economy need to pay greater attention to the international activities of workers and their institutions, that a critical economic geography must focus not only on how capital but also on how workers are remaking the economic landscapes of global capitalism, and that workers can successfully construct networks that allow global scales of praxis.

BUILDING INTERNATIONAL SOLIDARITY: USWA LOCAL 5668 GOES GLOBAL

For some 30 years the Kaiser Aluminum and Chemical Corporation plant near Ravenswood, West Virginia, operated with only one work-related fatality. In 1988 Kaiser Aluminum was acquired by Maxxam Inc. in a heavily leveraged takeover (*Forbes*, April 30, 1990). A year later, in an effort to reduce its debt, Maxxam sold the Ravenswood plant to a consortium known as the Ravenswood Aluminum Corporation (RAC), a company formed for the purpose of acquiring the facility by Charles Bradley of Stanwich Partners Inc., a group of international investors. Almost immediately RAC management instigated new work rules and a speedup of production, dissolved joint management–union safety programs, and combined several employment categories to eliminate nearly 100 jobs (Stidham, 1992). During the following 18 months five workers were killed and several others injured in accidents at the plant. According to the AFL-CIO, at the time this represented the worst accident record of any U.S. metals-processing facility (*AFL-CIO News*, July 22, 1991). The federal Occupational Safety and Health Administration (OSHA) calculated that lost work-day injuries at Ravenswood ran six times higher than the national average for manufacturing facilities and over three times the rate for aluminum smelters (*Charleston [West Virginia] Gazette*, December 21, 1991). Although RAC maintained the deaths were not the result of company negligence, the plant's USWA Local 5668 determined to make workplace safety a major issue in negotiations for the new contract, scheduled to take effect in November 1990. Local members also resolved to defend the union's pension formula, which the company intimated it wished to change.

For RAC, a fight over these contract issues promised to be long. Local 5668 has a history of militancy, having been one of only two locals to vote against ratification of the 1988 concessionary contract negotiated throughout the 11-plant Kaiser chain. Indeed, it was only the acceptance of the companywide agreement by a majority of Kaiser's USWA members that had ultimately forced Ravenswood workers to concede givebacks. However, the recent purchase of the plant by RAC augured a new era of labor relations. The 1990 contract would be the Local's first independently negotiated agreement. With the Local no longer obliged to follow the prescriptions of a national Kaiser contract, the plant's owners could no longer count on the willingness of the USWA's wider membership to concede givebacks to temper Local 5668's militancy.[4] During the several weeks immediately preceding the old contract's expiration on October 31, the company spent some $1.5 million fortifying the plant, hiring "security" guards, and advertising for "replacement workers." When the contract deadline passed without a resolution, company guards escorted the union workers out of one end of the plant while bringing their nonunion replacements in at the other. Whereas the company maintained the Local had failed to negotiate in good faith and that this was a strike, USWA officials said the company had locked them out to break the union.

During its first two months the dispute was, essentially, a local affair as union officials and RAC management excoriated each other in the local media. However, both the International union and the AFL-CIO soon began to get more deeply involved in the Ravenswood dispute. For the AFL-CIO the dispute quickly came to epitomize the weakness of unions in the face of U.S. labor law. During the union-busting drives of the 1980s companies frequently made use of a 1938 U.S. Supreme Court ruling that allows an employer to hire permanent replacement workers during a dispute over "economic" issues (*National Labor Relations Board v. Mackay Radio and Telegraph Co.*, 304 U.S. 333 [1938]). RAC's use of replacement workers came at the very time the AFL-CIO was gearing up for a major campaign to seek federal legislation outlawing such activities.[5] The AFL-CIO therefore determined to make the Ravenswood unionists a cause célèbre, a tactic that would require throwing its full weight behind the Local. For its part, the USWA's International office in Pittsburgh, Pennsylvania, felt the plant management had become entrenched in its bargaining position and believed the union needed to expand the geographic scale of the dispute if there were to be any hope of settling the new contract favorably (Chapman, 1992). The involvement of the International office appeared the best way to do this.

In December 1990, as International representatives were outlining plans for stepping up their campaign, a copy of an internal RAC audit conducted by the Price Waterhouse accounting firm arrived anonymously at Local 5668's union hall. This was a crucial event, for the document identified a number of RAC's creditors who had helped finance the plant's 1989 purchase. Sensing the significance of this information, Local officials passed the document on to the International union, which in turn informed AFL-CIO headquarters. At this point the AFL-CIO assigned a member of its research staff to look more closely into the RAC purchase, while IUD Director of Special Projects Joe Uehlein (who had been called in by USWA International Vice President George Becker) accessed the ICEF (now the ICEM) trade secretariat's computerized corporate database for information about the companies listed in the audit (Uehlein, 1992).[6]

As the USWA, the IUD, and the AFL-CIO began in the first months of 1991 to investigate RAC and the circumstances surrounding the plant's purchase, they unraveled a complex web of corporate connections that seemed to tie the 1,700 Ravenswood workers directly into the operations of a multibillion-dollar transnational corporation with truly global reach (see Figure 8.1). Their research revealed that some six months after his February 1989 purchase of the plant Charles Bradley had divided up ownership of RAC with several other investors. Retaining 20% for himself, he sold 27% of the company to local investor R. Emmett Boyle (who had subsequently been made CEO), 5% to local plant managers, and the remaining 48% to Rinoman Investments B.V., a Dutch company wholly owned and controlled by international investor Willy Strothotte. Delving more deeply into Rinoman, the union researchers mapped a trail of corporate ownership that appeared to lead from Ravenswood, West Virginia, to the financial hub of

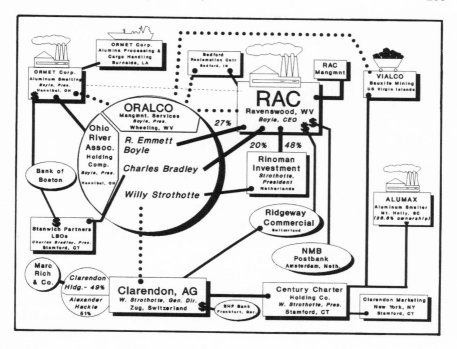

FIGURE 8.1. Diagram showing organizational structure of RAC.

Zug, Switzerland, and, specifically, to global commodities trader Marc Rich. Whereas the Steelworkers believed ownership in RAC to be personal for Boyle and Bradley, they became convinced that Strothotte's participation in the deal through his Rinoman Investments was merely a front for Rich and his group of companies (USWA, 1991).

Marc Rich and the International Connection

Through his various investments Marc Rich controlled in the early 1990s a global financial organization with operations in more than 40 countries. His companies were estimated to trade in excess of $30 billion worth of commodities annually. In 1983 Rich, his U.S. operating arm "Marc Rich and Co. International A.G.," and a business partner were indicted by the U.S. Department of Justice on 65 counts of tax fraud and racketeering, charges stemming from their alleged rigging of an illegal oil-pricing scheme, failure to pay at least $96 million in taxes, mail and wire fraud, and breaking the U.S. embargo on trading in Iranian oil.[7] In an attempt to make his legal problems go away, Rich sold 51% of Marc Rich International to an unindicted partner. Retaining 49% ownership, he then renamed the company Clarendon Ltd. and installed his old friend Strothotte as company president (*Philadelphia Inquirer*, December 5, 1991). Rich himself fled to Switzerland to avoid a possible 325-year jail sentence on criminal charges stemming from the indictments. These organizational changes were more cosmetic than substantial, however, for Clarendon continued to employ the same traders and financial experts as had Marc Rich International, retained almost entirely the same board of directors, and even used the same telephone numbers and addresses. The only major change was that Rich and his indicted partner were no longer listed as company directors (Copetas, 1985). Although a federal judge called this shuffling of corporate ownership simply a "ploy" to legally distance Rich from his U.S. operations, it was enough to allow Clarendon to continue operating in the United States and to expand aggressively into the aluminum-smelting industry worldwide, to the extent that by the early 1990s the company controlled more than one-third of the global independent primary aluminum and bauxite market (*Institutional Investor*, 1992).

The fact that Strothotte was the president of Clarendon was certainly an indication to the USWA that events in Ravenswood might be connected to Marc Rich. However, several other leads and corporate links provided clues to suggest the identity of the powers ultimately behind events at RAC. The Steelworkers' research showed that the plant's leveraged buyout was partial financed by a $260-million loan from Ridgeway Commercial (a Swiss affiliate of Clarendon) and by $140 million worth of revolving credit provided by Amsterdam's Nederlandsche Middenstandsbank, a bank that regularly did business with the Rich group of companies (*Philadelphia Inquirer*, October 23, 1991). Furthermore, Clarendon's president Strothotte was a co-owner of Oralco Management Services, the company responsible for managing the Ravenswood plant and several other operations allegedly indirectly controlled by Rich (*Forbes*, April 30, 1990). Clarendon also supplied RAC with most of the bauxite used in the smelting operation and purchased large quantities of the finished aluminum. Finally, the law firm that handled Bradley's purchase of the plant considered itself to be Marc Rich's general counsel (Horne, 1991). Although the USWA was willing to concede that the maze of corporate ownership, holding companies, dummy corporations, and secret partnerships made it difficult to prove conclusively that Rich was a part-owner of the plant, they argued that there was certainly more than sufficient circumstantial evidence to indicate he had been involved in the RAC buyout. Having become convinced that RAC was indeed part of the financial octopus that was Marc Rich's global commodities trading empire, the Steelworkers shifted the focus of their campaign from Boyle and the local plant management to Rich. The key to settling the dispute, they believed, now lay in identifying Rich's vulnerabilities and seeking to exploit them internationally. This represented a tangible refocusing of the scale of the Steelworkers' praxis from a concern with the activities of local management within the plant and the community to a concern with building national and international links with workers and other supporters beyond Ravenswood.

Elements of the Campaign

Soon after receiving the Price Waterhouse documents, the Steelworkers began devising a five-pronged domestic and international campaign to pressure Rich and his confederates (Chapman, 1992). First, the International union pursued the labor law aspect of the dispute with the National Labor Relations Board, because a favorable ruling that the dispute was a lockout rather than a strike would greatly strengthen the legal basis of the union's campaign. Second, it investigated the health, safety, and environmental aspects of the dispute with OSHA and the U.S. Environmental Protection Agency. At the behest of the IUD and the union, OSHA officials conducted several inspections of the plant during 1991, ultimately imposing $604,500 in penalties for 231 safety violations (USWA, no date). The USWA and the IUD used this litigation to pressure several of RAC's creditors to reassess their support for the company (Uehlein, 1992). Third, Local 5668 members and supporters engaged in local morale-boosting activities to maintain the locked-out workers' resolve in the face of what would ultimately be a lengthy dispute. Much of this focused around the local union hall and, later, "Fort Unity," a building constructed next to the union hall by Local 5668 members to serve as a focal point for rallies, meeting the numerous caravans of supporters from across the country and even the world—including a 300-vehicle caravan that brought over $100,000 in donations from union workers across the Midwest in October 1991—and providing food for locked-out workers as a way of maintaining solidarity. In addition, Local members operated an assistance center in a local shopping plaza to distribute strike funds, food and clothing, and financial advice. They also conducted informational picketing at the company's corporate offices in Connecticut, at the New York Stock Exchange, and at several university campuses and state capitols. Some even sought to embarrass Rich on the ski slopes near Zug, Switzerland, where he regularly played with other members of the international jet set.

Fourth, under the auspices of the AFL-CIO's Strategic Approaches Committee, the International union initiated a campaign to pressure over 300 end users to stop buying RAC metal or face the consequences of a national consumer boycott (Chapman, 1992). The USWA also persuaded a U.S. House subcommittee to investigate why the U.S. Mint had recently bought some $25 million worth of metal for coinage from Clarendon (*Forbes*, January 13, 1992). Fifth, the Steelworkers developed the international component of the dispute by stressing the links between Rich and RAC (see Figure 8.2). As Jarley and Maranto (1990) note, trade unions considering corporate campaigns to settle a labor dispute may adopt a number of different strategies. One such strategy involves escalating a campaign by alleging the company is a corporate outlaw, thereby potentially damaging its public image and encouraging government examination of its affairs. This became a crucial component of the union's international campaign against Rich. The

Working through a network of corporate fronts from his hide-out in Zug, Switzerland, international fugitive Marc Rich pulls 'strings' connected to R. Emmett Boyle, Chairman of the Ravenswood Aluminum Corporation (RAC) in West Virginia, USA.

Boyle knows about the U.S. Government's 65-count indictment charging Rich with *racketeering, conspiracy, tax fraud and trading with the enemy*.—and the $750,000 reward for his arrest.

Boyle tries to cover-up his links to Rich. But Boyle can't hide the ties that bind RAC to Rich.

They're on the record: Clarendon, AG, a Marc Rich Group company in Switzerland, provides RAC financing. Its General Director, Willy Strothotte (another Rich puppet), is an owner of RAC and partner of Boyle's.

On November 1, 1990, Emmett Boyle locked-out 1700 USWA metalworkers from their jobs without warning as their contract expired. They were experienced workers, averaging 24 years of service. RAC stole their jobs. It hired "permanent strike replacements." The community calls them *scabs*.

The United States National Labor Relations Board (NLRB) has filed a complaint against RAC for engaging in unfair

labor practices in violation of U.S. labor laws. That could cost RAC tens of millions of dollars in back pay awards to the union workers.

The locked-out Steelworkers, with the backing of the USWA and the world's free trade unions, will fight RAC, Rich and his puppets—like Boyle—as long as it takes, until justice prevails.

If you want to know more about what Marc Rich and Emmett Boyle are doing to the workers at Ravenswood and would like to help us stop them, please contact:

- UNITED STEELWORKERS OF AMERICA
  RAC INTERNATIONAL COORDINATED CAMPAIGN
  FIVE GATEWAY CENTER
  PITTSBURGH, PA 15222 USA
  TEL. 412/562-2442, FAX 412/562-2445

- INTERNATIONAL METALWORKERS' FEDERATION (IMF)
  GENEVA (SWITZERLAND) TEL. (022) 43 61 50

- INTERNATIONAL FEDERATION OF CHEMICAL, ENERGY AND
  GENERAL WORKERS' UNIONS (ICEF)
  BRUSSELS (BELGIUM) TEL. (32) 2 648 4316

UNITED STEELWORKERS OF AMERICA AFL-CIO/CLC   RACP 41

FIGURE 8.2. Anti-Marc Rich flyer distributed during the dispute.

Steelworkers surmised Rich was eager to ward off such unwanted government intrusion for two reasons (Uehlein, 1992). On a personal level, interest in his business activities might jeopardize any possibility of negotiating a settlement with the Department of Justice which would allow him to return to the United States and yet avoid imprisonment on the outstanding criminal charges he faced. Of greater significance, however, was the belief that by drawing attention to Rich's nefarious past and his links to the RAC dispute, the USWA could disrupt his international operations and so financially hurt him that he would use his influence with Strothotte and others to settle the dispute.

Having conducted their analysis of Rich's known holdings, by mid-1991 the IUD and the Steelworkers were ready to put the international element of the campaign into operation. Research had revealed that Rich had interests in Switzerland, Britain, France, Finland, the Netherlands, Spain, Czechoslovakia, Romania, Bulgaria, Russia, Israel, Venezuela, Hong Kong, and Australia, among other places. In June 1991 a delegation of IUD and USWA representatives traveled to Switzerland. Through the medium of the International Metalworkers' Federation and the ICEF trade secretariats, the U.S. delegation persuaded the Swiss metalworkers' union to organize several press conferences to lobby the Swiss parliament concerning its latent support for Rich and other such corporate operators.[8] Uehlein's links to the ICEF also enabled him to contact the Dutch bank workers' union and arrange, through them, a meeting with officials of the Nederlandsche Middenstandsbank (NMB), which had helped finance the original RAC purchase. This was a particularly important meeting because some weeks previously Boyle had announced his intention to buy out his partners using loans from the NMB. At their meeting with the bank's directors the U.S. unionists presented information showing that RAC's liability for fines and back pay, should the courts rule the dispute an illegal lockout, would leave the company in poor financial condition. Consequently, they suggested, loaning money to Boyle might be a risky proposition for the bank. Indeed, Boyle's buyout of his partners was never completed, the result, the USWA claimed, of the pressure the union had brought to bear, which "so embroiled [the NMB] in the dispute that they cut [Boyle] off" (Uehlein, 1992).

During the latter part of 1991 the U.S. union representatives returned to Europe periodically to develop further contacts with trade unionists and other sympathizers. In October, while leafleting at the London Metals Exchange, they learned that Rich was attempting to purchase the Slovakian National Aluminum Company in Czechoslovakia and even had a signed letter of intent to do so. Working through the International Metalworkers' Federation, the USWA made contact with the Czechoslovakian metalworkers' union OS KOVO to devise a strategy that would disrupt Rich's plans. The pressure subsequently mounted by the USWA through diplomatic channels in Washington, DC and by the Czechoslovakian unionists was sufficient to persuade President Vaclav Havel to intervene personally to prevent the purchase (Labor Research Association, 1992; *Business Week*, May 11, 1992; Moberg, 1992). Despite their successfully foiling Rich's plans to expand his metal-smelting operations into Czechoslovakia, however, the difficulty of organizing mass demonstrations and other activities an ocean away, combined with the enormity of the corporation with which they were dealing, convinced the USWA and the IUD that they needed to maintain a permanent presence in Europe (Uehlein, 1992). The establishment in December 1991 of a European office in Paris to coordinate anti-RAC and anti-Rich activities on a day-to-day basis enabled the Steelworkers to pursue the international aspect of their campaign more vigorously than had previously been possible.

For Rich, the apparent demise of Communism and the selling off of formerly state-owned enterprises held the possibility of new investment opportunities in Eastern Europe. Through their research the USWA and the IUD were able to identify several pending deals involving the Rich group of companies. Working through various trade secretariats, the Steelworkers' European coordinator made contact with a number of national labor federations and individual trade unions in the region. This geographic expansion of the USWA's corporate campaign allowed the union and its supporters to disrupt Rich's plans to pursue several business openings as well as generally to harass his operations throughout the continent. The IUF, for instance, organized a rally of some 20,000 trade unionists to protest Rich's proposed acquisition of a 51% stake in the famous Athénée Palace Hotel in Bucharest, Romania, a rally that, the USWA claimed, was instrumental in bringing about the government's intervention to prevent Rich from purchasing the hotel (Labor Research Association, 1992). Events in support of the locked-out Ravenswood workers were also staged in several other Eastern European countries, including Bulgaria and Russia.

Eastern Europe was the primary focus of this aspect of the campaign because it was there that Rich was seeking most aggressively to expand his operations. However, the USWA did not confine its activities solely to the former Soviet bloc. The Steelworkers were also active in Latin America and the Caribbean, locations where Rich has numerous investments and controls large reserves of bauxite, oil, and other commodities. Earlier in 1991 the IUD had contacted Jamaican Prime Minister Michael Manley, who formerly had worked with the USWA as a union organizer in the country's bauxite mines (for more on the USWA's role in organizing Jamaica's bauxite industry, see Harrod, 1972). Manley had himself made Rich's control of Jamaica's natural resources an issue in his 1989 electoral campaign (see *Institutional Investor*, 1992). At the USWA's urging Manley lobbied representatives of the Venezuelan government, where Rich had a deal pending. Simultaneously, working through the International Metalworkers' Federation, the ICEF, and several other trade secretariats, the USWA contacted trade unionists in Venezuela, who brought political pressure to bear on their government. Although Rich had been one of three final bidders on a multibillion-dollar aluminum smelter purchase, the protests organized on behalf of the Ravenswood workers and the light shed on Rich's business practices were sufficient to have him publicly dismissed from the bidding process by President Carlos Andres Perez (Uehlein, 1992). Following this success, in March 1992 the USWA also presented its case to the convention of the ICFTU and briefed representatives of the Confederation's Inter-American Regional Organization (ORIT), who printed up in several languages "Wanted" posters featuring Rich's face and resolved to disrupt his operations throughout Latin America and the Caribbean.

The End in Sight

By the spring of 1992 the Steelworkers had organized sustained anti-Rich actions in 28 countries (including the Netherlands, Britain, Canada, France, Venezuela, Romania, Bulgaria, Czechoslovakia, and Switzerland) on five continents (Uehlein, 1992). Plans were afoot for activities in May and June 1992 in Australia, Russia, Israel, Hong Kong, and Finland. Meanwhile, the domestic end user campaign had persuaded many RAC customers to cease purchasing aluminum from the company. This forced RAC to reduce production from some 24,500 tons per month before the lockout to 15,600 tons and helped shrink its revenues from $701 million in 1989 to $491 million in 1991, leading the company to default on $71 million in loans (RAC, 1992; *AFL-CIO News*, March 2, 1992; *Business Week*, May 11, 1992). Although Rich continued to deny involvement with RAC, by April 1992 the global campaign and the effects of the RAC boycott seemed, finally, to yield results for the locked-out unionists. Facing the possibility the dispute might drag on indefinitely in the face of the USWA's corporate campaign, and concerned about the financial losses this implied for RAC, Strothotte finally moved to end the lockout. Seizing control of the board of directors, he fired Boyle and ordered RAC management to resume contract negotiations with the USWA. During the next few weeks the details of a new contract were worked out, perhaps the most important provisions of which included the return to work of Local 5668 members with full seniority, the dismissal of the replacement workers, a strong union successorship clause, and wage and pension increases (USWA, 1992). For its part, the Local agreed to allow the company to combine some work classifications, with the result that approximately 200 jobs would be lost, mostly through attrition (Stidham, 1992). The union membership overwhelmingly approved the new three-year contract, and on June 29, 1992, about 20 months since they had been locked out, Local 5668 members returned to work (see Table 8.1 for a timeline of the most significant events during the dispute).

TABLE 8.1. Key Events in the Ravenswood Dispute

September 25, 1990	Negotiations for new contract begin. RAC and USWA exchange contract proposals.
October 31, 1990	RAC demands acceptance of the company's final offer; USWA counters and proposes extending current contract. RAC rejects extension, locks Local 5668 members out, and brings in replacement workers.
December 21, 1990	West Virginia State Board of Review rules dispute a lockout and awards workers unemployment compensation.
March 11, 1991	USWA initiates domestic consumer boycott campaign under auspices of AFL-CIO's Strategic Approaches Committee.
April 28, 1991	Seven thousand supporters of Local 5668 gather in Ravenswood for Workers Memorial Day observance for five workers killed on the job at RAC.
June 23, 1991	Local 5668 Negotiating Committee and IUD/AFL-CIO representatives make first trip to Europe to gather international support.
August 28, 1991	Stroh Brewing Company notifies USWA it will stop using RAC aluminum.
September 23, 1991	USWA and NLRB lawyers present evidence to administrative law judge that RAC illegally replaced Local 5668 members.
December 1991	Establishment of European office to coordinate anti-Rich activities.
December 4, 1991	U.S. House Government Operations Subcommittee opens hearings on U.S. Mint contracts awarded to Clarendon.
January 21, 1992	Anheuser-Busch notifies USWA it will stop using RAC aluminum.
January 28, 1992	USWA members return to Europe. Demonstration subsequently organized outside Rich's corporate headquarters in Zug, Switzerland.
February 4, 1992	Miller Brewing Company notifies USWA it will stop using RAC aluminum.
February 14, 1992	International Union of Food and Allied Workers' Associations trade secretariat organizes anti-Rich demonstration in Bucharest, Romania.
March 16–19, 1992	USWA briefs representatives of the Inter-American Regional Organization of the ICFTU on Rich's Caribbean and South American operations.
April 11, 1992	Strothotte gains control of RAC Board of Directors; subsequently fires Boyle.
April 29, 1992	Negotiations begin for new union contract.
May 27, 1992	Tentative agreement reached.
June 29, 1992	Local 5668 workers return to work.

Source: United Steelworkers of America (1992).

INTERNATIONAL LABOR SOLIDARITY
AND THE GEOGRAPHY OF THE GLOBAL ECONOMY

Learning From Ravenswood

Much has been made in recent literature (e.g., Shostak, 1991) about the need for unions to constantly adapt their modes of operating to meet new political and economic challenges. Arguably, one of the greatest contemporary challenges that workers and their institutions now face is the internationalization of production and the growing integration of the world economy. Although they encountered numerous difficulties, Local 5668 members and their supporters were able to confront these challenges by constructing an effective global solidarity campaign to ensure that the Ravenswood plant remained unionized. Their success provides some useful lessons for workers contemplating similar campaigns against transnational corporations.

Clearly, several aspects of the case study are specific to Ravenswood. For one, Marc Rich made an easy target for the locked-out workers to vilify, which strengthened their resolve to continue their corporate campaign against RAC. Many Local 5668 members conceived of the lockout as a personal battle of wills between Rich and themselves. In cases where workers are unable to personalize a dispute in this manner, organizing for action may prove considerably more difficult, especially when a company is successful in portraying its actions in taken-for-granted terms (e.g., "market forces") that are often accepted more readily as legitimate by those affected than are the actions of particular, clearly identifiable, corporate officials who can be labeled as malevolent. The Ravenswood workers were also fortunate to receive considerable financial support from their International union (to the tune of some $11.5 million) and elsewhere, support not always available to workers in similar situations. Furthermore, workers were able to draw on a powerful sense of place and community based upon work and personal relationships built up over many years to sustain themselves throughout the length of the lockout in ways that less stable work forces are perhaps not so easily able to do.[9] So deep was this sense of identity that during the year-and-a-half-long dispute only 17 (mostly younger) union members crossed the picket line to return to work. Finally, Local 5668 officials were extremely fortunate to receive anonymously a copy of the Price Waterhouse report, detailing much of RAC's corporate structure, which made the process of tracing the geography of ownership much easier.

Several broader insights may, however, also be drawn from the Ravenswood campaign. One of the most significant concerns the politics of coalition building. Community coalition building is often seen as being of paramount importance in organizing against a plant shutdown or efforts to bust a union. Indeed, Haas and the Plant Closures Project (1985: 35) have suggested that a broad community–labor coalition is "the single most important weapon against shutdowns." However, in such cases a spatially parochial notion of "community" is frequently adopted, one that defines community in terms of the boundaries of the particular locality itself. In effect, such a narrowly conceived vision of working-class community can isolate those opposing a shutdown because it confines action to only one geographic scale (i.e., the scale of the immediate locality). In the Ravenswood lockout, however, workers successfully extended their understanding of community to operate on a number of different scales. By expanding the scale of conflict geographically they were able to draw upon social resources that would not otherwise have been available to them and that provided a greatly enlarged resource base for mobilizing against RAC (see McCarthy & Mayer, 1977 who have argued that access to resources is a crucial determinant of social movements' abilities to organize). Equally, activities at different scales reinforced one another. Whereas the Local union members' resolve to remain on the picket line proved crucial for securing the time necessary to implement the global campaign, the influx of money, information, and emotional support from beyond the community and even beyond the country— many trade unionists came from South Africa and other foreign countries to meet with the locked-out Ravenswood workers, for example—provided essential resources to sustain the locked-out workers throughout the lengthy dispute. In addition, expanding the conflict beyond the confines of Ravenswood allowed the Steelworkers to draw into the dispute and pressure high-level corporate officials (Rich, Strothotte) who "didn't want to get their hands dirty with this stuff" (Uehlein, 1992).

A key part of efforts to expand the campaign geographically was the ability of the Steelworkers and the IUD to determine how RAC was organized and how the Ravenswood plant fitted into a larger corporate structure. Although they were indeed fortunate to gain access to the Price Waterhouse audit, the Steelworkers also gleaned much information from the ICEF's corporate database. Such research proved crucial to settling the dispute. As Joe Uehlein (1992) commented:

> It's very difficult to trace ultimately ownership and you have to be willing to go overseas to do research and network and get unions to help you. It's hard and it takes operating beyond the ordinary means, but it's clear that information feeds action. Without the information you don't know where to best hit or what kind of actions to organize. So you have to have the information. It's first the investigation, then the strategy, then the tactics.

Workers fighting corporate union-busting plans or shutdowns clearly need to generate and communicate up-to-date information, particularly in an era of widespread acquisitions and mergers. Given the nature of the global economy, the databases maintained by a number of international trade secretariats and the information made available via the Internet will undoubtedly play increasingly important roles in future solidarity campaigns and may, in fact, substantially transform the tactics used by workers in the future (see Waterman, 1993).

Having conducted their research, the Steelworkers also had to get their story out. Once again, they adopted a number of unconventional methods to do so. Certainly, mainstream media have traditionally provided outlets for unions and workers to inform a wider audience about particular labor disputes. However, there are problems associated with this approach, for it can result in unions becoming hostage to corporate-controlled media in which hostile editors, writers, reporters, and owners might distort, or even subvert, their message (Puette, 1992). Thus, although they made extensive use of mainstream media, together with several labor newspapers, the USWA also used electronic mail systems to inform workers and supporters around the globe directly about the campaign and to solicit information about Rich, a strategy that the International union headquarters in Pittsburgh has subsequently used in other campaigns, such as the 1994–1996 "cyber-campaign" it waged against the Bridgestone/Firestone company (Herod, 1998e).

Finally, Joe Uehlein of the IUD worried that the boycott may actually have been *too* successful, because it left RAC "a heavily damaged company and now we have to get customers back and perhaps we'd be better off had we not destroyed their customer base." Upon reflection, he suggested that a more effective strategy would have been to "escalate much more quickly on the international front ... and downplay the boycott somewhat." For Uehlein, it was the "Marc Rich piece which gave us our avenue for settlement. . . . This thing was won on the coordinated campaign and that's the key. It's beyond the picket line solely. It's the action which you engage in to pressure [the high corporate officials] that's key" (Uehlein, 1992). Such reflections suggest that to be successful workers need to be willing not only to coordinate activities internationally but to do so sooner rather than later.

Broader Issues for Theorizing Workers' Roles in Shaping the Geography of Global Capitalism

Moving beyond the specifics of the fight against RAC and Marc Rich, the Ravenswood campaign's success raises a number of broader theoretical issues that relate to how trade unions and other working-class organizations are reshaping the global economy. Three issues are of particular importance. First, as I have tried to argue throughout this volume, the case study presented above shows clearly that workers can play significant roles in molding the geography of capitalism by helping to determine patterns of capital flow and foreign direct investment, and thus of uneven development. The fact that workers, not just at Ravenswood but in other similar campaigns (see Moody, 1988; Moberg, 1989; Hecker & Hallock, 1991; and Bendiner, 1987, for examples), have been able to take on and beat corporations *at the global scale* necessitates a greater incorporation of workers in narratives about processes of globalization. Furthermore, such examples of workers' praxis affecting how the geography of uneven development is produced reinforce arguments about the necessity to theorize such uneven development not as emerging simply out of the internal workings, contradictions, and logic of capital, but to conceive of it as a highly contested process shaped by struggles both between and among capitalists and workers, struggles that are conditioned by the contingencies of history and geography.

Second, returning to Gibson-Graham's (1996) argument about the politics of representation, there has been a tendency by many on both the political right and the political left to present capital discursively as hegemonic at the global scale. The result of such a conceptualization is that capital is ultimately seen as "unfettered by local attachments, labor unions, or national-level regulation." From capital as global hegemenon "it is but one step," Gibson-Graham (p. 9) maintains, "to capital as absolute presence." Thus, capital becomes thought of both as unbeatable, particularly at the global scale, but also as able to pick up and leave one locality for another almost instantaneously. This feeds into discourses of rampant capital mobility that can be used to undermine efforts of workers—who are themselves assumed to be fettered by local attachments of kinship, language, religion, and so forth—to challenge corporate activities (see Herod, 1994). Yet, the success achieved by the Ravenswood workers and their supporters around the globe shows that global capital is not, in fact, hegemonic and can indeed be successfully challenged. Certainly, this is hard work for workers. But, through their actions workers are clearly able to produce spaces of opposition to capital globally. Rather than thinking of global capitalism as a billiard ball smooth system within which struggle against capital is ultimately pointless, the USWA's campaign shows that it *is* possible to open up and exploit cracks in the landscape of capitalism and thereby to create globally spaces of opposition to capital. Whereas capital's globalization is usually presented discursively as being more powerful than is that of labor, the Ravenswood campaign suggests this is not always the case and that a critical politics of labor internationalism and transnational solidarity must rethink the way global capital has been theorized and represented. Specifically, while those on the left have often urged workers to think globally and to act locally, in this case, I would suggest, we see a situation in which workers were successful in thinking locally (i.e., they were concerned primarily about the locking out of Local 5668 members by RAC) but in acting globally.

The third issue that emerges from the Ravenswood case but that is also an issue more generally in campaigns to develop solidarity between workers (especially internationally) is that of how space shapes efforts to build solidarity and, consequently, to theorize class relations. In cases such as that of Ravenswood in which capital mobility or continued investment in a particular locality is not an issue (e.g., in disputes over the right to organize or perhaps over wages), solidarity across space may be developed fairly readily as a means for workers in one location—who have nothing to gain from any loss their confederates may suffer—to support workers in another location. However, in other cases where corporations threaten to move investment and jobs from one location to another, one group of workers' loss can often be another's gain. Hence, while for North American and Western European workers the growth of global assembly lines has usually meant job loss, for those workers in less developed countries and/or newly industrializing countries it has often meant increased investment by transnational corporations.[10] Whereas in the first case workers may be more inclined to emphasize common class issues—one group of workers helping another gain union rights—in the latter they may more readily choose to organize around identities that are defined geographically, that is, they may vigorously defend and/or promote their own localities within the global economy to the detriment of workers located elsewhere. In cases of capital flight, then, it can frequently be much more difficult to develop solidarity across space as a means to defeat the efforts of capital to play different places against one another, and international labor organizations may face much greater problems in trying to coordinate global campaigns against transnational corporations. This raises important theoretical issues concerning how workers' practices are theorized, how workers use space, and how space itself may complicate class analysis (see Herod, 1991c).

Such issues of what some have termed the "space versus class" conflict and its impact on workers' abilities to develop solidarity across space have been addressed in an interesting manner by Johns (1998). In her study of the U.S./Guatemala Labor Education Project (an effort to develop solidarity between unionists in Guatemala and the United States) Johns makes a useful distinction between what she calls workers' "accommodationist solidarity" (of which she distinguishes several degrees) and workers' "transformatory solidarity." Specifically, she uses the first term to describe efforts to develop (international) solidarity across space that essentially accommodate capitalism to different degrees and in which spatial interests and the defense of particular economic spaces ("our region" versus "their region," for instance) dominate over class ones. Thus, she argues (pp. 256–257), "while using the rhetoric of internationalism and union solidarity, this type of solidarity organization seeks not to transform social relations but to accommodate them while reasserting the dominance of a particular group of workers within capitalism's spatial structures. By assisting workers abroad in organizing, this kind of solidarity hopes to take groups of workers out of

competition for jobs, reducing the incentive for capital flight" from one locale or country to another. The result of all this, she maintains, is a politics
that divides workers geographically, that lends itself to policies of collaboration with capital, and that "reinforces the spatial competition [i.e., the ability
to play places against one another] capital so loves." Thus, she continues, attempting to "rais[e] the wages and working conditions of Third World workers [through this kind of solidarity] has the appearance of working to eliminate uneven development, but in fact the goal of accommodationist
solidarity is the entrenchment of uneven development," such that the privileged geographical position of certain workers is maintained.

"Transformatory solidarity," on the other hand, is a type of solidarity in
which "universal class interests dominate over interests that are spatially derived and rooted." Such campaigns, she argues, are designed to "unite
workers across space in order to fundamentally transform social relations of
production into something more humane." Whereas workers engaged in
struggles articulated around the goals of accommodationist solidarity have
as their primary motive the defense of their own privileged spaces within
the global economy, Johns suggests that workers engaging in transformative
solidarity do not prioritize their own interests over those of workers in other
places but, instead, seek to challenge the very social relations of capitalism,
regardless of which particular places this may or may not benefit. In such
types of activity, it is class interests rather than spatial interests that are acted
upon.

Although in practice it may be hard to determine empirically in which
types of solidarity any particular group of workers is actually engaged at any
given time and place, I would suggest that Johns's distinction is nevertheless
useful theoretically because it does nicely indicate that workers' solidarity
activities can have different impacts upon the geography of globally uneven
development, depending upon the specific contexts within which they find
themselves and the particular goals they have in mind when participating in
such solidarity actions. Equally, when thinking about the role of international solidarity actions on the part of workers, it would be somewhat sim-
plistic to see such solidarity in "either/or" terms—i.e., that workers *either* engage in international solidarity actions as a response to globalization *or* they
resort to nationalistic defense of "their" particular spaces in the global economy.[11] As Lorwin (1929) long ago suggested, it is in fact often the case that
the growing integration of the global economy actually elicits simultaneously both nationalist and internationalist responses, frequently from the
same groups of workers. Hence, workers may choose to engage in transnational solidarity actions concerning some issues (which may or may not be
about transforming the relations of capitalism or of accommodating capitalism to protect certain spaces within the international division of labor, to use
Johns's terminology) while perhaps resorting to explicitly nationalistic

actions—which are implicitly about defending certain spaces from "outside" influences, however defined—concerning others (see also Wills, 1998b). How workers choose to interact with one another in the context of international competition between themselves, then, is a complex matter shaped by a host of things, not least of which is their geographical location. As is the case for most of us, workers do not live in a world of eithers and ors but, rather, in an often contradictory world of continuums and overlapping interests.

Such issues not only illustrate that workers should be seen as active geographical agents who can either exacerbate or reduce disparities in patterns of (global) uneven development through their choice of different kinds of praxis, but they also indicate the complexities of theorizing class actions within a geographic context.[12] Indeed, it is a sad fact that much theorizing about class—particularly with regard to the international activities of workers and unions—either has been remarkably nongeographic and aspatial or it has presented "class" and "space" as two separate categories necessarily in opposition to each other, with workers being assumed to organize *either* along class lines *or* along territorial lines. However, given that workers can act in different ways in different times and places, the goal, surely, is to see how spatial relations are embroiled in and shape class processes and how class processes are embroiled in and shape spatial relations. Thus, whereas many analyses have simply assumed that workers who organize around issues of place are experiencing a bad dose of "false consciousness," giving back agency to workers as makers of economic geographies allows us to see how they may seek to defend and/or enhance their own positions and spaces in the global economy—all the while doing so *as workers*—not simply because they have somehow been duped into doing so by capital but because they themselves see very real advantages to so doing (e.g., maintaining their own jobs rather than being made unemployed). If we are to think of class processes as shaping, and being shaped by, spatial relations, and if we assume that workers play some role in their own class creation (à la Thompson, 1963), then we must also allow that workers can have significant impacts on the shaping of the geography of capitalism.

GEOGRAPHIC SCALE, WORKERS' INTERNATIONAL SOLIDARITY, AND UNEVEN DEVELOPMENT

The global labor practices examined above generate crucial theoretical issues for understanding and analyzing processes of globalization. In particular, the Ravenswood case shows that, contrary to neoliberal rhetoric, workers can indeed play powerful roles in the international arena. Through its corporate campaign the USWA and its supporters successfully restricted

Rich's efforts to expand his operations in Eastern Europe and Latin America. Although Rich had intended to capitalize on the end of the Cold War by increasing his investments in Eastern Europe in particular, through their practice of solidarity the USWA and local unionists in the region were able to prevent him from acquiring several ventures. By disrupting his expansion plans and compelling him to alter a number of strategic business decisions, the USWA clearly played a significant role in defining the geography of Rich's international investments and the patterns of foreign direct investment in such countries as Czechoslovakia, Venezuela, and others. The activities of the ITSs and the Steelworkers' ability to shape international investment patterns, then, certainly highlight the need to reconsider the role played by workers and their organizations in the making of the geography of the contemporary global economy and in the processes of globalization that seem to be driving this geography. As I have argued throughout, in the production of the economic landscape in certain ways it is just as crucial for workers to reproduce themselves socially and generationally as it is for capital to do so, for workers clearly have a vested interest in ensuring that the geography of capitalism is made in some ways and not in others. Thus, the locked-out Ravenswood workers were keen to maintain their plant as a unionized, rather than nonunion, component of the local and global economic landscape.

But the practice of (international) solidarity also raises important theoretical questions concerning the geography of class struggle. Building solidarity across space—especially internationally—is not a straightforward matter but rather is fraught with complexities. It is a practice that is as much about geography as it is about class. Unions and workers must confront, and work within the context of, global landscapes that exhibit profound variations in living conditions, wages, types of occupation, position within the spatial division of labor, modes of political and legal regulation, and the like. Equally, whereas workers may sometimes use geography to reinforce common class positions (as when workers in physically separate communities organize across space to limit the whipsawing activities of capital), at other times they may use it to cut across class divisions (as when they unite with their employers and choose to defend "their" communities in the face of real or perceived threats coming from workers and employers located elsewhere [see Cox & Mair, 1988]). Workers, through their conscious decisions to engage in different types of (international) solidarity, then, may pursue strategies that reinforce common class positions or that may undermine them in favor of territorial allegiances. As such, their different types of solidarity actions can exacerbate patterns of uneven development, or they can help to reduce them.

The fact that workers may find different advantages presented to them by designing campaigns at one geographical scale rather than at another is important to bear in mind, for how workers choose to scale their activities when challenging a transnational corporation—that is to say, whether they develop strategies that are local, national, or international in scope—can clearly have significant implications for the evolution of the economic geography of capitalism. The choices workers make in deciding that one scale rather than another is the most appropriate for their goals are both shaped by, but will also shape, the geographic context within which they live their lives. Hence, on some occasions locally-oriented campaigns designed to pressure local managers of a transnational corporation may prove more suitable (e.g., by embarrassing them in the local community where they themselves live or by getting local government and its policing and regulatory powers involved), whereas on others appeals to national organizations and levels of government may be preferable (other AFL-CIO unions, the federal government), and on still others global campaigns may be determined to be necessary to secure strategic aims. Of course, frequently it is the ability to develop actions at all of these scales and to articulate connections between them that is the key to success. Thus, as we have seen in the case both of the Ravenswood workers but also of the Liverpool dockers (Castree, 2000), the ability to maintain disputes long enough that an international component can be developed often relies upon being able to develop solidarity at the local, regional, and national scales simultaneously that can sustain locally those workers involved in such disputes.

At the same time, however, it is also important to recognize that, although we are likely to see growing numbers of international solidarity campaigns as the global economy becomes ever more integrated, transnationalism on the part of workers may not just be about confronting corporations who are themselves transnational. We should not assume that international labor solidarity is necessary only against companies that themselves are transnational, but it may also be used to bring pressure on companies that operate largely within particular regions of the world. For example, in the early 1990s the United Farm Workers was involved in a dispute with the Chateau Ste. Michelle winery in Washington State. Due to its membership in the IUF, the UFW was able to call for aid from this trade secretariat, which in turn requested that Swedish unions refuse to transport or sell wines imported from Chateau Ste. Michelle. After a highly effective boycott that lasted seven months, the winery agreed not to interfere in the UFW's organizing efforts (Sanchez, 1995).

Likewise, we should not assume that it is always necessary for workers to implement a transnational solidarity campaign in order to take on a transnational corporation. This is particularly so if that corporation has several key plants upon which the rest of its production chain relies. For example, in the 1998 dispute between General Motors and the United Auto Workers, the ability of workers to shut down two crucial components plants in Flint,

Michigan, in a dispute over local working conditions effectively crippled GM's entire North American production for almost two months, forcing it to close or cut back production at some 146 plants and eventually costing the company several billion dollars in lost production, thereby making it one of the most expensive strikes in U.S. history (the geography of this dispute is chronicled in more detail in Herod, 2000). In this situation, the effects of workers' local actions were rapidly and widely transmitted throughout GM's corporate structure because of the company's reliance upon just-in-time production methods.[13] The inability of GM's vehicle assembly plants in the United States, Canada, and Mexico to secure in a timely fashion needed components from these two plants, and thus their inability to keep operating, illustrates that, under the right conditions, workers may exert great pressure on a globally-organized corporation through their local actions.

As paradoxical as it might seem, then, workers and unions may be able to confront successfully transnational corporations not only through engaging in actions designed to get workers in other countries to bring pressure to bear on such corporations' overseas operations—what we might think of as the traditional model of international labor solidarity—but also by bringing pressure to bear in quite local campaigns that focus upon a single plant or perhaps a handful of key plants within the corporation's overall global structure. The power that GM workers in two plants in Flint were able to wield over the corporation's North American operations because of their location within the production chain suggests that solidarity campaigns that are locally or regionally focused may, likewise, prove effective against transnationally organized corporations, should the right conditions prevail. Indeed, the strategy of engaging in highly selective local strikes (which may nevertheless then have national and international effects) is one that the UAW used consistently during the 1990s to force concessions from GM.[14] Again, this is not usually an either/or proposition, for such locally focused campaigns may depend for their success at least partially on a union's abilities to gain support (money to aid local strikers, for instance) from colleagues elsewhere. But, it is to say that workers may have open to them the possibility of focusing their actions at different scales, while the choices they make concerning at which scales they engage in praxis have important implications for how the economic geography of capitalism is made.

In summary, then, these insights suggest three sets of overall conclusions. First, as I have argued throughout, we should not succumb to the rhetoric of neoliberalism about the supposed absolute power of globally organized capital. Given the right resources and context, locally situated workers may indeed prevail in disputes with multibillion-dollar transnational corporations, either through very local campaigns or through campaigns that involve actions in many locations around the world. Second, contingent issues—such as whether a company is using just-in-time methods of production or, instead, more traditional Fordist methods—can play an important role in shaping which types of geographical praxis workers employ and the possibilities for success using these different modes of campaigning against their employers. In situations where one or two plants are making key components upon which the rest of the corporation's production chain relies, focusing on these few facilities may offer significant opportunities for success. On the other hand, in situations where many plants produce the same component and production may easily be switched among them should one be affected by a labor dispute, a more geographically widespread campaign of translocal solidarity may be necessary. Third, theorizing about class relations and class conflict needs both to incorporate much more centrally issues of geography and spatial praxis and to pay much closer attention to how workers may manipulate spatial relations—including geographic scales of organization—and landscapes to serve as sources of political power that they can use to shape class processes. Whether workers are able to achieve their geographic and political goals through local campaigns (as in the case of GM) or through engaging in transnationally articulated campaigns designed to connect together workers in different parts of the globe (as in the case of the locked-out Ravenswood unionists and RAC), or through campaigns articulated at other scales, will be shaped, ultimately, by the spatial contexts and contingencies within which they find themselves.

CHAPTER 9

International Labor Union Activity and the Landscapes of Transition in Central and Eastern Europe

Today, the deterioration of the economic landscape in Eastern Europe gives rise to concern and anxiety.

The East–West Cold War may be over. The North–South divide grows day by day. But if we can consolidate democratic trades unionism in Eastern Europe, if we can help create societies based on social and economic justice, if we can stop the banks and the multinationals from exploiting workers, if we can harness the energy and democratic commitment of Polish and Czechoslovakian and other workers to build solid unions of metalworkers, then this will help the task of metalworkers everywhere in the world.

—MARCELLO MALENTACCHI,
General Secretary, International Metalworkers' Federation[1]

The collapse of the Soviet Union and the end of the Cold War are, arguably, the defining events of the late twentieth century, if not the entire century. Indeed, so significant have been these events that they have led at least one U.S. State Department functionary to proclaim that the supposed triumph of Western-style liberal democracy and the market over Soviet-style totalitarianism and central planning have actually inaugurated the "end of history," as there are now no longer any great ideological conflicts that will serve as the motor of human social and political development (Fukuyama, 1992). Whether one subscribes to such a proclamation or not, one thing is undeniable: the collapse of Communism in Central and Eastern Europe has brought with it a dramatic set of consequences for workers and unions both within, but also beyond, the region. In particular, whereas the geography of trade unionism in this region had remained relatively stable during the four decades after World War II as prewar unions were captured by the various Communist parties of the region and integrated into the central state apparatus (see Herod, 1998d, for more on the role of unions under Soviet-style central planning; also, Godson, 1984), following the end of the Cold War this geography has become much more fragmented and heterogeneous. What was a uniform political space, at least in terms of its representation by the pro-Soviet World Federation of Trade Unions, is now a space of intense conflict and rivalry, a result both of local unions leaving the Federation and of the active policies of the International Confederation of Free Trade Unions, the World Confederation of Labour (WCL), and others to create alternative structures to that of the WFTU and the old Communist labor organizations.[2] Indeed, the past decade has witnessed an unprecedented flow of money, organizers, and information into the region as the ICFTU and other Western-oriented trade union organizations such as the AFL-CIO have sought to expand their influence in this vast, newly opened, space to which they have not had access since the 1930s.

In this chapter I examine how a number of international trade union organizations are attempting to come to terms with these changes and how, in turn, their activities are helping to reshape the geopolitical and geoeconomic order of post-Cold War Europe. As can be seen from the statements by the International Metalworkers' Federation's (IMF) General Secretary Marcello Malentacchi reproduced above, Central and Eastern Europe represents at once both a "problem region" and, potentially, a "keystone region" of great importance to organizations such as the IMF, the AFL-CIO, the ICFTU, the WCL, and others. This is of particular concern, given that some of the former Soviet satellite states of Central and Eastern Europe may ultimately become full members of the European Union and that a growing number of Western corporations are setting up operations in the region. Not only are many Western unions concerned to ensure the growth of strong and independent trade unions as part of age-old principles of solidarity that will help protect workers' rights in Central and Eastern Europe itself, but it is clear that many also want to ensure that the region does not become a low-wage competitor for their own members. Equally, many Central and East European unionists fear that, while foreign investment may help stimulate economic development, their countries will be used simply as a low-wage production platform to supply the rest of Europe with goods and services in a neocolonial relationship.

Although there may be some debate as to what extent organizations such as the AFL-CIO or other Western-oriented union bodies are simply attempting to defend their own privileged space within the global economy and to what extent they may be acting out of altruism toward Central and

East European colleagues—that is to say, whether particular organizations are engaging in "accommodatory" or "transformatory" solidarity, to use Johns's (1998) distinction, outlined in Chapter 8—there can be little debate that events in Central and Eastern Europe represent something of a geographical conundrum for workers both within and outside the region. In aiding workers in Central and Eastern Europe, organizations such as the International Metalworkers' Federation, the ICFTU, and others appear to be thinking in quite geographic and strategic—what we might call "geostrategic"—terms about the transition from centrally planned to market economies, and how the breakdown of the old geopolitical and geoeconomic blocs that dominated the European political and economic landscape during most of the last half century may affect workers both within the region and beyond. I argue here that these entities, though they may have different reasons for doing so, are in various ways playing a significant role in shaping the new geopolitical and geoeconomic order in post-Cold War Europe. The collapse of Communism may have marked the end of history in the manner in which Fukuyama has suggested—though this is doubtful!—but it has also unleashed a tremendous struggle to remake the geography of labor representation in Central and Eastern Europe.

The purpose of this chapter, then, is twofold. First, although there is a burgeoning geographic literature on the transition in Central and Eastern Europe, this has focused predominantly upon the geography of foreign investment (Murphy, 1992; Michalak, 1993; Buckwalter, 1995), economic restructuring (Pavlínek, 1992; Smith 1994; Bradshaw et al., 1998; Pickles, 1998; Meurs & Begg, 1998), the region's foreign debt (Gibb & Michalak, 1993), problems of implementing democratic political structures and of gaining legitimacy (Frankland & Cox, 1995; Staddon, 1998), and the implications of the transition for international geopolitics (for a good introduction to issues of transition and how it is conceptualized, see Pickles & Smith, 1998). Relatively little has been written that examines the new geographical relationships with which unions must deal during the transition to the market economy and the ways in which they are seeking to actively shape the economic and political geographies of the post-Cold War period. Yet, the end of the Cold War has brought with it an economic, political, and geographical transformation that is posing significant challenges for workers and trade unions throughout Europe (see Clarke et al., 1995, and Herod, 1998d, for more on issues affecting the trade unions in Central and Eastern Europe during the transition). In particular, the end of state control over the official trade union movements has seen the emergence of a multitude of new trade unions, labor organizations, and workers' groups of varying political beliefs, each vying for power, while the shift from a centrally planned system of production and allocation to a market economy is fundamentally changing the relationships between enterprise managers, the workforce, the state, and the trade unions. The first section of the chapter, then, serves to fill in some of the gaps evident in much of the geographic writing about the transition by examining some of the changes that have taken place since 1989 in Central and Eastern Europe as they affect workers and trade unions.

In the second part of this chapter I consider some of the activities engaged in by a number of international trade union organizations in the region. These organizations are playing an important role in shaping the economy and politics of post-Cold War Europe, at least in the field of labor relations and trade union politics. I use the analysis of such organizations' behavior to suggest three things which bear upon the broader thesis that I am exploring in this volume. First, the organizations that I examine in this chapter appear to be thinking quite geographically as they go about their work in the region. Indeed, understanding the specific geographic context within which such organizations find themselves provides insights into their activities and their geostrategic thinking with regard to the transition. This latter point itself suggests that a geographic perspective can add significantly to an understanding of unions' economic and political behavior. In fact, organizations such as the International Metalworkers' Federation appear to be thinking *glocally* (Swyngedouw, 1992)—that is, they are attempting to tailor their international activities to specific local circumstances and to design different programs for particular countries who have experienced different trajectories of "transition" during the 1990s rather than adopting a "one program fits all" model. Second, through such activities, then, these organizations are playing active roles in remaking the economic and political landscapes of the region.

Third, the importance of geographic scale is again emphasized. For entities such as the ICFTU or various international trade secretariats, expansion into Central and Eastern Europe represents an important extension of the scale of their operations and representation of workers, for it brings millions of workers under the same organizational umbrella as are those from many other parts of the world. At the same time, the issue of scale is also important because in most countries of Central and Eastern Europe there has been an intense struggle during the past decade over the scale at which political power should reside within the labor movement. Whereas national labor leaders and officials from many international labor organizations have often wanted power to reside at the level of the national unions in particular industries or the national labor confederation as a means to limit the ability of governments and capital to play off various parts of the region or even particular countries against one another, the leaders of local unions—naturally distrustful of centralization, given the history of the Soviet era—frequently wish to keep as much power and union finances concentrated at the local level. This tension between the local and the national or international has unleashed an important dynamic that has shaped the evolution of the political and economic landscape of the region.

CONSEQUENCES FOR TRADE UNIONS OF THE TRANSFORMATIONS IN CENTRAL AND EASTERN EUROPE

The collapse of the Soviet Union and the economic and political transformations that are currently taking place in Central and Eastern Europe represent a fundamental restructuring of the geopolitical order that dominated Europe for the past half-century. In this section I address some of the implications of these transformations for trade unions in the region as they struggle to redefine their roles and come to terms with the new economic and political forces that have been unleashed during what some see as a transition to the market economy (see Herod, 1998d and 1998e, for discussions of Cold War international trade union politics). In particular, the section examines the effects of privatization on the economies of the region, the rise of new trade union and workers' groups, the new roles such groups are attempting to define, and the influx of foreign capital.

Privatization

For many "reformers" the future success of the Central and East European economies has been seen to lie in the ability to break up and demonopolize the state sector—particularly the massive enterprises built up during the Communist period—and to encourage the formation of smaller-sized firms as a means to establish a system in which autonomous, decentralized production units can compete with each other and on the world market.[3] While privatization has been seen by many as the catalyst by which to bring about the transition to the market economy (Gaudier, 1991), the process has varied considerably among the region's countries. Privatization has been taken to mean different things by different people, including the sale of assets, the sale of shares, transfer of shares to workers, cutting off public funding to force enterprises to look to private sources, deregulation, the breaking up of monopolies and the sale of subsidiaries, and privatization of management through the replacement of state-appointed managers by private ones, while the rate at which it has occurred has varied widely throughout the region.[4] Similarly, even what it means to talk of initiating a "market economy" through privatization continues to be a source of some dispute. Thus, in the West the term tends to refer to the ways in which narrowly defined economic institutions operate, whereas in Central and Eastern Europe it has tended to refer to the dispersion of political power and the breakup of the state monopoly over economic resources (Piore, 1992). Privatization has also raised serious questions and posed problems for workers and trade unions in the region, in particular with regard to the codification of property rights, compensation to former owners of businesses seized after World War II, how state-run enterprises are to be valued, who will control such enterprises after privatization, and whether restructuring should precede or postdate privatization (i.e., should restructuring be carried out by the state prior to privatization, or should the new private owners restructure the enterprises after purchase?). This latter question, particularly, has proven to be a source of major conflict, as many of the old *nomenklatura* continue to use their former positions of influence in the state sector to acquire economic and political power in the market economy sector.

Although there is frequently a tendency in the West to see Central and Eastern Europe as a fairly homogenous bloc that is moving as a whole inevitably toward a market economy—an approach that Burawoy (1992) has called "transitology"—in reality the variety of mechanisms adopted to break up the state sector means there is actually great divergence in experience. In the Czech Republic, for instance, small- and medium-sized enterprises (mostly services, retail stores, and small cooperatives) were sold at auction under the auspices of district privatization commissions appointed by the central government, whereas privatization of large enterprises was carried out through a system in which citizens could buy vouchers to be used to bid for shares of state enterprises. In Poland, by way of contrast, large enterprises were principally privatized through a process of capitalization in which they were first converted into government-controlled joint-stock companies and then shares of stock were subsequently transferred to third parties. Small and medium-sized enterprises, though, have usually been privatized by liquidation and the direct sale of assets. In Romania, some 30 profitable small-sized enterprises were among the first to be partially privatized through the auction of assets in 1992, although the government later launched a larger privatization campaign in which citizens were entitled to receive free of charge 30% of the capital of former state-owned companies, which was to be held in a number of mutual funds (called Private Ownership Funds) (IMF, 1993).

Whereas some governments (e.g., in Hungary and Poland) were early proponents of privatization, others (e.g., in Bulgaria, Slovenia, and Romania) were slower to develop privatization programs. Indeed, as Poul-Erik Olsen (1995) of the International Metalworkers' Federation has commented about the early days of the transition, "In Romania and Bulgaria everybody spoke about privatization but, in fact, very little happened." Equally, whereas in Hungary, Bulgaria, and the Czech Republic privatization in the early days of the transition was often little more than a legal change in ownership, in Poland and Slovenia it more frequently went hand in hand with attempts to bring about a rapid restructuring of enterprises.[5] Moreover, although legislation authorizing privatization may have been passed on paper, in many instances the procedures for actually carrying this out were not clearly established. Even in the Czech Republic, where the government of former Prime Minister Václav Klaus was one of the most fervent proponents of neoliberal policies, trade unionists in the mid-1990s were still complaining that the process of privatization had taken on the aura of a somewhat artificial "cryptoprivatization," that even after nearly a decade of transition there was still "zero mechanism established for the functioning of any capital trade," and that much privatization had been little more than speculative activity of the part of banks and portfolio investors (Falbr, 1996). Indeed, many of the newly democratized trade unions actually played quite active roles in pressuring the government to speed up its privatization program in the belief that it would allow for more effective economic reform policies to be implemented—the Czech power workers' union, for example, roundly criticized the Klaus government for its failure to dismantle the energy monopoly České Energetické Závody and to privatize a number of power plants in the early 1990s (De Luce, 1993; Herod, 1998d), whereas in Poland Solidarność has been a forceful advocate of the privatization of state enterprises (see Ost, 1989, 1995).[6] While official policy in the region during the 1990s may have been to push for privatization, the absence sometimes of the necessary economic and political institutions, such as banks that can provide venture capital, stock exchanges, taxation and insurance policies, and so forth, means that the process of privatization has been problematic.

The process of privatization, then, has been geographically uneven throughout the region, with different consequences for different groups of workers. Nevertheless, workers are facing a number of common challenges that go beyond the specifics of their particular country's privatization process. Foremost among these is the rise in unemployment levels as many enterprises have been sold off and/or shut down because they are ill prepared to compete according to "free market" principles. Rising levels of unemployment have gone hand in hand with a dramatic collapse in real wage levels associated with often rapid and substantial price increases in food, housing, transportation, and consumer goods to their "true" market levels. Many workers have also lost access to state-provided social services upon which they have traditionally relied (such as universal and free education and health care, family allowances, uniform old-age pensions), services that are either being discontinued due to cutbacks in government expenditures or are themselves being privatized (as with the rise in private medicine, growing numbers of employee-funded, contribution-defined pension plans, etc.). Combined with the loss of "existential security" (the triumvirate of job, income, and housing security formerly provided by the state), many workers are finding they are increasingly unable to provide for themselves and their families, a situation that has fueled social unrest and sparked strikes and demonstrations. Such unrest has particularly been directed against the widening income gaps between the majority of workers and the relatively small class of affluent entrepreneurs, themselves often former Communist managers of recently privatized enterprises.

Privatization is also beginning to bring about dramatic shifts in the industrial structure and economic geography of the region, shifts that have been quicker and deeper in some areas and sectors than in others. The beginning of a turn away from heavy industry in some regions and the shift to consumer durables' production, together with the growth of subcontracting, the breakup of large collective units of production, greater market penetration by high- and low-quality imports, and the despecialization and respecialization of many enterprises, is having disruptive effects on millions of workers and their communities as old factories have to retool and new ones must be built. In addition, thousands of small, private new enterprises have been established in the wake of privatization. Equally, thousands of old (and new) establishments have also shut down. This is causing problems for unions trying to organize these workplaces because frequently they cannot keep up with the rapid pace at which some of these businesses open and close, so that many businesses have come and gone before the unions are able to organize them. Most unions have also reported difficulty in winning bargaining rights from invariably hostile employers (even when the workforce overwhelmingly favors unionization), especially in the smaller private firms now being established (IMF, 1994). Thus, although they have generally supported privatization, many unions fear that the economic and social dislocation being wrought by it may expose workers to the worst excesses of capitalist development—leading many to refer to the region as the "Wild East"—unless the unions have the freedom and ability to organize workplaces and unless they can prevent the total withdrawal of the state from the provision of social services and enforcement of proworker protective legislation (Mureau, 1995b).

Trade Unions and Labor Relations

The transition from centrally planned economies to market economies and the end of state control over the trade unions have lead to dramatic changes in the roles played by unions and in the way in which labor markets function. During the Communist period trade unions primarily served as the conduits or "transmission belts" by which directions concerning production were transmitted from central planners to workers on the shop floor and (theoretically at least) information about shop floor conditions was made known to the central planners and party officials. Unions had a dual role, which was to mobilize and discipline the workforce for production and "socialist emulation" while also representing members' legal rights and interests in the face of management (i.e., state) arbitrariness (see Herod, 1998d). In practice, however, the former role usually took precedence, and the unions' primary consideration was the promotion of party and government policy. Certainly, there were variations on this model. Both East Germany and Czechoslovakia, for instance, were relatively industrialized societies prior to the Communist period, and both had well-established pre-Communist social democratic traditions. Consequently, the Communist parties and trade unions in these countries were more inclined to follow relatively egalitarian wage policies, especially during the 1950s and 1960s, and to adopt policies that were designed to protect workers in the factory than were the unions in countries such as Bulgaria and Romania that were predominantly agrarian societies at the end of World War II. In these latter countries, there was a much weaker pre-Communist tradition of trade unionism, and the unions tended to act in the postwar period primarily as agents of modernization and enforcers of industrial discipline on workforces more attuned to the rhythms of agricultural production than to those of the factory floor (Pravda & Ruble, 1986). Nevertheless, a number of commonalties did run through the role played by Communist trade unions and their practice of industrial relations.

The Soviet model of unionism imposed after World War II was one with a very centralized structure of union control in which key union positions were appointed by party officials (Clark, 1966; Herod, 1998d). As Pravda & Ruble (1986: 4) put it, "trade unions remain[ed] closely subordinated to the party yet organizationally distinct from it, [a] distinction [that] afford[ed] a degree of administrative latitude but exclude[d] policy neutrality, let alone union autonomy." Only at the enterprise level did unions enjoy even a limited degree of local control, although the legal structure severely limited what they could legitimately negotiate with management (Nagy, 1984). Furthermore, unions tended to adopt an organizational structure in which all those employed in a particular sector were eligible for membership, even managers. Although most of the region's countries had both sectoral and enterprise-level contracts, there was little collective bargaining of the sort that Western unions would recognize, and collective agreements rarely included provisions covering wage rates or basic working conditions such as working time, leave, and so forth (IMF, 1994). Disputes between workers and management in the various enterprises were often settled by referral to higher authorities for mediation or arbitration rather than by industrial action, so that resolution was supposed to be through amicable negotiation and not adversarial collective bargaining. Indeed, strikes were seen as failures of the system rather than legitimate union weapons for improving the condition of the membership (Pravda & Ruble, 1986).

The fact that the state played the role of employer and manager of enterprises also meant that there were few employers' organizations of the sort that exist in most advanced capitalist economies. In their place existed a network of trade associations and chambers of industry, commerce, or economy—themselves usually little more than extensions of the ministry of trade—that had very little to do with the conduct of labor relations (IMF, 1994). Although some limited efforts to develop less state-controlled collective bargaining were made in a number of countries in the 1980s (particularly Hungary and Poland), these had few tangible effects, largely because the state continued to retain control over the principal elements of the employment relation (e.g., wage determination) (Héthy, 1991).

During the post-Communist period this situation has changed in dramatic ways. Principally there has been a transformation in the relationship between the unions and the state such that the state no longer controls the trade unions in the way it formerly did. Two models of independent unionism have emerged: the growth of new unions now allowed under the law (the "Polish model") and the taking over of formerly Communist-controlled unions by reformist elements and their restructuring along democratic, autonomous lines (the "Czech model") (Héthy, 1991; Brewster, 1992). In addition, some unions are still dominated by the old *nomenklatura* but have declared their independence from the state institutions (see Herod, 1998d, for more details). These developments have had two major sets of consequences.

First, the ending of state control over the union movement is bringing with it a new role for unions. Although some countries lag behind others, in all there have been efforts made to lift legal barriers to collective bargaining, to bring about the separation of government, trade unions, and employers' associations, to exclude political parties from being labor relations actors, to provide legal recognition of unions' rights to strike (although this is still severely limited in some countries), and to create national-level tripartite institutions that, while recognizing that collective bargaining should be free of government interference, simultaneously seek to bring unions, employers, and government representatives together to address matters of "national interest" such as inflation and minimum-wage rates (Héthy, 1991; Musil, 1991).[7] As a result, unions are increasingly taking on the role of representatives of workers' interests and rights in dealings with government officials and the still small number of employers' associations. Furthermore, the withdrawal of the state from the unions' activities has led many to adopt very decentralized confederative organizational structures. While this represents an attempt to prevent the centralization of power reminiscent of the Communist era and to preserve local autonomy, it is hampering the ability of the national leadership of many of the recently democratized unions to address nationally issues relating to wages, work conditions, and the like. In the former Czechoslovakia, for instance, the national chamber of trade unions (ČSKOS) that replaced the old Communist body in 1990 initially adopted a voluntaristic model patterned after the AFL-CIO in the United States, although the successor chambers to it in the Czech and Slovak Republics (the Českomoravská Komora Odborových Svazů [ČMKOS—Czech-Moravian Chamber of Trade Unions] and the Konfederácia Odborových Zväzov Slovenskej Republiky [KOZ SR—Trade Union Confederation of the Slovak Republic]) have subsequently sought to develop stronger centralized control, so as to present a more unified front to the employers and, perhaps more importantly, the government (Falbr, 1996).[8] Nevertheless, in a union such as the Czech Metalworkers' Federation OS KOVO some 75% of union dues collected still remain with the local union at the enterprise level, with the remainder being divided among regional and national organizations (OS KOVO, no date). Similar proportions are reported throughout the region (IMF, 1994; Olsen, 1995). Such a "localist" structure has the potential to lead to a veritable geographic patchwork of wage rates and working conditions that could provide companies the opportunities to play workers in different plants and regions against one another in a race to the bottom, a situation that U.S. workers have increasingly faced during the past decade as employers have broken up national agreements in several industries (see Herod, 1991a). At the same time, such struggles over the scale at which power will reside create tensions between national officials and local leaders and members with which unions must deal.

Second, there has been a proliferation of new unions under the revised labor laws. In the former Czechoslovakia, for example, the 1991 Charter of Fundamental Rights and Liberties stipulated that "trade union organisations shall be established independently of the State" and that "it shall be inadmissible to limit the number of trade union organisations" (Article 27, 2, quoted in IMF, 1994: 23–24). As part of this attempt to replace Communist-era labor law in which control was highly centralized, it was decreed that as few as three workers could come together to charter a trade union. Although the government presented this as a way of ensuring worker democracy and choice of representation, many saw it as a way of weakening unions, a key goal of the Klaus administration in its quest to introduce neoliberal market mechanisms to the country (Uhlíř, 1996). In 1997 efforts by the ČMKOS to have legislation passed allowing only the majority union to conclude collective bargaining agreements were defeated (WCL, 1997c)—a defeat that was seen either as helping managers to undermine the power of organized labor or as defending union pluralism, depending upon the point of view of the unions affected. In the Slovak Republic, on the other hand, legislation was passed in 1996 that did, in fact, give the majority union the monopoly of concluding collective agreements, a decision lauded by the majority unions but decried by the minority unions (such as those affiliates of the Christian-oriented World Confederation of Labour) as an attack on pluralism (WCL, 1997b). Similar legislation encouraging the proliferation of unions has been passed in other countries. For example, in Romania a 1993 law permitted as few as 15 workers to form a union.[9]

Although the growth in the number of independent trade unions will ensure workers' rights to organize unions of their own choosing, it is nevertheless proving problematic as workers become divided among a plethora of old and new unions, which are often very small in size (Mureau, 1995a). In addition, some employers have taken advantage of these laws to encourage breakaway groups of workers to form their own unions with which the employers then negotiate, a practice that threatens to weaken the bargaining power of organized labor (IMF, 1992a). Something of a paradox has existed in this regard. Although the old unions may have been tainted in the eyes of many workers because of their activities during the Communist period, during the early 1990s they were often the only ones with operational structures in place and financial assets upon which workers could draw. The new unions, on the other hand, frequently have been short of cash and sometimes have had to rely upon donations from Western unions to remain functioning. Thus, whereas many workers have shied away from the old unions because of their Communist past, they have also been reticent to join the new unions, which are often seen as ineffective. Additionally, in some countries (such as Poland), companies—particularly those in the foreign investment sector—have successfully, though illegally, forbidden newly recruited workers from belonging to any trade union, and such workers must undertake not to organize a new union, something to which many agree due to fear of losing their job in an era of employment shrinkage, inabilities to seek legal redress through the courts, and the authorities' tacit condoning of such practices (Wójcik, 1998).

The situation for many unions has been exacerbated by efforts in some countries to establish at the plant level "works councils" that have the right to consult with management on certain issues independently of the union (Thelen, 1991).[10] Such councils, while allowing for labor participation in some work-related affairs, are formally separate from the unions. Although collective bargaining at the plant and branch level is still legally within the purview of the unions, the works councils do have the right to consult with management on a range of issues, including safety rules, the introduction of new technologies and work methods, personnel decisions, and welfare arrangements. Whereas worker representatives on these councils are often union members, in no country is there a legal requirement to be so. Thus, although many government and management officials have pushed such councils as bodies for representing workers within the workplace, many union officials fear they are simply being established as a way for managers to avoid having to deal with the unions directly (Uhlíř, 1994). For example, at the U.S.-owned United Technologies Automotive plant in Hungary it was management that established that plant's works council and appointed its members. When a group of workers attempted to form a union in 1995 to elect members to the council, they were summarily fired (ICFTU, 1998). Furthermore, some national union leaders have argued that works councils foster "plant egoism" (given their legal obligations to work with management for the good of the plant) in which the parochial interests of particular plants may be placed before broader union interests, a situation that can lead to problems of fragmentation for the national union—especially in the kind of decentralized union structure outlined above. Others suggest that, although in Germany the power of the national unions and their centralized structure mean they have usually been able to co-opt such works councils to the advantage of workers, in Central and Eastern Europe the present weakness of many unions means that the councils are often more readily co-opted by managers and unable effectively to defend the interests of workers.

Not only are the roles of the unions changing, but so is the way in which labor markets operate. A number of characteristics concerning the operation of labor markets during the Communist period can be identified (Freeman, 1992). Perhaps foremost among these was the tendency for workers to be hired at enterprises on the basis of the need to meet *politically* determined production quotas, regardless of the actual labor costs involved. The shortage of raw materials in many countries also encouraged enterprises to hoard labor so that workers would be on hand when components and other production inputs did arrive, whereas the shortage of money meant that enterprises frequently remunerated their workers with nonmonetary compensation (e.g., housing allotments, special permits to buy scarce goods, vacations in enterprise or union-owned resorts, etc.). Because wages were determined centrally by the planning agencies, they bore little relation to productivity or availability of skills. In addition, enterprises frequently paid their workforce with add-ons, so that sometimes as little as a quarter of a worker's wages was actually made up of base pay. Finally, to a greater or lesser degree, labor mobility was also restricted as a means of retaining skilled workers in particular enterprises and regions.

The political and economic transformations currently taking place, however, are dramatically changing the way in which labor markets operate in Central and Eastern Europe. In turn, this is causing consternation and confusion for many workers and trade unions. The primary concern undoubtedly has been the rise in unemployment rates, as many enterprises have laid off workers or closed down completely. This has been combined with growing levels of worker migration, particularly from regions with a disproportionate share of heavy industries, which have been among the hardest hit by the transition. Although some migrants have relocated within their home countries, the lack of employment opportunities and housing has forced many to move further afield, principally to more developed Central and East European countries such as the Czech Republic and Hungary, to a number of West European countries (particularly Germany, Austria, France, Sweden, Norway, Denmark, and the Benelux nations), and to Canada, the United States, and Australia. This is causing a number of problems. In the metals industry, for instance, migrants are disproportionately men aged between 20 and 40 who are relatively well educated and skilled (IMF, 1992b). Not only is this having significant effects on civil and social life (women and children are frequently left behind by such migrants, either temporarily or permanently), but it is depriving many enterprises and communities of their most skilled workers. Additionally, since such workers also represent a source of cheap labor, some West European unions have expressed fears they may exert a downward pressure on wages in the West. In Germany, agreements have already been established with several Central and East European governments in which these countries' firms can use their own imported specialized labor (subject to an upper limit) while operating in the German market. Even within Central and Eastern Europe, the influx of low-wage foreign workers has also been a problem in (relatively) higher-wage countries such as the Czech Republic and Hungary, which have been the destinations for thousands of migrants from further east.

The withdrawal of the state from activities in which it formerly participated, combined with the concomitant restructuring of the unions and their adoption of new roles, is changing the way in which wages and other benefits are determined. In particular, there have been efforts to decentralize the wage determination process, shifting it from the realm of the central state to the enterprise or plant level. Many labor rights that were formerly guaranteed under the law are now being devolved to the realm of collective bargaining to be settled on a case-by-case basis, thereby raising the specter that some workers will lose many basic legal rights if they fail to secure them in their contracts or if employers simply ignore these contracts (as many are doing). Although unions are pushing for effective collective bargaining to be adopted throughout the region, in many instances the lack of employers' associations or the unions' failure to convince employers to bargain collectively and in good faith, especially in the smaller private firms, makes this problematic—especially in view of the fact that in several countries of the region during the 1990s there was no legal obligation for employers to bargain collectively.[11] This has had several consequences. Not only does it mean that throughout the region the procedures and mechanisms for collective bargaining have often operated on something of an ad hoc basis with few formal guidelines, but unions frequently have found it difficult to get access to basic information from the company that is necessary for realistic bargaining to occur.[12] Given that there is usually no legal obligation to do so, most employers have steadfastly refused to disclose information that Western unions are often provided as a matter of course. Furthermore, the lack of legal requirements to bargain collectively has also frequently caused confusion concerning which union has the right to lead negotiations for, and sign agreements on behalf of, the workers (IMF, 1994). This has sometimes resulted in the employer choosing to bargain with the union that it perceives to be most favorable to management goals. Relatedly, whereas at the plant level it is usually fairly obvious which union is most favored by the workforce, at the national level this often is not the case and sometimes leads the government (which is still often the enterprise owner) to make a political decision to negotiate with some unions and not with others in a particular industry.

In most Central and East European countries the lack of employers' associations in many sectors means that collective bargaining is primarily focused at the level of the individual enterprise or plant, which, while providing local unions freedom to negotiate agreements sensitive to local conditions, means that such unions often are unable to bring much national political and economic power to bear on the employers/state. Although agreements can be negotiated at higher levels if an employers' association exists, many employers and government officials have fought against regional or national contracts because they fear that these would give unions greater political and economic leverage. Furthermore, employers have frequently been successful in getting workers to sign *individual* contracts with them by offering higher wages and better benefits to those who do not join the union, a practice that undermines trade unions' power (Mureau, 1994). In addition, a number of large enterprises in several countries are also using wage tariff systems established by the government according to skill levels and length of employment with the company (IMF, 1994) rather than determining wages through collective bargaining. Still other employers simply impose conditions of remuneration unilaterally on their workers.

There is also significant variation in how unions in Central and Eastern Europe seek to settle disputes and the legal environment within which they operate with regard to conflict resolution. In Bulgaria, for example, the Law for Settlement of Collective Labor Disputes has provided for negotiation and arbitration of industrial disputes, with parties allowed to resort to arbitration after a minimum of two weeks of negotiations. Although the right to strike has been enshrined in the law, the employer must be notified in writing ahead of time of the union's intention to strike and of the expected duration of such a strike. Relatedly, spontaneous "work stoppages" (the equivalent of "wildcat strikes" in the United States) have been limited to one hour's duration by law. In Romania the law calls for obligatory mediation and conciliation before a union can strike, while the 1991 Law on the Settlement of Collective Trade Disputes forbade strikes over issues that require adoption of a government decree to be enacted, such as increases in the guaranteed minimum wage. Workers have also been judicially prevented from striking to protest violations of the collective bargaining agreement, thus making it virtually impossible to legally defend their interests through industrial conflict—although, as the profusion of industrial action in Romania during the 1990s showed, this has not necessarily stopped workers from striking. In Poland, the right to strike was granted to all workers in 1991 with the exception of those in the public sector. A strike may be initiated only after a vote has been taken and only if employers are given five days' notice of such a planned action. In Hungary, new labor laws made no legal obligations for labor and management to conclude an agreement, and strikes aimed at changing the provisions of the collective agreement before its expiration have been prohibited. Strikes with "unconstitutional aims" (however defined) have also been declared illegal.

Clearly, then, there is a fairly wide variation in the legal protections offered workers with regard to their ability to strike. There is also a considerable variation in the propensity of workers to use the strike as a weapon to settle disputes. In the early 1990s, the unions in the Czech Republic, Slovakia, and Hungary were often reticent to strike and preferred instead to operate through other means—such as public demonstrations—to achieve their goals, although this began to change during the late 1990s as workers engaged in a number of significant showdowns with their respective governments (Czech railroad workers, for example, engaged in an important nationwide strike in February 1997, which was widely seen as the harbinger of a more militant stance by Czech unions [Pehe, 1997]). In Poland, Bulgaria, and Romania, on the other hand, strikes (often illegal) have been used more frequently and in a more overtly political manner, with strikers' goals often geared more toward changes in government policy than to workplace issues (IMF, 1994). Evidently, as the transition progresses, the models of unionism and industrial relations adopted throughout the region, and the implications these have for workers and patterns of economic and political development, are likely to vary quite considerably as each country seeks to develop its own road to the post-centrally planned economy (see Herod, 1998d). At the same time, however, there are a number of similarities in the experiences of workers and unions across Central and Eastern Europe, not the least of which is the loss of employment and the difficulties of coming to grips with a changing economic and political landscape that is emerging after nearly half a century of central planning.

Foreign Investment

Many Western corporations see considerable business opportunities in Central and Eastern Europe's apparent transition to the market economy.[13] Not only does the region represent a significant potential market of some 400 million people (if the republics of the former Soviet Union are included) that is chronically short of consumer durables and other services, but large reserves of natural resources, a plethora of very skilled manual workers in some countries, relatively low wages, the willingness of government officials and workers often to ignore or relax work safety and environmental regulations, and the proximity to Western Europe have attracted considerable interest among foreign investors. Consequently, there has been a slow but steady flow of investment into the region from European Union countries in particular. Equally, strapped for cash, many governments and enterprises have been keen to attract Western capital—which they see as a means to jump-start their economies—and have pushed joint ventures, direct acquisition of enterprises, franchising, and other forms of collaboration with Western corporations. While they face many problems in doing business in the region—not least of which is the question of how to repatriate their profits in hard currency—some Western corporations have taken the plunge and invested substantial quantities of capital in the region. Initially this was largely geared toward the purchase of state-owned enterprises, as Western companies sought to establish subsidiaries in the region. More recently, however, in Hungary and Poland particularly, much recent foreign direct investment (FDI) has been oriented not toward purchasing former state-owned enterprises but to establishing new "greenfield" projects, many of which are designed to produce not for the local market but for export to the European Union. For their part, much of the early investment by Western governments was in the form of credits that went toward rescheduling debt payments, a phenomenon that has largely continued.

Although many Western investors have sunk large amounts of money into the region, the volume of FDI has been rather low relative to early expectations. Between 1991 and 1996 the countries of the region (including the former Soviet Union) received a net inflow of $42.2 billion in FDI—a figure that represented only about 4% of the transition economies' gross domestic product, compared to inflows of about 6% of GDP for Latin America and 13% for the East Asian developing economies (International Monetary Fund, 1997: 106). Investment began to pick up during the mid-1990s—some $22.2 billion and $18.2 billion of new investment was made in 1997 and 1998, respectively (Coolidge, 1999)—though, as many countries effectively ran out of state-held assets to sell to foreign investors, levels began to drop off toward the end of the decade. Generally it has been the more industrially developed economies of Central Europe (the Czech Republic, Hungary, and Poland) and the Baltic states—and particularly those countries such as Hungary and Estonia that chose a privatization method that allowed major sales to foreign investors—that have attracted the lion's share of FDI. Between 1989 and 1998, Hungary received a net influx of $17.4 billion in FDI, Poland $23.2 billion, and the former Czechoslovakia $11.8 billion (the bulk of this in what is now the Czech Republic), compared to Romania's $3.4 billion and Bulgaria's $1.2 billion (Coolidge, 1999).[14] The dominant streams of investment have been in manufacturing and goods production, but this investment emphasis varies geographically. In the Czech Republic investment has primarily been in manufacturing and transport, whereas in Hungary the leading recipients have been manufacturing and the financial sector. German and Austrian companies have been major investors, although corporations from other West European countries, the United States and Canada, together with a number of Asian nations, have also invested to a lesser degree. German and Austrian firms have invested particularly in the German-speaking areas of the Czech Republic, Poland, and Hungary. Japanese firms, though, were relatively slow to invest in the region, and most early Asian investment came from South Korea and the People's Republic of China (Michalak, 1993).

The growth of Western FDI is having significant impacts upon labor markets and patterns of production and consumption. Such impacts have implications for workers and unions, three of which warrant particular consideration. First, the introduction of Western-style management and labor relations is forcing many workers to learn new ways of working and is transforming the manner in which the "regimes" (Burawoy, 1985) of labor control operating in the workplace are articulated. In turn, this has raised fears that foreign investors will try to introduce work methods that workers in their home countries have been successful in prohibiting but that the nascent labor organizations in Central and Eastern Europe may yet be too weak to prevent (see Uhlíř, 1994). Given the lack of other indigenous traditions of industrial relations upon which they can draw, particularly traditions based on experience with market economies, many are concerned that the unions may find it difficult to resist such employment practices. Furthermore, the struggle to remake the geography of trade union representation and labor practices is being shaped by the uneven geography of economic restructuring as it plays out across the industrial landscape. In those parts of Central and Eastern Europe in which restructuring is proceeding with relatively little unemployment, unions and workers appear fairly open to adopting Western-style trade unionism and the market economy. On the other hand, in regions in which restructuring and privatization are bringing with them high levels of unemployment—such as in the heavy manufacturing Ostrava region of the Czech Republic—many workers and union officials are less receptive to such ideas, are often more sympathetic to the old Communist unions, and have opposed (through strikes and other actions) many proposed "reforms" and the breakup of state enterprises. Through such activities—both successful and unsuccessful—workers are helping to shape divergent paths of local economic development in the post-command economy era (I address this issue in more detail in Herod, 1998d), though the uneven geography of transition has also unleashed myriad conflicts among different sections of various unions and among national and local union officials.

Second, many worry that the low wages and lack of work safety provisions in the region will encourage a branch plant economy to develop that services Western Europe in a semicolonial relationship. Such an economic relationship would have implications not only for workers' rights but also for the type of economic development that will take place in the region, development that will be determined to a large degree by foreign decision makers (Uhlíř, 1995). Many union officials see in such an arrangement a future for Central and Eastern Europe that more closely resembles that of, say, Mexico, with its *maquiladoras*, than that of the Western European economies to which they aspire.

Third, many workers in Western Europe have begun to raise questions about how the growth of Western investment in Central and Eastern Europe will affect their own wage rates, job security, and working conditions. The availability of very skilled, low-wage labor in the Czech Republic, Hungary, and Poland, in particular, together with these countries' efforts to join the European Union, mean that higher-wage workers in countries such as Germany and Austria may face increasing competition for work in the future (Senft, 1995).[15] Central and Eastern Europe may thus increasingly come to serve as a geographical safety valve for Western employers (see Harvey, 1985c) who can use the threat of relocation to undercut demands for wage increases, improvements in working conditions, or even unionization itself. Furthermore, if Central and East European countries do eventually gain full membership to the European Union, West European employers could also freely and without legal restriction import lower-cost labor from the east, which would allow them to use ethnic, linguistic, and cultural differences to divide their domestic workforces.

Having outlined a number of ways in which the political economy of Central and Eastern Europe has been changing since 1989, and particularly the impacts of such changes upon workers and their organizations, in what follows I examine how a number of labor union organizations—both international and national in scope—have been quite active in Central and Eastern Europe in the realms of economic policy and union politics. What I seek to highlight here are two things. First, an examination of such activities illuminates the active roles played by these organizations in remaking the political and economic geographies of the region. As such, the activities recounted below provide further evidence of the ability of organized labor to operate at a global scale and to play an active role in shaping the geography of global capitalism. Second, in carrying out their activities such organizations have had to deal with a variety of different spatial concerns and, accordingly, have often had to tailor their strategies to particular local situations. The spatial context within which they have been operating has, in other words, forced them to implement programs in one way in one part of the region and in another way in a different part—that is to say, it has forced them to operate glocally. Such a recognition once again emphasizes the significance of geographical scale. Thus, although many in the West tended to think of Central and Eastern Europe as a relatively homogeneous single region when it was part of the "Soviet bloc," the great variety of situations before and after 1989 suggest that such a simple scalar categorization was (and is) problematic. Strategies that may have appeared to be implementable at one scale (e.g., a union's singular "Central and East European *regional* strategy") may be quite inappropriate when the variety of local and national conditions to be found across the region is considered, a situation that suggests that multiscalar policies are necessary that recognize national differences among countries but also local differences between plants and areas within the same country.

TRANSNATIONAL LABOR ACTIVITIES AND THE COURSE OF THE TRANSITION IN CENTRAL AND EASTERN EUROPE

The end of the Cold War has unleashed a dynamic struggle to inscribe a new historical geography of trade unionism onto the world space-economy. Perhaps the most significant element in this struggle has been the loss after 1989 by the pro-Soviet WFTU of much of its Central and East European membership as many unions and national confederations left and became affiliated with the ICFTU, a situation that led even the Federation's own representatives to concede that in the early 1990s its "very existence . . . was questioned" (Zharikov, 1995; see Herod, 1998e, for more details). For its part, the ICFTU has been quite active in Central and Eastern Europe since the early 1990s in seeking to promote both a transition to a market economy and the growth of Western-style trade unionism. In December 1990, for example, it created a Coordinating Committee on Central and Eastern Europe composed of representatives from affiliates both inside and outside the region, together with representatives from various international trade secretariats, the European Trade Union Confederation, the Trade Union Advisory Committee of the Organisation for Economic Cooperation and Development, and the International Labour Organisation. Chaired by Richard Falbr, President of the Czech ČMKOS, the Committee held several meetings to discuss policy toward the region and to develop programs for the exchange of information and technical expertise, training of union personnel, and the establishment of in-country educational centers to be run by local trade union officials. The ICFTU has also developed cooperative arrangements with trade unions in most of the countries of Central and Eastern Europe, including those that were not directly under Soviet control such as Albania (where it has links with BSPSh, the Union of Independent Trade Unions of Albania) and the countries of the former Yugoslavia. By early 1999 it had affiliated with national union centers representing some ten-and-a-half million workers in Bulgaria, Croatia, the Czech Republic, Estonia, Hungary, Latvia, Lithuania, Moldova, Poland, Romania, the Slovak Republic, and the former Yugoslavia (Oulatar, 1999).[16] ICFTU officials have also established an office in Moscow and made visits to discuss the situations in Belarus, Ukraine, Georgia, Uzbekistan, Kyrgistan, Azerbaijan, and Kazakhstan (ICFTU, 1996).

Primarily, the ICFTU has represented its efforts in the region as being designed to ensure the emergence of a regulated market economy while also trying to facilitate the growth of strong and democratic trade unions to protect workers' rights.[17] To further this goal the ICFTU has held seminars and conferences (in Hungary and Romania) on the social dimensions of structural adjustment policies and on countering the negative consequences of privatization and economic deregulation, together with workshops in Bulgaria and Hungary, which focused on comparative experiences with various social pacts, and in the Czech Republic, where the topic of discussion was labor migration within Europe. More broadly, the ICFTU has been keen to establish a Social Clause to be operated by the World Trade Organization and other international trade agreements that would guarantee freedom of association, the right to collective bargaining, a minimum age for employment, nondiscrimination and equal remuneration in employment, and a prohibition on forced labor. The ICFTU has also supported affiliates' efforts to include clauses protecting workers' rights in the generalized systems of preferences operated by various countries in their trade agreements.

The World Confederation of Labour, made up principally of Christian trade unions, has also been increasingly active in Central and Eastern Europe. Since 1989 it has affiliated a number of national trade union confederations and workers' groups, including the National Federation of Workers' Councils in Hungary, the Lithuanian Federation of Labour, the All-Ukrainian Union of Workers' Solidarity (VOST), the Krestanska Odborova Koalice (KOK—Christian Trade Union Coalition) in the Czech Republic, the Slovak Christian labor confederation NKOS, the Confederation of Democratic Unions of Romania (CSDR), and the Alfa Cartel-CNS, which claims a membership of some 1.2 million members and is the second-largest national labor center in Romania. The WCL supports the expansion of the European Union eastward and has developed plans to encourage the national Christian union centers of Central and Eastern Europe to join the European Trade Union Confederation as soon as their countries are admitted to the Union (WCL, 1998a). As a Confederation, the WCL has also been working, through seminars, conferences, and other educational forums, to encourage unions in the transition economies to develop their structures and to organize new members (particularly in small- and medium-sized companies), and to help them operate in a market economy. As part of this strategy, in 1998 the WCL restructured its European Section by combining it with its Coordination Committee on Central and Eastern Europe so that the Section could address issues of the transition in a more fully integrated manner (WCL, 1998b). It has also opened a liaison office in Bucharest, Romania, to help coordinate with its Secretariat office located in Brussels and to provide workers in Central and Eastern Europe a means to gain access to West European WCL affiliates. The Confederation has also encouraged many of its national affiliates to develop direct union-to-union programs with Central and Eastern Europe—by way of example, in 1997 the Dutch Christian trade union center Christelijk Nationaal Vakverbond (CNV) ran a seminar in the Slovak Republic as part of a program to help the Slovak NKOS railway workers' section develop collective bargaining procedures (WCL, 1997a).

The WCL also has a number of organizations similar to the international trade secretariats that work in particular economic sectors and that are affiliating new members in the region and attempting to establish training and other programs. In 1997, the World Federation of Industry Workers (FMTI—Fédération Mondiale des Travailleurs de l'Industrie), for example, started a seminar series at the WCL Bucharest office that has been attended by representatives from affiliates in Bulgaria, Romania, Hungary, the Czech Republic, and Lithuania, the main goals of which have been to inform union leaders about the structures and activities of trade unions in a market economy, to strategize about issues of collective bargaining at the enterprise and branch level, and to set up training programs (WCL, 1999; FMTI, 1997). FMTI has also worked with a number of miners' and chemical workers' unions on health and safety training (WCL, 1998c). In 1998 the World Federation of Clerical Workers (WFCW) Solidarity Fund helped pay for the Croatian clerical workers' union to open a new office in Zagreb that will provide, the Federation hopes, a better physical presence from which to service and recruit members (WCL, 1998d). The WCL's International Federation of Textile and Clothing Workers (IFTC) has also been active in Croatia, cosponsoring a seminar for women workers with its Croatian textile workers' federation affiliate TOKG in Pula, Croatia, in March 1998, together with seminars on "Social Dialogue and Relations between Central and Eastern Europe and the European Union" (in 1994 in Vienna), "Privatisation" (in 1995 in Dubrovnik), and "Safety and Health in the Textile and Clothing Branches" (in 1997 in Timisoara, Romania) (WCL, 1998e; IFTC, 1997). The IFTC has also affiliated local unions in Hungary, Romania, Bosnia, and Poland (the Light Industry Department of Solidarność affiliated in 1992), and signed a "Co-operation Protocol" with unions in Croatia and Bosnia that would take local unionists to training seminars in Belgium and the Netherlands run by affiliates in those countries. For its part, the International Federation of Trade Unions of Transport Workers (FIOST—Fédération Internationale des

Organisations Syndicales du Personnel des Transports) has developed a number of training programs for its affiliates in Bulgaria, Hungary, and Romania dealing with privatization, social legislation, trade union structures and financing, and organizing (FIOST, 1997). Other of the WCL's "International Trade Federations" have likewise become increasingly active in Central and Eastern Europe in the past few years.

The AFL-CIO, too, has been active in the region, principally through its Free Trade Union Institute (FTUI).[18] Created in 1977 to serve as the AFL-CIO's regional organization for Europe, at first the FTUI was designed to help democratic trade unions in Spain and Portugal as they struggled against the military regimes in those countries, though in the early 1980s it began to take a greater interest in Central and Eastern Europe, providing underground support to Solidarność in Poland during the period of martial law. After the revolutions of 1989 the Institute developed contacts and close working relationships with anti-Communist forces in the trade union movements in several countries in the region, providing moral, financial, and technical support. Much of the support for these activities came from the National Endowment for Democracy, which was established by the Reagan administration to inculcate pro-U.S. values abroad and for which the FTUI served at the time as the coordinating body for the AFL-CIO.[19]

Two issues relating to changes in the region have been of primary concern to the Institute. The first is the potential for Central and Eastern Europe to become a region of unfettered and anarchic free market capitalism. Such a development would both threaten the living standards of workers in the countries of the former Soviet bloc but could also serve to undermine the wages and working conditions of West European workers. Consequently, the AFL-CIO has come to characterize the process of the transition in Central and Eastern Europe as one in which the principle struggle being waged "is now between the defenders of humanity and the peddlers of unfettered capitalism" (former AFL-CIO President Lane Kirkland, quoted in FTUI, 1994: 5). Such a struggle is important, the Federation maintains, not just for the rights of workers in Central and Eastern Europe but also as a way of maintaining living standards for U.S. workers. "Free trade has become a mantra," Kirkland argued, "that conveniently blinds its advocates from the harsh world," a mantra that has resulted in "American workers . . . los[ing] millions of jobs over the past decade to the forces of flagless capital and free trade idolatry." As a result, he suggested, "international solidarity is an absolute necessity" for both U.S. workers and the AFL-CIO, who must commit themselves "to providing independent trade unions abroad as much support as possible to ensure their survival and to help strengthen them as effective democratic institutions so that the tyranny of the market will have no place in a civilized world."

The second of the FTUI's concerns has been that of the resurgence of

Communist parties and former Communist parties in the region, especially given the electoral successes of Communists and former Communists in several Central and East European countries, as well as in former Soviet Asian republics such as Kazakhstan and Uzbekistan. Such a resurgence is of concern not just from an ideological point of view but also because it threatens to hamper the process of economic transformation that the FTUI sees as crucial to the future of the region. As the Institute (1994: 8) has suggested:

> While communism itself is unlikely to return, and outright repression is being felt in only some countries, the strong resurgence of ex-communist parties and political groupings has brought life back to the nomenklatura elites and ex-communist institutions that dominated political and social life up until 1989. The consequences are already being felt. The networks that these individuals and institutions form are struggling to gain control of economic assets, government posts, the right to limited broadcast media channels, and other privileges. Economic and political reform are being stifled, impeding the development of both a private economy and a civil society. While serious reform is needed to transform the bankrupt economies of the old order, under the control of these interweaving and competing networks of party, government, and trade union officials, privatization is often closed and corrupt. This type of "reform" locks out workers and free trade unions from the reform process and leaves them even more vulnerable to restructuring and layoffs.

As the AFL-CIO has interpreted the situation (FTUI, 1994: 6–11), the main reason for the failure of market reforms to work the way they "should" has been the inability of the " 'intelligentsia-based' democratic parties" to develop strong ties with the "worker-based independent unions." The former's focus on market reform and their frequent disdain of the independent unions, whom they have seen as impediments to reform and restructuring, have often led them to ignore "the crucial importance of free trade unions, both as a representational and mediating institution, to a democratic society and to a market economy." The independent, non-Communist unions, on the other hand, "have no serious political ally." The solution to such problems, the Institute maintains, is the development of strong, independent trade unions that must be supported as "a democratic bulwark against dictatorial trends." The failure to enact "necessary" economic and political reforms, and the operation under the direction of the old *nomenklatura* of "a crony capitalism that relies less on free markets than on mafia-like networks," present for the AFL-CIO a situation in which "continued economic hardship and social dislocation will foster political instability . . . [and] ultranationalist ideologies that [will] threaten the region with an expanding circle of violence and ethnic conflict."

It is in response to this situation that the AFL-CIO has attempted to play a most active role in remaking the political geography of Central and Eastern Europe. The key to this strategy—as the AFL-CIO presents it—has been the FTUI's attempts to "bring together free trade unionists and democratic activists in a common agenda" (FTUI, 1994: 7). For the FTUI, the independent unions in the region are a crucial element in the creation of a liberal democratic, free market, civil society. The triumvirate of "democracy, civil society, and market reform . . . [will] all falter," officials suggest, "without their most forceful advocate, the free trade unions." As part of its strategy to strengthen the non-Communist unions, the Institute has conducted education and training programs in most countries throughout the region, focusing particularly upon the operation of market economies and the role of unions therein, organizing strategies, collective-bargaining procedures, and health and safety. Much of this has been carried out by FTUI staff from the Institute's field offices in Bucharest, Kiev, Moscow, Sofia, and Warsaw. The Institute has also provided material aid to help establish centers such as the Lithuanian Center for Labor Education and Research and the Ukrainian Institute for Labor Education and Research, and has provided funding to allow local unions to publish newspapers (FTUI has supported the Podkrepa union's Printshop and Union Publications center in Bulgaria, for example). It has also provided resources so that unions in the region may also learn from one another. For instance, AFL-CIO officers began running a program in the mid-1990s that enabled unions in Bulgaria and Romania to study the operation of Solidarność's Consulting and Negotiating Bureaux (BKNs). However, the goal clearly is to emphasize the benefits of Western-style (particularly U.S.) trade unionism so that the unionists in the region may "learn about the fundamental basis for free trade unions, the history of international and American labor movements, and specialized material on safety and health, collective bargaining, organizing, and election campaigns" (FTUI, 1994: 14).

Several other Western labor organizations have also expanded their operations and/or area of interest into Central and Eastern Europe since 1989. A number of individual unions have established direct bilateral relationships with their counterparts in the region. The German metalworkers union IG Metall, for example, worked closely with Solidarność in Poland in late 1995 as the latter was negotiating collective-bargaining agreements (Senft, 1995). IG Metall has also developed close linkages with the Czech labor movement, both the national ČMKOS Confederation and individual unions, particularly the Czech Metalworkers' Federation OS KOVO. Indeed, representatives from OS KOVO have been invited to attend collective-bargaining sessions involving IG Metall and its German employers. In 1995 IG Metall appointed a coordinator for the Czech Republic, the Slovak Republic, and Slovenia to encourage regional cooperation and mutual assistance between German unions and those in these three countries, while the Czechs also have a seat at the works councils of German companies owning plants in the Czech Republic. Likewise, U.S. unions such as the International Association of Machinists, the American Federation of Federal, State, County and Municipal Employees, the United Food and Commercial Workers, the United Mine Workers of America, and the Service Employees' International Union have conducted seminars in the region with various unions and have developed ongoing working relationships with some.

Of the trade secretariats, the International Metalworkers' Federation has been one of the most active in Central and Eastern Europe, seeing its main objective as that of facilitating the emergence of "a strong trade unionism capable of asserting itself as a negotiating power" (Mureau, 1995b) that will be able to defend the rights of workers and prevent them from being subjected to the worst excesses of capitalist development. To do so, the IMF has initiated a number of projects with metal unions in the region. Perhaps the clearest example of the IMF's involvement in reconstructing trade unionism in Central and Eastern Europe is the Federation's sponsorship of a number of training seminars designed to familiarize the region's workers with the roles played by unions in market economies. Initiated in response to requests for information from trade unionists in the transition economies, the seminars have covered a wide range of topics. These have included: the basics relating to trade union structure in a number of West European and North American labor movements; how to put together collective bargaining agreements; how to operate welfare and benefits programs; the provision of technical knowledge about pension plans and other welfare schemes; how to locate information on Western transnational corporations; enforcement of health and safety regulations; implementation of workers' education schemes; the implications of "lean" and just-in-time production; unions' roles in the conversion of military production to civilian production; how to read company balance sheets; negotiating and organizational skills; and aspects of Western labor law that might be of interest to unionists in Central and Eastern Europe (Olsen, 1995). As part of this program the IMF has translated a number of its manuals and handbooks relating to union structure and operation into several of the region's languages.

Initially the IMF focused primarily on the former Soviet satellite states of Central Europe—in 1994 alone the Federation conducted some 40 seminars in Bulgaria, Romania, the Czech and Slovak Republics, Slovenia, and Poland, for example. In late 1994 the Federation also sent a delegation to Macedonia. Although it initially adopted a cautious attitude with regard to the countries of the former Yugoslavia for fear that becoming too embroiled in the Balkan conflict might compromise its ability to work in the area after the end of the war (Mureau, 1994), the Federation did subsequently affiliate metalworkers from Macedonia in June 1995, setting up a pilot scheme to gather information on problems local unionists face. It also had some informal contact via third parties with opposition labor groups in Serbia and Kosovo (Olsen, 1995). As a result, by 1999 the Federation had affiliated 11 unions in the Balkans and sees this region as crucial in the future, as unions try to reconstruct themselves after nearly a decade of war (Lapointe, 1999). While in the early 1990s staff from the IMF's Geneva office conducted these seminars, by the mid-1990s the Federation was relying upon a number of its Western affiliates to implement these programs—for example, Finnish unionists have gone to the Slovak Republic to train local unionists—and was also training Central and East European unionists, who have now developed their own training structures and largely run such programs themselves. Adopting something of a rolling-front attitude toward the region, the Federation began in the late 1990s to reduce its own activities in the Czech Republic, the Slovak Republic, Hungary, Poland, and Slovenia (where it no longer conducts educational programs) and started developing new programs further east in Russia, Ukraine, Belarus, and Moldova. As part of this strategy, in June 1999 the Federation affiliated two unions in Belarus, one in Russia (the 1,360,000-member Miners' and Metallurgical Workers' Union of Russia), three in Ukraine, and one in Moldova (IMF, 2000a). Subsequently, the Federation also affiliated the Automobile and Farm Machinery Workers' Union of Russia and the Engineering Workers' Union of Russia, which have a combined membership of some 700,000.

Despite much success, the IMF's efforts have, however, run into a number of difficulties. In particular, Federation officials feel that the absence among Central and East European unions of a genuine collective-bargaining tradition or limited experience in protecting workers' interests through collective bargaining have been a major handicap, a situation that has led to frustrations because unions often have not had sufficient technical expertise to adequately understand company finances and macroeconomic analysis, and they have not always been sufficiently familiar with the content of (rapidly changing) new laws (Mureau, 1995b). Moreover, even after a decade, in some countries there are still relatively few employers' associations with whom unions can negotiate, and wherever state ownership continues to be a dominant feature of the economy the role of the employer is not always clearly defined. In addition, some terms and concepts that are second-nature to Western trade unionists are unknown to Central and East European unionists and workers (IMF, 1994).[20] This situation is sometimes further complicated by the fact that the IMF's own staff and the representatives from its Western affiliates come out of different union traditions and may have different appreciations of such terms themselves. Rivalries between, and among, new and "reformed" unions have also sometimes made it difficult to coordinate IMF educational seminars. Indeed, such has been the intensity of this rivalry that, on several occasions, the Federation has been forced to develop separate programs for new and reformed unions.[21]

The political instability in the region has also presented problems for the IMF. In the early 1990s, in particular, rivalries and high worker expectations about the ability of unions to rapidly increase their standards of living produced a high turnover rate of union officials that, in turn, made it more difficult for IMF officials and representatives from the Federation's Western affiliates to develop consistent long-term working relationships with some unions. This has been combined with serious problems of corruption on the part of many government officials that has made dealing with some state agencies difficult. Nevertheless, the IMF hopes that by projecting an image of cooperation it can encourage the independent unions throughout Central and Eastern Europe to work together on a number of issues common to them all—such as the implications of privatization—and prevent the fragmentation and multiplication of unions in the metals sector that, the Federation fears, will weaken workers' collective bargaining power (Mureau, 1994). As a practical way of encouraging such cooperation, in the mid-1990s the IMF asked its Czech and Slovak affiliates, for example, to send delegations to Bulgaria and Romania to share information on their privatization experiences.

In general, the IMF has been careful not to advocate particular models of labor relations but has preferred to let the region's unions make their own way. According to Poul-Erik Olsen, former Director of the IMF's Eastern Europe Education and Working Environment group, the Federation's position has been one in which "We do not say 'this is the model for your development' but that 'we will assist you in your development.'. . . We try to discuss with them what is required by the union in that particular country in that particular situation . . . and where we can step in to help" (Olsen, 1995). However, at the same time, the Federation has been actively seeking to impress upon its new affiliates the importance of developing strong national structures (Mureau, 1994). The tendency for both new and reformed unions to adopt a very decentralized, locally-oriented brand of trade unionism has led IMF officials to worry that such a model, while pluralist in nature, will hinder efforts to develop strong national-level organizations capable of addressing issues resulting from privatization and of confronting multi-establishment employers. In the eyes of the IMF, a strong, nationally-oriented, branch federation system of union organization not only will provide a structure in which the same union represents workers in different plants of the same company, but it will also help to break down the old Communist union structures in which the unions were largely organized on the basis of local Party political jurisdiction rather than enterprise structure. One way the IMF has sought to do this is by providing information to the national union leaderships on the proportion of union dues which the Federation's own Western affiliates spend locally, regionally, and nationally, and the roles that such affiliates' regional and national organizations play in wage bargaining, health and safety issues, legislative lobbying, and the like. In turn, the national leaderships have used this information to try to persuade local unions of the need for some greater degree of power and union funds to be placed in the hands of the national unions.

The Federation has tried to pay particular attention to the great variation of economic conditions, histories, political situations, local cultures, and needs across Central and Eastern Europe in a process that might be described as the "glocalization" (Swyngedouw, 1992) of its programs. For example, whereas the unions in the Czech Republic have tended to work more independently of the IMF and have consulted with the Federation more on issues related to privatization, the IMF has been much more active in developing links with workers in the Slovak Republic, particularly concerning the issue of trade unions and electoral politics, issues about which the Federation's Slovak affiliate (OZ KOVO) became particularly concerned with the election of Vladimír Mečiar in 1994. Indeed, as the IMF's Poul-Erik Olsen (1995) has commented, in recognizing the wide variation in conditions that exist across the region, the Federation has tried to devise "programs that are really tailor-made." Thus, he has suggested, "it is quite obvious that Romania and Bulgaria with their type of privatization and their type of problems . . . are facing different problems than are Hungary, Poland, and the Czech Republic. . . . This means priority will be given to different things."

Likewise, as time has passed so, too, the issues in which unions are interested have changed, such that whereas "early on it was privatization and restructuring," by the mid-1990s it was "more on what is the role of the unions in the various countries" (Olsen, 1995). Hence in Poland, for example, the Federation's early seminars with Solidarność focused on general issues, whereas by 1995 it was running individualized seminars for union officials in each of the four main industrial sectors (steel, aerospace, automobiles, and machine building) that Solidarność represents. Additionally, in designing its activities the Federation has been cognizant of the emerging geopolitical situations in various parts of Central and Eastern Europe. Consequently, it has organized a series of events solely for the so-called Vishegrad countries (originally Hungary, Poland, the Czech Republic, and the Slovak Republic, which were later joined by Slovenia), which, with the exception of the Slovak Republic, all have long industrial histories—unlike other Central and East European countries such as Romania or Bulgaria—and which, perhaps more importantly, are also the countries (with the exception of the Slovak Republic) that any future expanded European Union is likely to look to incorporate first. Similarly, in seminars on the conflict and labor situation in the Balkans, the IMF has been keen to include metalworkers from Bulgaria and Romania so as not to "split the region . . . [since] we have no interest in having any division between the unions [in the area]" (Olsen, 1995).

Indeed, in May 2000 the Federation organized a conference in Macedonia for 31 metal unions from Western, Central, and Southeastern Europe on how to develop cross-border and cross-ethnic cooperation in the region (IMF, 2000b).

In response to concerns by its Western affiliates that lower workplace safety and environmental standards in Central and Eastern Europe might serve as an incentive for employers to shift investment from Western Europe and North America, the IMF has conducted a number of seminars concerning these matters, addressing the issue of environmental degradation via that of workers' health and safety. It has also worked with unions in Japan concerning work conditions in Japanese-owned facilities, such as Suzuki's plant in Hungary. However, it has encountered some difficulties in this area, not least of which is a fairly low awareness among many Central and East European workers of the importance of health, safety, and environmental issues, such that the secretariat's seminar leaders have found that many workers are quite willing to work in substandard conditions if this means they can attract foreign investment, retain their jobs, and improve their wage and benefits packages (Mureau, 1994). Nevertheless, IMF seminar workers have tried to stress the importance of health, safety, and environmental issues for workers' quality of life, and by the mid-1990s most felt that some progress had been made in this regard (Mureau, 1995b). As part of its workplace health and safety education effort, the Federation has also attempted to develop programs geared specifically to women workers and women trade unionists, given that in the metals industries in Central and Eastern Europe women make up approximately one-third of the workforce and typically experience the worst working conditions and lowest remuneration rates.

In addition to such direct contact with unions in Central and Eastern Europe and the Federation's encouragement of its Western affiliates to closely cooperate in these training programs, the IMF has been keen to ensure that corporations adopt various codes of conduct regarding their operations and that a social charter guaranteeing basic labor rights regarding freedom of association, the right to organize and bargain collectively, and the right to strike be adopted as part of any future trade agreements involving the region (IMF, no date b).[22] As part of this desire, the Federation has encouraged its affiliates to pressure their own governments and a number of international agencies concerning issues of the region's foreign debt, economic growth, and balanced development. One campaign the IMF has been working on is its effort to pressure the World Bank and the International Monetary Fund to adopt less austere policies with regard to Central and Eastern Europe and to "ensure that they pursue investment policies which do not reduce workers' living standards and employment opportunities but encourage pro-worker economic development" (Mureau, 1995b). However, IMF officials recognize that the narrow constituencies of individual unions, and even the Federation itself, makes it difficult for them to influence institutions such as the World Bank and the International Monetary Fund regarding investment policies toward the region. Consequently, Federation officials have determined to work on such matters primarily through the ICFTU, which, with its membership made up of national trade union centers rather than unions in particular crafts or industries, is much more broadly based, both sectorally and geographically, than is the IMF.

UNIONS AND THE SPACES OF TRANSITION

Whether the countries of Central and Eastern Europe are experiencing a transition to a Western-style market economy and civil society or to some other form of economy and society—and how the trajectory of transition appears to be varying significantly across the region (Pickles & Smith, 1998)—is a subject of much academic, political, empirical, and ideological debate (e.g., see Stark, 1996). What is not a matter of debate, however, is the fact that the International Metalworkers' Federation, the ICFTU, the AFL-CIO, and other trade union organizations have been very much involved in remaking the political and economic geography of post-Cold War Central and Eastern Europe. Already by 1992 the IMF had come to represent nearly two million metal workers in Central and Eastern Europe and was actively engaged in helping to rebuild the region's trade union movements (IMF, 1992b). Subsequently, the Federation opened a regional office in Budapest to coordinate its activities in the region, a practice replicated by several other trade secretariats. Likewise, by the end of 1998 the ICFTU had affiliated national centers that represented some ten-and-a-half million workers across the region (Oulatar, 1999), while the WCL and its international trade federations similarly have signed up affiliates among national centers and individual sector unions.[23] Certainly, much about the future remains unclear. Many of the nascent trade unions in the region are very weak, and the loss of benefits, reduced living standards, and general hardships have made some workers nostalgic for the days of Cold War stability. This raises questions about the path that unions will take in the future and how this will both reflect, but also shape, the trajectory of "transition" in various countries. Additionally, a very real possibility also exists that, as the unions of the former-Soviet Union become affiliated with Western organizations such as the IMF, they may come to dominate them because of their large size. This, in turn, might lead to shifts in the internal balance of power of such organizations. Indeed, it was partly out of such concern that the IMF, for example, was initially somewhat cautious in its dealings with these unions, particularly given the political instability in Russia. However, after making some contacts in the mid-1990s with groups of instrument makers in the Moscow region via its Norwegian affiliate (Olsen, 1995), by 1999 the Federation had begun to step up its activities in the successor states as it affiliated several unions and began publishing a Russian-language newsletter (Lapointe, 1999). Furthermore, the activities of the IMF, ICFTU, WCL, and others in Central and Eastern Europe have provoked questions from some quarters about whether the added financial and organizational burdens on these organizations will impact their ability to conduct programs in other parts of the globe, especially the less developed countries. Although so far this does not seem to have been much of a problem, this may not always be the case.

Stepping back from the specifics described above, the activities of unions both from within and from outside the region raise a number of significant conceptual issues for geographers and other social scientists. Perhaps one of the most significant of these relates to how events in Central and Eastern Europe have been represented discursively. Specifically, analyses of the restructuring of the economies of Central and Eastern Europe and of the rescripting of geopolitical discourses in the post-Cold War period have tended to focus on the activities of capital and the state. Thus, capital has typically been seen as the active agent of change that is "opening up" the region to new market possibilities while states have been seen as facilitators of this through their encouragement of privatization via the selling off of various state-owned enterprises and the establishment of the regulatory frameworks necessary to allow capital markets to operate. Unions and workers, on the other hand, have often been represented as passive social actors who must serve as the bearers of such transformations (in the form of higher unemployment rates, for example) and who frequently stand in the way of "market reform," even though, as we have seen with the case of the Czech power workers' union and Poland's Solidarność, many Central and East European unions have been vociferous advocates of, and players in, the process of privatizing state enterprises—which they see as a way of stimulating economic development and, perhaps more importantly, of breaking the political control of the old *nomenklatura* (Volkov, 1992; Ost, 1989, 1995). As evidenced by the activities of unions described above, workers and their organizations are playing significant roles in the processes of economic and political transformation in the region. Organizations such as the AFL-CIO and IMF are spending large amounts of time and money working with unionists in Central and Eastern Europe in their efforts to adapt to the transition, to operate in market economies, and to develop mechanisms for effective collective bargaining. Likewise, the unions within Central and Eastern Europe are helping to shape the evolutionary paths of local economic development through their acceptance of, or opposition to, government policies of privatization and enterprise restructuring, and through their abilities to negotiate higher wages, work rules, job preservation agreements, and the like (see Herod, 1998d).

    As in other chapters, so, too, with regard to the activities of various in-
ternational labor organizations, it is clear that not only does a geographic
perspective provide insights into the mosaic of economic and political conditions, cultural values, labor laws, union organization, and types of labor
markets operating throughout the region, but it can also help to explain the
thinking and political praxis of social actors such as trade unions. The IMF
and the German IG Metall, for instance, perceive and treat Central and Eastern Europe in geostrategic terms. The region's geographic proximity to
high-wage Western Europe, and the possibility that the boundaries of the
European Union may one day be extended to include countries such as Poland, Hungary, the Czech Republic, and others, mean that Western capital
may come to use Central and Eastern Europe as a "spatial safety valve" by
threatening to move production to, or to import labor and commodities
from, the region. Indeed, Central and Eastern Europe's very proximity, and
the greater likelihood that labor markets and production links will become
more highly integrated between European countries in any future expanded
European Union than they will be between Europe and other parts of the
world, make such threats more potent weapons that capital can use to undermine workers' wages and working conditions in Western Europe than
are threats to move to, say, Southeast Asia. For organizations such as the International Metalworkers' Federation, then, ensuring the emergence of
strong unions in the region is key to their ability to defend workers' rights,
both in Central and Eastern Europe but also beyond. It is in this sense that,
echoing General Secretary Malentacchi's comments quoted at the beginning
of this chapter, Central and Eastern Europe represents both a region of great
"concern and anxiety" for the IMF and other bodies but also a region of
great possibilities, the emergence of which as a geographic bloc of strong
trade unionism, IMF officials believe, "will help the task of metalworkers everywhere in the world." By expanding the geographical scale of their activities and membership, entities such as the International Metalworkers' Federation and the ICFTU hope to limit the ability of capital to use Central and
Eastern Europe as a space through which they can engage in whipsawing
activities. Equally, through their own membership of such organizations, unions in Central and Eastern Europe hope to prevent their countries from being tied into a neocolonial political, economic, and geographical relationship with Western Europe, with all the implications this brings with it for the
future spatial transfer across Europe of wealth (on the concept of the geographical transfer of value and its impacts upon the uneven development of
capitalism historically, see Hadjimichalis, 1984).[24]
    Finally, the fact that, through seminars and education programs aimed
at bringing to them the experiences of operating in a market economy, Central and East European unions are adopting for themselves some elements
of trade union praxis developed in other countries also raises significant

questions about the transmission of ideas across space. Whatever else it may be, the diffusion of information is an inherently geographical process that calls into question how knowledge is geographically contextualized and how trade union practices developed in one place may be adopted, modified, and remade in other places (see Wills, 1998a). Furthermore, given the variety of experiences and conditions across Central and Eastern Europe, developing new models of trade unionism in the region will require an explicitly spatial sensitivity. Ideas that work in one situation or context may not work elsewhere. Geographic context is important. Blanket, aspatial solutions will not work. This is something about which the IMF and other organizations appear to be intensely aware. Indeed, as the IMF's General Secretary Malentacci has commented:

> It will take some years before new models of labour relations emerge in Central and Eastern Europe. They may be partly based on models from other countries and integrate some of their characteristics but, in the end, their configuration will lie in the hands of the change makers themselves who will have to build up their own models, using their innovative abilities while taking into account their historical and cultural traditions, their specific needs and aspirations. Systems cannot be simply transferred from one country to the other but the experience and knowledge gained during [IMF] seminars will be valuable assets in the reconstruction process and the search for a new role in societies at the crossroads. (quoted in IMF, 1994: 3)

Clearly, the IMF and other labor organizations are thinking explicitly about issues of geographic context, international economic and political geography, place, and space as they carry out their activities in Central and Eastern Europe. Equally, it is clear that understanding the IMF's and other unions' behavior, and the behavior of trade unions more generally, requires a geographic perspective.

CHAPTER 10

Labor Geographies
A Conclusion and a New Beginning?

In this book I have attempted to examine how the geography of capitalism is made at various geographical scales and, specifically, what the role of workers and their organizations is in this process. By showing how groups of workers have been able to shape the geography of capitalism, both in co-operation with, and in opposition to, various segments of capital, I want to suggest that even when workers have spatial goals that coincide with those of capital and that may actually be beneficial to certain segments of capital, they pursue these spatial goals as *active* geographical agents making space as part of their own very real political and economic goals in ways that they themselves perceive as being advantageous. Likewise, the fact that different groups of workers often pursue different spatial agendas means that the resolution of their geographic praxis will bring with it different outcomes for the landscapes of capitalism. Accepting in this manner that workers play active roles in making the geographies of global capitalism at various scales is important both conceptually and politically because it allows us to think about the spatial aspect of political praxis, for if we recognize that space and society are imbricated the one within the other, then social action is also spatial action, such that any effort to implement progressive social change will, by definition, invariably imply geographical change. Likewise, changes in the geographic relationships and structures within which people live their lives may help bring about more equitable social organization, for example as when mass transport is constructed to allow low-income people who do not have access to private cars to get about the urban landscape so that they are no longer spatially entrapped in inner-city ghettos.

In this concluding chapter, then, I want to step back from the specifics of each of the examples upon which I have drawn in order to make a number of broader remarks about how the geography of capitalism is made and what labor's role is in that process. Several issues strike me as being important to think about. Perhaps the most significant of these is the issue of how space can be used by both capital and labor—and by different segments within those two categories—to further political and economic agendas. The ability to shape a landscape in one way rather than in another can have significant implications for the balance of political power between different social groups and can play a central role in political struggle. For example, in the case of the New York garment workers outlined in Chapter 3, their success in defending a particular space in the built environment from the incursions of certain types of global financial capital was a central plank in their ability to defend jobs in the industry. Likewise, efforts by the AFL-CIO and its allies in Latin America and the Caribbean to physically transform the urban environment within which workers lived—in essence, to restructure the spaces of their lives—were intimately tied up with the politics of the Cold War, whereas dockers' abilities to carve out regulatory spaces surrounding East Coast ports within which certain work practices would be proscribed were a key part of their political strategy for dealing with the effects of containerization.

The control of space and the ability to shape the landscape's physical form, then, are clearly both reflections of, and sources of, power. Often, such control is exercized along class lines. However, frequently such efforts to shape the geography of capitalism in particular ways cut across class lines and result in the formation of geographical coalitions articulated around spatial, rather than class, interests. Analytically this is significant, for approaches that are not sensitive to issues of space and geography—and here I am thinking principally of various brands of orthodox Marxism and neoclassical economics—or which implicitly conceptualize space and class as being produced independently of one another, only to be brought together in some sort of "interaction" subsequent to their formation, have usually interpreted such coalitions either in terms of "displaced" class struggle (in which class conflicts are "projected" onto the landscape rather than being fundamentally about how the landscape is itself constructed) or as examples of false consciousness on the part of workers. However, in so doing such aspatial approaches fail to recognize that these types of coalitions—and the defense of place that they entail—may have deeply felt significance for people that shapes how they act on certain occasions and in certain places and, indeed, shapes the way in which classes as social categories are themselves formed. Thus, workers may act to defend class-defined spatial interests or to defend spatially-defined class interests, and do so in ways that, depending upon the context within which they find themselves, may coincide or collide—that is to say, these various sets of class and spatial interests may reinforce one another or may contradict one another.

Indeed, such spatial conflicts have been quite evident on a global scale with the recent protests against the efforts of the World Trade Organization and the International Monetary Fund to facilitate "free trade." Whereas many of the North American union members who were at the 1999 anti-WTO demonstrations in Seattle, Washington, for example, have come to see neoliberal globalization as an attack on their standards of living, many workers in developing countries see such protests as little more than an effort to limit capital flight from the "developed world" and to maintain the historical dominance of the global north. Such conflicts of geographical interest among the global working class, then, highlight some of the problems of developing class-based "proletarian internationalism"—the traditional cry of the political left—in the context of a highly unevenly developed global space-economy. At the same time, however, organizing around spatial interests may also produce unexpected coalitions and coalescences of interests. Thus, anti-WTO protests saw U.S. environmentalists and labor unionists— two groups that have often been at odds with each other but who see the WTO as a vehicle for undermining U.S. environmental and labor laws— uniting on several issues, prompting the now-famous slogan "Teamsters and turtles, together at last."

Within this broad question of how space and landscapes can be manipulated for certain political goals is, clearly, the issue of geographic scale. In particular, there appear to be at least three issues of importance here. First, we can see from the examples examined that the geographical scales at which social life is organized can be manipulated in different ways at different times. On some occasions social actors (workers, employers, environmentalists, pro- or antichoice abortion activists, the homeless) may choose to organize themselves at one particular scale, whereas at others they may find it more appropriate to their needs to organize at quite different scales. Likewise, whereas different types of problems may sometimes lead workers in different times, places, and industries to adopt similar strategies with regard to, for example, expanding or reducing the scale of bargaining, workers facing similar problems (e.g., a management attack on working conditions) may opt to pursue at different times and in different places radically different spatial strategies with regard to the production of scales of contract bargaining. Much of what determines these decisions, of course, will be shaped by the changing geographic contexts within which such social actors find themselves. As conditions change, so must the production of geographic scale be constantly renegotiated, for a particular "scalar fix" will not always serve the same needs as when first established. Thus, as regional economies wax and wane, particular groups of workers may find it more or less advantageous to bargain locally or nationally to secure their goals. Similarly, as the political and regulatory environment changes, so might this impact the scale at which social actors choose to organize. Hence, for instance, in the case of the United States, passage of the Taft–Hartley Act in 1947 and the criminalizing of secondary boycotts and sympathy strikes led many unions in the 1950s to begin making the national, rather than the local, union leadership the sole contracting agent with employers as a way of protecting the individual local unions that made up the national union from lawsuits now made possible under U.S. labor law (Moody, 1988).

This, then, raises a second set of issues with regard to the scale at which social life is organized, namely, how political praxis may crucially revolve around the ability to build linkages across various geographical scales or, alternatively, to uncouple processes and actors operating at one scale from those operating at another. In the case of labor, an obvious way in which this plays out is in the construction of linkages of solidarity between workers located in communities which are many miles from each other—something that the probusiness architects of Taft–Hartley recognized and sought to hamper. Building networks of solidarity across space—perhaps to limit the capacities of corporations to whipsaw communities against one another—often involves workers in one community using the resources of various national or international organizations to which they belong to facilitate developing contacts with workers located elsewhere, as in the case of the Ravenswood, West Virginia, workers. Equally, social actors' capacities to expand (or to contract) the geographical scale of a political or economic dispute to bring in (or to exclude) other actors who may have resources that can be of use in a particular situation can also significantly shape how a particular dispute or struggle evolves. Thus, workers involved in a local contract dispute may appeal to groups outside their community for help as a way to bring pressure to bear upon recalcitrant local employers. Alternatively, they may consider it beneficial to cut themselves off from broader geographical structures and relationships—such as a geographically uniform national master contract—if local conditions are more preferable than are national or international ones. At the same time, employers' success in limiting the influx of resources—money, organizers, information about conditions elsewhere—from national or international organizations into a local campaign may give them an advantage.

Third, we should not assume that social actors who are in conflict with each other (workers and employers, for instance) will necessarily choose to organize at the same geographic scale. Indeed, the cases examined in the previous chapters suggest that whereas sometimes it makes more sense for social actors to organize at the same scale as their opponents—a transnational labor campaign to confront a transnational corporation, for instance— at other times it is more fruitful to organize at different scales. Organizing at one scale to shape what happens at another can be key to achieving political and social goals. Thus, returning again to the example of the 1947 Taft–Hartley act, although the Act was a federal piece of labor law, it allowed individual states to pass right-to-work statutes in which individual workers could opt out of joining the union in a union shop workplace, a provision that has helped create what Davis (1986: 129) has called a "union-resistant geography of American industry" in much of the southern and western part of the United States. Although recent evidence would suggest that the geography of industry in the U.S. South is perhaps not as union-resistant as it once was (Herod, 1997c), the point is that probusiness Republicans, allied with antiunion southern Democrats, were able to use their control of the federal-level organs of government after the 1946 elections to shape the possibilities for union organizing at the state and local level, with myriad consequences for the economic and political geography of the country.

While the implementation of Taft–Hartley and right-to-work legislation is an example of one group of social actors using their control of national-scale institutions to shape the possibilities for action by subnational-scale institutions, we should not assume that it is always necessary to construct larger scales of organization than one's opponents to achieve success. Indeed, it is often the case that moving to more circumscribed scales of organization ("going local," as it were) can be a more effective strategy than is trying to outmaneuver an opponent by organizing at a scale larger than the one at which they are themselves organized. In the case of the Local 5668 workers in Ravenswood, for example, the local union's ability to break out of a concessionary national master contract so as to be able to negotiate locally allowed them to adopt more aggressive bargaining positions with their employer. Depending upon the circumstances, then, "going local" may enable workers to target particular strategic parts of, say, a transnational corporation in ways that may not be possible if workers are forced to operate within the strictures of larger-scale (e.g., regional or national) structures of organization.

This last point, about the purchase that organizing at the local scale can sometimes give workers facing corporations that may be organized at much broader scales, is particularly significant for understanding the process and politics of neoliberal globalization, especially with regard to how global processes and actions may shape local contexts but equally with regard to how local organization can shape global processes. Thus, much neoliberal rhetoric maintains that, as globalization proceeds, local customs and practices must give way to the dictates of globally organized capital. Certainly, this is the perspective of the WTO.[1] Unleashing the power of global capital, it is suggested, will sweep away local impediments—however defined—to the emergence of a fully integrated, super-conductive planetary capitalism (see Herod et al., 1998, and Toffler, 1980). Put another way, the shrinking of relative distances between various parts of the planet that modern forms of transportation and telecommunications have brought about, it is argued by some, is bringing with it a world in which borders no longer matter, in which "nothing is overseas any longer" (Ohmae, 1990: viii), and in which, seemingly, the significance of geography and the local are erased.

Yet, this rhetoric, while politically powerful as a way of undermining opposition to neoliberal globalization through seeking to portray no alternative to it, does not represent the reality of things. Instead, I would suggest, it presents only a partial reading of the nature of contemporary global capitalism. Indeed, it could equally be argued that the shrinking of relative distances between places actually makes geography (in the sense of the differences between places) and local idiosyncrasies more, not less, significant politically. As capital, commodities, and people now have the ability to whiz around the globe at seemingly breakneck speeds and in ways not previously possible, the differences between places—what makes up their "localness"—actually may become more important in the decision-making processes of social actors. As David Harvey (1989: 294) has suggested: "As spatial barriers diminish so we become much more sensitized to what the world's spaces contain." Hence, in the days when capital could only be transferred across the planet (in the form of letters of credit) at the speed of a swift horse or a sleek ship, small differences in profit rates between different places were fairly insignificant, because by the time it took to get capital investments to such places the market may have changed dramatically. Today, when trillions of dollars, pounds sterling, and yen can be shunted with a computer keystroke between markets at the speed of light down fiber-optic cables, minute differences between places can be acted upon by investors. In this regard, local political action—a strike or slowdown in production here, workers agreeing to work overtime there—can fundamentally shape global flows of capital.

Concomitantly, the fact that the planet is more interconnected than at any time in history means that the consequences of actions are now transmitted faster and wider than ever before. Through the medium of television and the Internet events on the other side of the planet can now be watched as they happen, forcing politicians, human rights activists, workers, business people, union leaders, generals, and others to respond to them in much shortened time frames. Equally, the repercussions of an event in one particular place may be felt more quickly elsewhere than ever before, a situation that, arguably, makes events in those places more significant than when repercussions spread more slowly across the landscape. In the 1998 dispute between the United Auto Workers and General Motors, for instance, the autoworkers union was able effectively to cripple GM's North American operations within days by striking at just two plants in Flint, Michigan (see pages 219–220 in this volume and Herod, 2000, for more details). Because the two struck plants sat at crucial intersections within GM's highly integrated organizational structure, the industrial action by just 9,200 UAW members in a single community led to the laying off of 193,517 GM workers at 27 of

GM's 29 North American assembly plants and 117 components supplier plants in the United States, Canada, Mexico, and even Singapore. [The dispute ultimately cost GM production of some 500,000 vehicles and an after-tax loss of an estimated $2.3 billion in the second and third quarters of 1998.] As this example and others (such as the New York garment workers) show, the power of the local is not necessarily eviscerated in a global economy, and one particular scale of organization (i.e., the global) does not "trump" all others. Rather, the power of actors organized at one scale to shape what happens at others (the global shaping the local, the local shaping the global, for example) will depend upon the particularities of each case.

Of course, if unions are to use such "localist" campaigns to target strategic points in the production chain successfully, they will usually need to know something about how the different parts of a corporation are linked together, both functionally and spatially. Discovering such information is as much a cartographic project as it is anything, requiring a sensitivity to the geographical structuring and organization of any corporation against which such campaigns are to be mounted. Already, in the highly integrated North American auto industry, such an example of what we might call "guerrilla cartography" is yielding important information for unionists. Hence, members of the International Research Network on Autowork in the Americas (IRNAA), a group of critical scholars working on labor issues in the industry, are currently attempting precisely to map such links as part of a broader project—the "Mapping Supplier Chains" project—that is examining the emerging structure of the automobile industry in various countries. The network sees as one of its primary tasks to map "the changing contours of this supply chain, from vehicle assembly plants, to in-house suppliers of major components including drive trains, to outside suppliers responsible for the manufacture of integrated systems, to commodity suppliers and job shops contracted to deliver parts and small components to other plants in the chain."[2] Such spatial information will provide useful insights into what are the vital control points of various auto producers' production chains that labor activists may then use to pressure management on certain issues. As production chains in many industries have been stretched out across the globe in the past few decades, such cartographic activism on the part of workers will likely play an increasingly important role in labor politics. The days when capitalism was largely organized on the basis of production at single plants within which the entire production process took place have long since gone, and mapping supply chains stretching around the world is now usually a necessary first step to action.

Space and geography, then, clearly make a difference to workers, but workers also make a difference to space and geography. This raises a number of interesting and important questions concerning the politics of space and the politics of organizing, specifically how workers seek to organize in space, with space, against space, around space, and across space. Indeed, Southall (1988: 466) has argued that unionization is, in fact, a process of "coming together" by workers that is inherently one of "organizing over space." Put more simply, what I suggest here, then, is that, if they are to be successful, workers and their organizations will have to address the question of what kind of organizing is appropriate for what kind of space. For example, in the United States the typical model of organizing manufacturing workplaces has incorporated within it a certain number of quite specific spatial assumptions (see Green & Tilly, 1987), these being principally that organizing takes place within national space-economies, that it focuses upon identifying union "hot shops," and that the primary mechanism of organizing is the picketing of large centralized facilities that have regular shift changes and only a few entrances (i.e., that can be serviced by a small number of organizers who may hand out leaflets and other literature to workers—who are easily identifiable by their dress—as they leave the premises). However, while such assumptions may be reasonable for organizing large manufacturing facilities, they are less likely to match the realities of many service-sector workplaces, such as offices or retail locations, where it may be more difficult to identify hot shops, where there may not be regular shift changes, where workers may work in smaller groups and in closer physical contact with management, where service-sector workers may work at multiple job sites (as in the case of janitors, service-call employees, bike messengers, taxi drivers, etc.), and where the need frequently for workers to dress in a professional manner may make them difficult to identify as they leave work.

As several authors have noted (e.g., Bronfenbrenner, 1993; Savage, 1998; Berman, 1998), organizing service-sector workers, then, may require new models of organizing that incorporate within them quite different sets of spatial assumptions and that necessitate different types of spatial strategies than those typically used in organizing manufacturing workers. Thus, in the case of the "Justice for Janitors" (J4J) campaign in Los Angeles during the 1990s, janitors, their supporters, and the Service Employees' International Union (SEIU) adopted an innovative strategy of "geographical unionism" (Wial, 1994) that involved moving their struggle for recognition and improved wages and benefits beyond the private spaces of the multiple buildings within which they worked and into the public spaces of the streets. Rather than seeking to negotiate union contracts with each individual janitorial service, which were themselves contracted by the buildings' owners—a strategy that would have been more traditional but one that would also have been problematic, given that there are dozens of companies who contract with building owners to clean multiple work sites scattered throughout the city—J4J instead used street events and other public actions to pressure both contractors and building owners to create standard-ized wage and benefit packages that would be put in place in various geo-graphically defined districts within the city. Within such districts, workers would be remunerated at a standard level. By changing its spatial strategy, J4J avoided the problem of having to organize on a workplace-by-workplace, company-by-company basis in an industry in which only a small number of workers may be employed in any particular work site or by any particular company. Similar tactics have been used by unions in other indus-tries in which workers may have attachments to the industry though not necessarily to particular employers (e.g., waitressing—see Cobble, 1991).

Such innovative spatial tactics have also been used quite successfully in the fight against garment sweatshops, both within the United States and abroad—as U.S. television personality Kathie Lee Gifford found out to her detriment in 1995 when the National Labor Committee (NLC) publicized the fact that a line of clothing that had been designed by her and that carried her name was being made in sweatshops in Honduras and New York City. Through a series of press conferences and a campaign to make consumers aware of this situation, the NLC was able to use the world's media to spot-light worker abuses and so effectively to shift the geographical focus of la-bor struggle in the garment industry from the private sphere of the sweat-shop floor (a space in which it is easy to hide abuses) to the public sphere beyond the factory gate, in the process expanding the scale of anti-sweatshop activity from the individual plants to a much wider global audi-ence.[3] The furor that was created when the story hit the global media high-lighted one of the fashion world's dirty little secrets and led President Clinton to announce a plan to ensure that L.L. Bean, Liz Claiborne, Warnaco, Phillips-Van Heusen, Nike, Tweeds, Patagonia, Nicole Miller, Karen Kane, and Lucky Brands clothing would begin placing disclaimers on their gar-ments to assure buyers that children were not exploited in their manufac-ture. In similar fashion, the Union of Needletrades, Industrial and Textile Employees' (UNITE) Stop Sweatshops Campaign has used the power of consumers to pressure retailers, who set prices within the industry, to en-sure that garment manufacturers follow laws concerning wages and work-ing conditions (Johns & Vural, 2000). Equally, the United Students Against Sweatshops (USAS) has used similar tactics to pressure various universities and professional sports teams to make sure garments sporting their logos are not made in sweatshops. Global media and consumer purchasing power have similarly been used to highlight the plight of Indian children involved in the carpet industry. By changing their focus from the spaces of produc-tion to the spaces of consumption, then, unions such as UNITE and other groups such as USAS have been quite successful in bringing the conditions found in garment sweatshops to the attention of the public, who, through their purchasing power, may be able to effect change in the industry.

Different types of spaces, then, may require different types of organizing strategies. But equally, different types of organizing may mean different things for the way in which the spaces of capitalism are made. Returning to the distinction made by Johns (1998) concerning different types of solidarity actions, it is obvious that the practice of "international solidarity" can have quite different implications depending upon its goals (see also Salt et al., 2000). Thus, efforts to organize across space may be implemented as part of a strategy to protect certain privileged spaces within the global economy. Workers who fear that their communities might experience capital flight to lower-wage areas may, for example, engage in developing solidarity across space with workers in such low-wage regions to raise the latter's wages as a way of limiting their appeal to potentially relocating capital. This kind of "accommodationist" solidarity, as Johns terms it, will have very different implications for the geography of capitalism than will solidarity campaigns that seek not to protect certain spaces but to challenge more fundamentally the social relations of capitalism, regardless of the implications of this for different places (what Johns calls "transformatory" solidarity). Whereas in both of these scenarios workers may be engaged in developing solidarity across space, the implications for the making of the geography of capitalism of their different reasons for doing so will be quite different—in one, patterns of uneven development may be exacerbated whereas in the other they may be reduced or, at the very least, reproduced in different ways.

Analysing workers' spatial positions may perhaps also give us some general insights about the types of organizing in which they are likely to engage or that may be more likely to succeed in particular circumstances. This is not to say that we can develop some taxonomic list that will allow us simply to read off what will be the actions of workers in different situations, but it is to say that certain geographical situations within which particular groups of workers find themselves may lend themselves more often and easily to certain strategies than others. For instance, workers who are perhaps less class-conscious and identify their interests more along spatial (e.g., national or regional) lines may be less likely to engage in solidarity across space in cases where they seek to benefit from the flight of capital investment from another community to their own than are workers who are more class-conscious and who recognize how corporations may use space to whipsaw communities against each other. Likewise, workers may be more likely to develop solidarity across space to help each other in cases where each group has nothing to lose by doing so or is not in direct competition for continued capital investment in their community. In a situation, then, in which a corporation is planning to close down a facility in one community and move those jobs elsewhere, workers in the receiving community may be little inclined to engage in solidarity actions with the community losing the jobs. On the other hand, in a situation in which there is no likelihood of investments and jobs being shifted from place to place but in which the corporation is impinging upon the rights of workers in its various plants located around the country or world, worker solidarity may be more readily engendered.

Certainly, in the case of the United States recognition of the geographical context of workers' organization in the 1880s fundamentally shaped the nature and subsequent development of the country's labor movement. Earle (1992) has argued, for example, that the geography of support and nonsupport for the 1886 general strike was significant in shaping the subsequent structure of the American Federation of Labor (AFL). Through an examination of the geographical spread of the strike, Earle shows that support was primarily concentrated in a few large northeastern cities and that, in geographical terms, the strike was not very general. The spatial lessons learned by unionists from this pattern of support, Earle has suggested, shaped the thinking of those who came together in late 1886 to found the AFL. In particular, he argues, the founders of the AFL quickly came to realize that the U.S. working class was divided along geographical lines and that support for unionism was concentrated in a small number of locations in the industrial heartland. The failure of the general strike to mobilize workers across large sections of the country led AFL leaders to conclude that their own interests were best served by a narrow craft-based organization focused upon their self-identified constituencies in the urban northeast and midwest. Furthermore, in light of the kinds of state repression which were unleashed after the 1886 Chicago Haymarket massacre, Earle maintains that these leaders preferred an organizational structure that was geographically decentralized so as to keep any industrial violence localized, thereby insulating the wider union federation from retaliation by politicians intent upon damning the entire labor movement for the perceived sins of a few. Thus, he notes (1992: 381), there is a "remarkable paradox [in that] sectarian trade unionists organizing a general strike on behalf of the entire working class in May [1886 had] within eight months withdraw[n] into the sectarian philosophy of pure and simple trade unionism," a paradox that can be explained, at least partially, on the basis of the geographical lessons learned from the spatial structure of the general strike.

Clearly, in such a case the geographical context in which workers found themselves was one of the key things that shaped their actions and the activities of labor union organizers in the late nineteenth century. This lesson from the nineteenth century brings with it a set of considerations as we enter the twenty-first century. In particular, as Hyman (1999) has suggested, contemporary trends in capitalism—at least in the global north—are dramatically changing the nature of work and the social and spatial relationships between home and the location of paid work. Whereas traditionally union strategy has often been based on the assumption that workers are full-time employees who work for a specific employer, often in a long-term relationship (the classic example here being the "mass" worker in mining, manufacturing, or transport), contemporary changes in the way in which capitalism is organized have brought with them more contingent (part-time/temporary/short-term) work, government make-work schemes, more self-employment, the growth of "white-collar" work, telecommuting, and generally a looser relationship to the labor market and the point of production than has historically been the case. Such changes in the way in which capitalism functions have resulted in a situation in which "the spatial location and social organization of work, residence, consumption and sociability have become highly differentiated" and the typical employee today "may live a considerable distance from fellow-workers, possess a largely 'privatized' domestic life or a circle of friends unconnected with work, and pursue cultural or recreational interests quite different from those of other employees in the same workplace" (Hyman, 1999: 3). Such developments are changing the spatial basis upon which labor unionism has traditionally relied, and are leading to a new "fragmented" geographical structure to workers' lives (see Herod, 1991b) that unions and other workers' organizations will have to address if they are to be successful.

Thus, although it is something of a stereotype (though not wholly!), the spatial organization of capitalism upon which mass unionism in the nineteenth and much of the twentieth centuries was built, and upon which workers (particularly males) constructed their identities as workers, was one in which the paid workplace and place of residence were often spatially proximate and in which working-class identity in one sphere was reinforced in the other. Spatial relationships helped to reinforce collective identities, since people who worked together invariably lived close to one another and often socialized together. However, with the increasing spatial fragmentation under contemporary capitalism of the places of paid work from the places of residence, together with the growing feminization of paid work, the traditional identities upon which collective union action was historically based are being undercut and new identities are emerging, identities often constructed around spaces of consumption rather than those of production. As Hyman (1999: 3) has suggested, this spatial "disjuncture between work and community (or indeed the destruction of community in much of its traditional meaning) entails the loss of many of the localized networks which strengthened the supports of union membership (and in some cases made the local union almost a 'total institution')." Moreover, some have suggested that such spatial and other changes have undermined the identity of the "collective worker" and spawned a growing "individualism" on the part of workers. Although this is something of a simplification, Hyman does suggest that the "eclipse of the 'mass worker' whose institutionalized solidarities were reinforced by the broader networks of everyday life does mean that the possibility and character of collectivism are today very different when work and everyday life are increasingly differentiated."

In such a changing spatial and social context, many observers have suggested, if workers are to be successful in securing their goals, then traditional industrial or craft-based unionism may have to give way to what some have referred to as "social movement unionism" (see Waterman, 1998, 1999), in which the focus of union attention is expanded beyond the strict confines of the workplace to include the broader community (including the spaces of consumption). Implicitly, such social movement unionism incorporates within it quite different spatial assumptions than either craft or industrial unionism, for it seeks to overcome artificial spatial divides between the workplace and the broader community to expand the scale of collective action, in much the way that, for instance, the UNITE's Stop Sweatshops Campaign has used the power of the broader purchasing public to pressure indirectly garment manufacturers. Although it may take time for union bureaucracies to retool the ways in which they operate, social movement unionism would certainly provide a different geographical model (the "new beginning" referred to in this chapter's title) in which the spatial fragmentation of workers' lives (e.g., between home and paid workplace) could be addressed.

In summing up, then, in this volume I have tried to examine how the geography of capitalism affects the ways in which workers live their lives and, concomitantly, how the ways in which workers live their lives affect the geography of capitalism. Evidently, if they are to be successful, unions and other working-class institutions need to be sensitive to the geographical contexts within which they operate. Workers have little choice other than to live within landscapes which may sometimes enable, and at other times may constrain, their social actions—they can no more live "beyond" geography, beyond the spatial relations of society, than can capital. Workers clearly have spatial imaginations upon which they act and which shape their political practices.

In exploring some of these issues, then, I hope not only to have raised questions about how the geography of capitalism is made but also questions about how we think of "labor" as a category. Frequently "labor" has been viewed as an undifferentiated, hegemonic category, with workers' class position being assumed—in the last instance—to explain everything about why they do what they do in the realm of social and political praxis. When all is said and done, such approaches implicitly suppose that workers will recognize that they share certain unifying characteristics that will ultimately outweigh all other distracting allegiances such as those of place, nation, ethnic identity, gender, and so forth.

Recently, however, "nonessentialist" approaches to social analysis have begun to question how the triumvirate of the social categories of race, class, and gender are perhaps not as internally coherent as was once assumed. For example, whereas early feminist work emphasized the commonalities within the category "woman," nonessentialist approaches have sought to examine how race, class, and gender intersect, such that what it means to be a wealthy white woman is invariably quite different from what it means to be a poor black woman. Indeed, such differences may be so great as to overwhelm any common bonds of sisterhood such women might feel. In the case of workers, historical experience has often shown that racial, gender, and other identities may overwhelm any sense of class solidarity that workers may share. Any analytical approach to understanding how workers act in particular situations, then, requires an understanding of the intersections of these different identities. As I have tried to emphasize in the cases examined, however, any nonessentialist approach must also include an appreciation of how the geographic context within which workers live their lives— that is to say, their geographic rootedness and spatial sensibilities—can lead different groups of workers to adopt different strategies and to pursue different agendas at different times in different places. Many approaches to understanding social praxis have regarded space as a conceptual annoyance to be theorized away. The reality, though, is that social praxis cannot be understood without an appreciation of the spatial context within which it takes places. Put more bluntly, the triumvirate of "race, class, and gender" should really be a quadrumvirate of "space, race, class, and gender."

Recognizing the significance of space in workers' political praxis is, then, important theoretically and politically. If workers are to exert political power, they must learn to effectively negotiate the landscapes within which they live their everyday lives. This seems particularly pressing at a time when longstanding models upon which organized labor has traditionally based its power appear not to be working so well. Union membership is at a historical low in many industrialized countries. At the same time, many workers in nations undergoing industrialization are facing myriad problems as they attempt to form unions and to use them as vehicles to attain their objectives in an increasingly neoliberal world. This is not to say that unions and other working-class institutions do not retain influence; as we have seen, workers and their organizations continue to shape the landscapes of global capitalism. But as a host of social, economic, technological, and political developments—presented to us under the name "globalization"—tear asunder the spatial relationships upon which the certainties of the past were based, workers must respond. If they are to secure their goals and play their roles in shaping the future, workers must build new organizations and reinvigorate old ones in ways appropriate for the new geographies of capitalism. Such a reconstitution of the world's labor movements will be, by necessity, a fundamentally geographical project.

Notes

Chapter 1

[1]Although Krugman and his compatriots appear to have "rediscovered" geography and have attempted to incorporate more of a spatial sensibility in their analyses of economic systems, I would argue that the view of space they adopt is still one that does not really appreciate the ways in which space can be actively shaped to pursue particular social and political ends.

[2]Whereas environmental determinist arguments of the early 20th century had taken space largely to be a naturally given thing that determined social development, the approach adopted by most Marxist geographers during the late 1970s and 1980s saw space as socially produced. This managed to avoid the excesses of environmental determinism by suggesting that, although spatial structures might shape patterns of social behavior (as the environmental determinists had earlier argued), these spatial structures were themselves *socially* produced and so might be transformed as social relations were transformed. In such a conceptualization, then, space was not conceived of as the immutable "God-given" explanator of the environmental determinists but, instead, as a social product that itself could be remolded.

[3]I have provided a summary of much of this empirical work elsewhere (see Herod, 1998a).

Chapter 2

[1]During the 1990s there was much discussion about the supposed transition from a Fordist regime of production oriented around "just-in-case" (JIC) production to a post-Fordist one oriented around "just-in-time" (JIT) production. Specifically, whereas JIC production relied upon vast quantities of components being produced ahead of time and stockpiled for eventual use, JIT production methods assume that components arrive at assembly plants as close as possible to the time that they are needed, a system that dramatically reduces the amount of capital an auto assembler

270

needs to have tied up in components at any one time (see Herod, 2000). While JIC methods allowed assemblers to use components suppliers located several days' drive away, under JIT the need to ensure that components arrive at the assembly plant just in time for use means that many U.S. automobile manufacturers have switched to using suppliers located within a short distance of their assembly operations, leading to the emergence of a new geography of production in the industry (e.g., see Mair, Florida, & Kenney, 1988).

[2]See Sayer (1984) for more on the differences between "necessary" and "contingent" social relationships.

[3]For more on the connections between academic geography and imperialism, see Hudson (1977), Harvey (1984), and the collection of essays in Smith and Godlewska (1994).

[4]Thus, imperial apologist and commercial geographer L. Dudley Stamp (1937: 65), for instance, compared the productivity of textile workers in the damp northwest of England rather favorably with those of sunny northern India, noting that in the early part of the twentieth century in Cawnpore (now Kanpur) "nine men were still required to work a [spinning] mule of 800 spindles where, it was said, only three would be necessary in a Lancashire mill."

[5]For a basic survey of literature in this human ecology tradition, see Theodorson (1961).

[6]Examples of such approaches include Berry et al. (1976), Dicken (1971), Dicken and Lloyd (1990), Greenhut and Hwang (1979), Isard (1956), Krumme (1969), Latham (1978), McNee (1960), Rees (1974), Smith (1966, 1970), and Walker (1975).

[7]For more details, see Harvey (1972), Massey (1973, 1995 [1984]), Gregory (1978), and Smith (1979).

[8]Wolff and Resnick (1987) provide a good overview of differences between neoclassical and Marxist approaches to economics.

[9]Reviews of the most significant theoretical debates within behavioralism are provided by Carr (1983) and Hayter and Watts (1983).

[10]Examples of such approaches include Greenhut (1956), Keeble (1968), Krumme and Hayter (1975), Lever (1975), Townroe (1975), and Stafford (1991).

[11]This, of course, is a very different kind of "new" economic geography from that now being promoted by the economist Paul Krugman, which is very much in the locational modeling tradition. For a critique of Krugman's "new economic geography" see Martin (1999) and Martin and Sunley (1996).

[12]For examples of early Marxist work on multinationals, see: Hymer (1976, 1979), Peet (1987), and Taylor and Thrift (1982, 1985). For work on urbanization under capitalism, see Lefebvre (1970, 1976 [1973], 1991 [1974]), Harvey (1978, 1982, 1985a, 1985b, 1989), Tabb and Sawers (1978), Walker (1981), Scott (1988), and Katznelson (1992).

[13]See Massey (1978), Walker (1978), Anderson (1980), Peet (1981), and Smith (1979, 1981) for more on the debate over the fetishization of space.

[14]Harvey (1973), for instance, touched upon Lefebvre's ideas (although critically) in *Social Justice and the City*.

[15]Although he recognizes this contradiction, Lefebvre tends to equate the articulation of space's use value with the political actions of the state, which organizes space through its power to control urbanization and the like.

[16]Harvey (1982: 388–390), for example, provided a devastating critique of neo-classical location theory as practiced by two of its leading lights, Walter Isard (1956) and August Lösch (1954). Specifically, he showed how assumptions of equilibrium in the landscape rely themselves upon the assumption that capitalists cannot gain anything by relocating elsewhere or by investing in situ to make their own production more profit-able than their competitors. This, in turn, leads to a situation in which competition for relative locational advantage stops, such that landscapes are no longer dynamic but are, instead, ossified at particular points in time with average profit rates across the space-economy in balance, though at zero percent. "This is an extraordinary result," pro-claimed Harvey (1982: 389), for "it means that competition for relative locational advan-tage on a closed plain under conditions of accumulation tends to produce a landscape of production that is antithetical to further accumulation."

[17]The main protagonists in this debate were Beard (1995), Martin et al. (1993, 1994), Massey (1994), Massey and Painter (1989), and Painter (1994). For a summary of this debate, see Herod (1998a), especially pages 25–27.

[18]It should also be pointed out that low wages, low levels of union representa-tion, and low strike rates are not necessarily indicative of low levels of class struggle but can also reflect the power of capital and the state to suppress some forms of worker praxis (such as organizing), while the suggestion that "work stoppages are a quite direct reflection of the level of underlying worker resistance to capital" (p. 124) ignores the fact that a strike is merely one expression of such opposition (relatedly, see also Herod, 1997c, which argues that, in fact, by some measures the southern United States has higher levels of class struggle than the national average). As James Scott (1985) has shown in a different context, resistance to capitalist social relations can also be manifested in other, less dramatic, ways: absenteeism, slowdowns, delib-erately producing defective products, and so forth. Equally, it is doubtful that mem-bership in a union is per se necessarily indicative of high levels of class struggle. Whereas some unions are very militant, others are often bureaucratic and politically conservative.

[19]Of course, this does not mean that capital and labor will see eye to eye on all things, but it is a recognition that on some issues they may feel they have common interests.

[20]In a similar vein, I am sure that most capitalists do not think of their locational strategies in terms of their creating a "spatial fix" à la Harvey (1982), even if that is, in fact, precisely what they are doing. To quote Foucault (himself quoted in Dreyfuss & Rabinow, 1983: 187) in this regard, "People know what they do; they frequently know why they do what they do; but what they don't know is what what they do does."

[21]Given that regions are dynamic social entities, this raises methodological problems of defining where one region ends and another begins. However, such methodological problems should not be confused with ontological questions con-cerning whether a region actually exists. In other words, contrary to the opinions of those who draw inspiration from Kantian idealism and who see regions as little more than figments of the imagination that can be manipulated by the individual re-searcher to serve his/her needs (cf. Hartshorne, 1939, and Hart, 1982), I would argue that regions are indeed real social entities (i.e., they do exist as material constructs) even if methodologically we may have a hard time delineating their exact bound-aries at any particular time.

[22]There is also a more technical meaning of the term "scale" that refers to the ratio between the distance on the Earth's surface and how that distance is represented on a map. In the discussion here, I am not referring to this more technical notion of scale but, rather, to the scales by which we organize our world (local, regional, national, global, etc.).

[23]For example, to what extent does the right to privacy protect one from the instrusions of government within the home? In several states in the United States and in other countries, many sexual acts between consenting adults are illegal, even when conducted within the "privacy" of one's own home. Whereas many progressives would presumably argue that such an intervention by the state should be defended against on the basis that it may infringe on the privacy one might reasonably expect to enjoy within the confines of one's home, at the same time most would (I hope) argue that the "privacy" of the home should not be so absolute that, say, murder may be committed with impunity just because it occurs within one's home. Equally, the "privacy" of the home is often impinged upon when the state seeks to regulate industrial homework, leading to the question of at what point the "home" becomes the "factory" (see Herod, 1991b). In other words, the boundaries between public and private space as they relate to the home may be—and perhaps should be—quite porous with regard to certain issues and not at all porous with regard to others. Which of those activities the state may choose to regulate at home, of course, is the subject of political struggle. For more on historical efforts to define the boundaries between public and private spaces in this regard, see Marston (1990).

[24]Travel-To-Work-Areas are frequently used by economists, urban planners, geographers, and others as measures of the spatial extent of a particular city's labor markets.

[25]Smith (1990 [1984]: 135) suggested, for instance, that it would be possible to "use the dialectic of differentiation and equalization to derive the actual spatial scales produced by capital."

[26]In this regard, see the Afterword in the second edition (1990) of *Uneven Development*; see also Smith (1989, 1992, 1993, 1995, 1996).

[27]Some of the principal writings in this regard are: Agnew (1993, 1997), Brenner (1997, 1998), Cox (1993, 1996), Delaney and Leitner (1997), Herod (1991a, 1995, 1997a, 1997b), Jonas (1994), Kelly (1997), Leitner (1997), Miller (1997), Smith (1992, 1993, 1995), Staeheli (1994), and Swyngedouw (1992, 1996, 1997).

[28]I have put "local" in quotation marks because being able to define strikes as local affairs was precisely the growers' goal. Thus, while there were multitudes of strikes across the state of California that were often sparked by the same sets of issues related to wages and working conditions, the growers were keen to represent them all as local problems.

[29]Herod (1998a) provides a review of work in this tradition in geography; see also the collection of essays in Herod (1998b) for other examples.

Chapter 3

[1]Examples of such literature include Kanter (1995), Ohmae (1990, 1995), Bryan and Farrell (1996), Toffler (1980), World Bank (1996), Gingrich (1995), and Gates et al. (1995).

[2]Indeed, in Chapter 8 of this volume I look precisely at how members of a local union in the United States developed transnational networks and a global campaign as a way of matching the global scale of organization of a corporation with which they were engaged in a contract dispute.

[3]In 1995 the ILGWU and the Amalgamated Clothing and Textile Workers' Union, the two unions representing the bulk of unionized garment workers in the United States, merged to form the Union of Needletrades, Industrial and Textile Employees (UNITE).

[4]Several feminist writers within geography (e.g., Hanson & Pratt, 1995) have criticized the historical division within the discipline between "economic" geography (which is often seen to be the realm of men) and "social" geography (which has often been seen to refer to women's realms). In reproducing this division here, I do not mean to endorse it but merely to point out that historically there has been such a division and that the ability of workers to shape the "social" geography of the city through their residential choices has usually been seen to be greater than their ability to shape its "economic" geography (taken here to refer to the location of particular economic activities such as industrial production).

[5]By way of comparison, in 1984 the Lower Manhattan area around Wall Street had a total of 130 million square feet of office space, the City of London had some 60 million square feet, and the entire county of Los Angeles had some 118 million square feet (*Business Week*, July 23, 1984).

[6]As first passed in 1954, Section J-51-2.5 of the Administrative Code of the City of New York provided tax benefits to property owners to upgrade apartments that they rented out in their homes. The 1975 revision redirected those tax benefits to conversion activities rather than upgrading and to large buildings rather than small ones. The revision changed the thrust of Section J-51-2.5 to benefit large property owners and professional developers, rather than small property owners as originally envisioned (Zukin, 1982).

[7]The city determined the characteristics of a loft building most likely to be converted into office space as containing a central elevator core, windows on two sides, and at least 10,000 square feet per floor. Only 3 million of the 8.5 million square feet of apparel production space in loft buildings in the Garment Center lacked these characteristics and was therefore unsuitable for office conversion.

[8]Made up of the Mayor, the City Council President, the Comptroller, and the presidents of the five boroughs of Manhattan, Brooklyn, the Bronx, Queens, and Staten Island, the Board of Estimate functioned at the time as the city's legislative body, controlling taxation, expenditures, and planning. It was found unconstitutional in 1989 because, by giving each borough equal representation through its borough president, it overvalued the influence of less populated Staten Island at the expense of the more populated boroughs such as Queens.

[9]Already, for instance, to reduce expenditures on rents many companies have relocated from southern Manhattan to office space in other New York City boroughs such as Brooklyn and even to parts of northern New Jersey (such as Jersey City, Hoboken, and Weehauken) that have been experiencing a boom in office construction during the past two decades.

[10]It should also be noted that the ILGWU has played a significant role in shaping the geography of housing in Manhattan, beginning in the 1920s with a series of worker housing projects such as the 2,820-unit development (often referred to as

"Penn South") located south of the Garment Center between West 23rd and West 29th Streets and between Eighth and Ninth Avenues (see Vural, 1994, for more on the ILGWU's housing efforts in New York City; see Chapter 7 of this volume for more on trade union housing cooperatives).

[11]Such a belief is illustrated in the comments of New York State Assemblyman (now U.S. Representative) Jerrold Nadler at a public meeting to discuss the establishment of the special zone: "I don't know how anyone reads the future. I do know that there is a difference between being laid off in 20 years or in 20 months. I do know there is a difference between absorbing a job loss of 100,000 or 200,000 over a 20- or 25-year period or a much shorter time" (Nadler, 1987).

Chapter 4

[1]In the industry, the term "East Coast" is usually taken to include all the ports of the North Atlantic, South Atlantic, and Gulf coasts. In the case of the ILA, the term "national" is usually taken to mean the East Coast from Maine to Texas, since the dominant dockers' union on the West Coast is the International Longshore and Warehouse Union, which split from the ILA in 1937 after disagreements with leaders in New York (see Larrowe, 1955, and Kimeldorf, 1988, for more details). Although from a geographical point of view the term "coastwide" is probably more appropriate, its use would be problematic because "coastwide" is usually used in the industry to refer to each of the North Atlantic, South Atlantic, and Gulf coasts separately (i.e., a "coastwide" agreement for the North Atlantic). Unfortunately, the situation is complicated by the fact that less frequently the term is also used to refer to all the ports from Maine to Texas. This is further complicated by the ILA's geographical structure. On the East Coast the union is broken into two districts: the "Atlantic Coast District" (which includes every port in which the ILA represents dockers north of Cape Hatteras, North Carolina, including all Canadian ports along the Atlantic seacoast) and its "South Atlantic and Gulf Coast District" (which includes every port in which the ILA represents dockers south of Cape Hatteras along the Atlantic and Gulf coasts, together with several locals in Puerto Rico). However, although the ILA's "Atlantic District" extends from Virginia into Canada, Canadian locals bargain separately from those in the U.S. East Coast industry. The ILA's two other districts are its "Great Lakes District" and its "Pacific Coast District" (which in actual fact only includes a handful of ports in the state of Washington), which bargain separately from those in the East Coast industry. For the sake of clarification, in this volume I use the term "North Atlantic" to include ports from Maine to Virginia; "South Atlantic" to include ports from North Carolina to Tampa, Florida; "Gulf" to include the remaining ports to Brownsville, Texas; and the terms "national," "industrywide," or "East Coast" to refer to these three coasts collectively.

[2]Cargo generally falls into one of two categories. General cargo consists of items such as machinery, steel products, textiles, paper, meat, sugar, liquor, vehicles and their associated parts, coffee, and bananas. This is in contrast to bulk cargo, which includes petroleum, grain, mineral ores, and molasses. Containerization is aimed at speeding the handling of general cargo, although there has also been mechanization of bulk cargo loading and unloading.

[3]This should not be taken to mean that the history of technological innovation under capitalism is always a monolithic linear progression of labor force deskilling (and this is where Braverman's [1974] analysis is incorrect). Rather, whereas technological innovation may cause deskilling in one sector or industry, concomitantly workers in other sectors (or even, for that matter, in the same sectors) may be *enskilled* (see Willis, 1988, and Jones, 1982). Hence, the rise of computer-aided manufacturing has necessitated that workers learn new skills, and the acquisition of these new skills can empower workers. Likewise, although containerization has reduced dramatically the number of dockers required on waterfronts, the skill required to operate the large gantry cranes and to maneuver containers through the air and onto/ off ships means those dockers who remain in the industry are less easy to replace during strikes, thereby giving them greater bargaining power.

[4]For more on union leaders' corruption, see New York State Crime Commission (1953), Lamson (1954), Larrowe (1955), Hutchinson (1970), Jensen (1974), and Kimeldorf (1988).

[5]See the statement by Alexander Chopin (1963), NYSA Chairman, before the U.S. House Committee on Merchant Marine and Fisheries, hearings on HR 1897, HR 2004, HR 2331, bills to amend the 1936 Merchant Marine Act, 88th Cong., 1st Sess., 1963, in which he states (p. 880) that "we [the NYSA] proposed that the union join with us in reevaluating present restrictive work practices of more than half a century, which have become outdated by new developments in the steamship industry." One example of such traditional practices was those engaged in by locals in the Chelsea section of Manhattan. Whereas most New York locals allowed management to redeploy four members of the gang from the dock into the cargo hold as needed, the Chelsea locals on Manhattan's Lower West Side barred this practice. Although International officers agreed to standardize across the port the practice of allowing employers to redeploy gang members in such a fashion, Chelsea dockers conducted a number of wildcat work stoppages in opposition to the change (*Journal of Commerce*, December 22, 1964, and January 5, 1965).

[6]For example, Thomas Gleason (1955), at the time ILA General Organizer, stated that the union was "not against the advances of machinery or mechanization in the industry [but] we are going to protect our people. . . . Something [will] have to be done to protect the men on the jobs."

[7]The Taft–Hartley Act allows the president to call an 80-day "cooling off" period during labor disputes that threaten the "national security," however defined. During the cooling-off period both sides must continue to work under the provisions of their previous contract and are supposed to work out a new contract. Only one such cooling-off period may be invoked during any particular labor dispute.

[8]Full shipper loads are containers filled with cargo beneficially owned by one shipper or consignee. LCLs are containers that either are not full or are full but consist of the cargo of more than one shipper/consignee (i.e., they are consolidated loads). Often the designation LCL is used in conjunction (and interchangeably) with LTL, a term referring to a road or rail trailer that contains either less than a capacity load or a cargo of more than one shipper/consignee.

[9]The provision stated that (Anon, 1973: 8–9): "[w]here an employer member of NYSA supplies a container, which is the property of such member, to a consolidator for loading or discharging of cargo in the Port of Greater New York, it will

be stipulated that such container must be loaded and unloaded by ILA at long-shore rates."

[10]The Teamsters subsequently challenged the right of the NYSA and the ILA to decide between themselves that the ILA had jurisdiction over this work (see Herod, 1998c, for more details).

[11]Prior to this, containers generally had been carried on the decks of conventional break-bulk ships.

[12]Containers fully loaded by a single shipper, containers arriving in New York for shipment but loaded by ILA members in a different port, and containers traveling to or from points outside the 50-mile limit were all exempt from this provision. The basis for choosing such a distance developed during the 1957–1968 period as a result both of contemporary LTL movement patterns and of the dispersed dock facilities in the Greater New York region. The 50-mile radius, centered on Columbus Circle in Manhattan, was deemed to best represent a basic understanding of the geographic limit of port activities.

[13]The evasion clause read:

> Should any person, firm or corporation, for the purpose of evading the provisions of [the 50-mile rule], seek to change such pattern by shifting its operations to, or commencing new operations at, a point outside said agreed upon geographic area, then either party may raise the question whether said point should be included within the geographic area, and upon agreement that the purpose of the shift in its operations was to evade the provisions of the [50-mile rule], then said point shall be deemed to be within the said geographic area for the purpose of these rules. (*Journal of Commerce*, January 14, 1969)

[14]Given that a large number of NYSA members operated in other ports, the union was able to insist on this provision because it was dealing with many of the same employers up and down the East Coast.

[15]Before being able to pass the penalties and extra costs of stripping and restuffing onto the consolidators, the ocean carriers would have to file new tariffs, a potentially lengthy process since these would have to pass muster with the Federal Maritime Commission. Consequently, in the meantime the carriers themselves would have to bear the cost of such stripping and restuffing (see Jacobs, 1973).

[16]The NYSA and the ILA did not attempt to claim under the Rules the right to regulate the small number of containers owned by manufacturers or consignees themselves (see Federal Maritime Commission, 1987: fn. 9).

[17]After a strike in 1975, the 30-day warehouse exemption was eliminated as it related to export cargo and tightened with regard to import cargo (*American Trucking Associations, Inc. v. National Labor Relations Board* [1984]).

[18]Although the restriction was not strictly limited to consolidators with facilities less than 50 miles from the port [Section 1(e) of the Rules merely stated that carriers were not to supply containers to "any consolidator or deconsolidator"], in 1980 the U.S. Supreme Court subsequently interpreted the prohibition as applying only to those consolidators whose terminals were located within the 50-mile area (see *National Labor Relations Board v. International Longshoremen's Association* [1980], and Federal Maritime Commission, 1987: fn. 5).

[19]A qualified shipper was defined in the Rules as a "manufacturer or seller hav-

ing a proprietary financial interest (other than in the transportation or physical con-solidation or deconsolidation) in the export cargo being transported and who is named in the dock/cargo receipt." A qualified consignee was defined as a "pur-chaser or one who otherwise has a proprietary financial interest (other than in the transportation or physical consolidation or deconsolidation) in the import cargo be-ing transported and who is named in the delivery order" (*National Labor Relations Board v. International Longshoremen's Association* [1980: 514]).

[20]In *Container News* (1969) such rehandling costs were estimated at approxi-mately $250–$300 per 20-foot container. John Bowers, Executive Vice President of the ILA, in a written response to questions posed during hearings conducted by the House of Representatives' Subcommittee on Merchant Marine in 1983, stated that the goal of the Rules was not to encourage double handling of containers. Bowers ar-gued that the ILA required all export cargo subject to the Rules to be brought to the piers loose, whereupon it would be handled by ILA labor. Containers of import cargo would be stripped once at the piers and the contents delivered loose to the consignees. Double handling would only be required if the Rules were violated and an export container not stuffed by ILA labor were delivered to the port (see Bowers, 1983)

[21]Though they generally opposed what they viewed as union "feather-bedding" practices, many stevedores in various port employers' associations supported the ILA on this point because they, too, wanted to keep work at the piers rather than see it migrate to inland warehouses. (Had work migrated inland, the stevedores, who are the ones who employ the dockers and are contracted by the steamship companies to load/unload their ships, would also have lost out to the consolidators.) Other mem-bers of port employers' associations supported the ILA out of a fear that, if the courts overturned the Rules, the docks would continually be tied up by lengthy strikes as the ILA searched for other ways to preserve work.

[22]Significantly, similar conflicts have arisen in the British Transport and General Workers' Union between dockside workers and inland TGWU members (*Journal of Commerce*, June 16, 1972).

[23]In New York, for instance, ILA members had historically brought import cargo from a ship's hold and placed it on the tailgate of a waiting truck but had not moved the cargo into the truck's interior. In Philadelphia dockers had merely brought cargo to the truck's side, whereupon the driver had loaded the trailer. Yet another system existed in Hampton Roads, Virginia, where dockers had moved cargo from the ship's hold all the way into the interior of the truck trailer (American Trucking Associations, no date; Groom, 1965).

[24]The relevant citations are: for Consolidated Express, *Balicer v. International Longshoremen's Association and New York Shipping Association*, 364 F.Supp. 205 (1973); for Twin Express, *Balicer v. International Longshoremen's Association and New York Shipping Association*, 86 LRRM 2559 (D. NJ 1974). The NLRB also subse-quently determined that the Rules were illegal in the ports of Norfolk, Virginia, and Baltimore, Maryland.

[25]Secondary boycotts against employers who are not party to the collective bar-gaining agreement are prohibited by sections 8 (b)(4)(B) and 8 (e) of the National Labor Relations Act (29 U.S.C. 158 [b][4][B] and 158 [e]). The present form of these statutes was enacted as part of the 1959 Landrum–Griffith Act (73 Stat. 542) and was

intended to eliminate the type of collusive boycott known as "hot cargo" clauses, under which an employer relinquishes the freedom to handle or provide goods and services to other businesses whom the union has designated. Section 8 (e) of the NLRA provides in part that

> [no] labor organization [or] any employer [shall] enter into any contract or agreement, express or implied, whereby such employer ceases or refrains or agrees to cease or refrain from handling, using, selling, transporting or otherwise dealing in any of the products of any other employer, or to cease doing business with any other person, and any contract or agreement entered into heretofore or hereafter containing such an agreement shall be to such extent unenforceable and void.

Despite this provision, the U.S. Supreme Court has ruled that collective bargaining agreements that have secondary impacts on neutral employers who are not party to the primary agreement are nevertheless lawful, as far as labor laws are concerned, providing the union has no forbidden secondary purpose to affect the employment relations of the neutral employer and that the collective bargaining agreement constitutes a primary bona fide work preservation practice (*National Labor Relations Board v. Pipefitters*, 1977).

[26]See Appendix to *National Labor Relations Board v. International Longshoremen's Association*, 447 U.S. 490 (1980), for the full text of the Rules, as amended in 1974.

[27]This conflict and its discursive representation are recounted in more detail in Herod (1998c).

[28]See *National Labor Relations Board v. International Longshoremen's Association*, 447 U.S. 490 (1980), p. 491.

[29]See *National Labor Relations Board v. International Longshoremen's Association*, 473 U.S. 61 (1985).

[30]The Rules were questioned under Sections 15, 16, and 17 of the Shipping Act. Section 15 stipulates that all agreements regulating carrier tariffs must have the Commission's prior approval. (The ILA and the NYSA claimed the Rules were entitled to an automatic labor exemption from this requirement because they were the result of a collectively bargained agreement under the National Labor Relations Act and that, consequently, the FMC lacked jurisdiction over them.) Section 16 makes it unlawful for a carrier to "give any undue or unreasonable preference or advantage to any particular person, locality, or description of traffic . . . or to subject [them] to any undue or unreasonable prejudice or disadvantage" (46 U.S.C. 815, as amended [1982]). Section 17 of the Act states that carriers engaged in foreign commerce must "establish, observe, and enforce just and reasonable regulations and practices" relating to freight transportation (46 U.S.C. 816, as amended [1982]).

[31]For more detail, see *Council of North Atlantic Steamship Associations v. Federal Maritime Commission*, 672 F.2d 171 (D.C. Cir., 1982), *cert. denied*, 459 U.S. 830 (1982), and *New York Shipping Association v. Federal Maritime Commission*, 854 F.2d 1338 (D.C. Cir. 1988), *cert. denied sub nom. International Longshoremen's Association v. Federal Maritime Commission*, 57 U.S.L.W. 3486 (US January 23, 1989).

[32]A. C. Novacek, Executive Vice President of Seatrain Lines Container Division, testified that, although Seatrain had several times considered merging its stripping

and stuffing operations at a single waterfront terminal located in Port Newark, the deciding factor that led the company to close its off-pier terminals had been the implementation of the Dublin Agreement (Novacek, 1973).

[33]They were able to do this because Florida is a right-to-work (RTW) state that does not require all members of a unionized facility to be members of the union. Whereas in ports such as New York the ILA had won the right to represent dockers and all employers had to recognize this, in southern ports where RTW legislation exists there are many nonunion operators. Indeed, the fact that every coastal state from Virginia to Texas is a RTW state has posed great problems for the ILA's efforts to develop national strategies to deal with issues such as containerization, an issue that is dealt with in more detail in Chapter 5.

[34]One main difference between the East Coast and West Coast version of the 50-mile rule was that in West Coast ports the rule was only applied to members of the employer's association (the Pacific Maritime Association) and any other company with which they subcontracted. Unlike in East Coast ports, in West Coast ports there was no restriction against ocean carriers providing their containers to other, non-PMA consolidators (*Journal of Commerce*, June 26, 1989).

Chapter 5

[1]Scotto is quoted in *Journal of Commerce* (March 22, 1965: 24); Gleason is quoted in Gleason (1955: 878).

[2]Much of the information for this analysis comes from various newspaper reports and press releases, particularly the *Journal of Commerce*, a trade newspaper. Unless specifically warranted, I have not included these as references in the text. Interested readers can consult Herod (1992) for more detailed references.

[3]Mers (1988: 136) makes mention of the animosity toward the New York leadership felt by some southern dockers even many decades later.

[4]As one commentator has said in this context, "People in the South in general hate New York, so why would shippers be any different?" (Storey, 1991).

[5]For more on this history and these practices, see New York State Crime Commission (1953), Rosenbaum (1954), Larrowe (1955), Hutchinson (1970), Jensen (1974), and Kimeldorf (1988).

[6]The struggles between the two unions to represent waterfront workers were themselves largely struggles over issues of geography. Before a vote could be taken to determine which union workers preferred as their representative, the NLRB had to decide what was the appropriate geographical voting unit. The old union argued for a single unit stretching from Maine to Texas, since this would allow it to include many of its supporters outside New York. The new union argued for a smaller unit focused on New York, where it received most of its support. In the end, the NLRB ruled that the appropriate unit was indeed the smaller one (see Herod, 1997d, for more details).

[7]The May 1954 election had been ordered by the NLRB because of complaints by the AFL that the earlier one had been rife with fraud and intimidation of AFL-ILA supporters (see Herod, 1997d).

[8]Statement by Louis Waldman, attorney for the ILA-IND, quoted in *Journal of Commerce*, November 27, 1956.

[9]This was a legitimate fear because, although it was resoundingly beaten in a third representation election held in New York in October 1956, the IBL had plans to challenge the ILA-IND a fourth time. The ILA-IND's fear of this fourth challenge prompted union leaders to seek a peace proposal with the Seafarers' International Union and the International Brotherhood of Teamsters (both of whom had earlier backed the IBL) that would settle a number of jurisdictional disputes and so eliminate much of the IBL's support (*Journal of Commerce*, January 7 and January 14, 1957).

[10]This situation was further complicated by the fact that while negotiations were still taking place in the North Atlantic, in the South Atlantic and Gulf ports a regional agreement had been reached based on a formula developed in New Orleans that, although lower in overall monetary terms, would actually provide a higher wage increase than in New York. Port of New Orleans dockers were offered a three-year agreement that provided a 31 cents an hour monetary boost, of which 27 cents was payable in wages, whereas in New York the employers had offered dockers a package worth 40 cents an hour, but of which only 14 cents would be applied to a raise in the basic wage rate (*Journal of Commerce*, January 31, 1957). This disparity led Bradley to insist that the NYSA match the southern ports' wage offer or face further delay in settling the New York contract, something the Association ultimately did with much reluctance.

[11]Although the master contract provided for uniform contributions, it did allow each employers' association to offer different welfare and pension benefits. The contract also allowed the ILA-IND to cancel the agreement at the end of the second year if the Independent were challenged in a NLRB election by any rival union (*Journal of Commerce*, February 18, 1957).

[12]Secondary boycotts are disputes in which one local union that is not a signatory to a particular contract strikes in support of another that is. By developing master contracts implemented through the International union but to which local unions are signatories, however, unions can evade such legal proscriptions because different local unions are now signatories of the same regional or national contract and so can strike in support of one another, assuming that any strikes relate to the provisions of the master agreement. Because of the 1947 Taft–Hartley Act's prohibition of secondary boycotts, then, between 1947 and 1957 ILA members in, for example, Philadelphia risked litigation and penalties if they went on strike in support of dockers in, say, Hampton Roads who were governed by a separate contract that applied only between the union and the employers in Hampton Roads—although in practice the union often ignored this prohibition. Only after a master agreement had been negotiated was the union safe from such litigation.

[13]The craft-based AFL and the industrially organized CIO had merged in 1955.

[14]The union sought a GAI equivalent to 1,600 hours' pay. The employers, however, were only prepared to concede a guarantee based on 75% of a docker's average annual earnings during the preceding two years, a formula that for most would fall below the union's proposal (*Journal of Commerce*, December 16, 1964).

[15]DiFazio (1985) provides a fascinating ethnographic account of how some

Brooklyn dockers on the GAI have spent their time when there was no work for them to do.

[16]Philadelphia dockers secured a GAI of 1,300 hours. In Baltimore dockers were unable to force their employers to concede a GAI, while in Boston ILA officials came to an oral agreement to discuss a GAI after the port returned to work (Department of Labor, 1980). (Dockers in Baltimore and Hampton Roads finally negotiated GAIs during talks for the 1968–1971 contract.) West Gulf employers agreed to establish minimum gang complements consisting of 18 dockers on bagged cargo, 16 on general cargo, and 14 on containerized and prepalletized cargo. Despite the ILA's objection, the employers gained the right to count the gang boss and drivers as part of the regular complement rather than have them included as additional gang members, as the union wished. In the South Atlantic dockers had pressed for a minimum of 18 dockers on a general cargo gang (exclusive of gang boss and drivers) to match the levels in New Orleans (which set the pattern for the East Gulf), although they, too, ultimately settled for terms similar to the West Gulf (*Journal of Commerce*, March 3, 1965).

[17]The members of the CONASA were the NYSA, the Steamship Trade Association of Baltimore, the Philadelphia Marine Trades Association, the Boston Shipping Association, the Hampton Roads Shipping Association, and the Rhode Island Shipping Association (RISA). The NYSA had not previously been authorized to bargain for Providence employers, and the CONASA agreement brought RISA into the master contract for the first time. The CONASA voting structure, based on port size and other considerations, was as follows: New York controlled 40% of the votes; Boston 10%; Providence 5%; and Philadelphia, Baltimore, and Hampton Roads 15% each (Dickman, 1973).

[18]LASH is a system in which freighters carry preloaded barges that they drop off or pick up in their ports of call, thus relieving the freighters themselves of having to wait in port to be loaded and unloaded with cargo.

[19]Unlike in the North Atlantic, where benefits were worked out on a port-by-port basis, the SAENC contract covered all ports in the region under a single GAI agreement. The 1,000 hours was a compromise: the union had wanted 1,400 hours, the employers a GAI set at 700 hours (*Journal of Commerce*, March 7, 1972).

[20]This was particularly so because the CONASA argued that the Rules' nullification in these ports indicated they were untenable in others. The union, on the other hand, maintained that since they had only been specifically outlawed in these three ports they were still valid elsewhere.

[21]The federal Taft–Hartley Act allows individual states to pass right-to-work legislation outlawing the union shop (in which workers hired at a unionized facility are required to become union members within a certain time period or risk being fired at the discretion of the union). Such laws have encouraged the growth of nonunion operations and made it harder for unions to organize workplaces.

[22]The participating employers' groups were the NYSA, the CONASA, the West Gulf Maritime Association, the Southeast Florida Employers' Association, the Mobile Steamship Association (MSA), and the South Atlantic Employers' Negotiating Committee. Of these groups, the SAENC and the MSA took part in management caucuses and were consulted on contract issues but, unlike the other employers' groups, did not participate in direct across-the-table negotiations (*Journal of Commerce*, May 28,

1980; also Dickman, 1980: 33). Only the New Orleans Steamship Association did not attend the talks.

[23]Although basic wages were the same throughout the industry, the cost of shipping a 40-foot container through New York was estimated to be about $250 higher than shipping through Baltimore because of the former's higher GAI expenses. In North Atlantic ports all dockers received six weeks' vacation pay annually, whereas dockers in New Orleans, for instance, received vacation pay based on the number of hours they had worked during the preceding year (*Journal of Commerce*, January 17, 1984).

[24]Many commentators argued that Gleason's strategy was designed to favor North Atlantic ports by ensuring that South Atlantic and Gulf ports had to fund a coastwide GAI. If the GAI remained a local provision, they argued, northern ports and employers would be disadvantaged because containerization had displaced many more jobs in those ports, and so dockers would be more likely to draw from any GAI fund. Making the GAI coastwide would at least ensure that southern employers were likewise burdened with funding the scheme.

[25]As was the case with any coastwide GAI, many southern dockers felt that the JSP disproportionately benefited northern ports. Ports in the South were less heavily containerized and more likely to handle noncontainerized bulk goods (e.g., agricultural wares) than were northern ones. Thus, dockers in northern ports were more likely to be displaced by containerization—and so more likely to draw funds from the JSP—than were southern dockers, yet the assessments southern employers had to pay into the JSP made them less competitive in the face of nonunion labor, a problem northern ports did not really face because there were few, if any, nonunion operations.

[26]When first established, the Carriers Container Council was called the Container Carriers Council. The Council was initially created as an unofficial, informal body (a part of the NYSA) after the creation of the Rules on Containers. First mentioned in the 1972 master contract agreement, the Council acted as a body to make known the various positions and interests of container carriers in collective bargaining. From 1977 on, the Council was a key player in collective bargaining negotiations.

[27]The abolition of these ports' GAIs was not expected to have much impact on the lives of most dockers, since ports in this region of the coast were fairly active.

[28]This is illustrated by comments made by ILA General Organizer (and later International President) Thomas Gleason, who stated that the union was "not against the advances of machinery or mechanization in the industry [but] we are going to protect our people. . . . Something [will] have to be done to protect the men on the jobs" (Gleason, 1955) and that "when the men first saw it [containerization] they didn't want to work it *but some of us* convinced them that this was progress and that we should do it as long as we shared in it" (Gleason, 1980b, emphasis added).

Chapter 6

[1]In the United States the National Union of Labor was in fact established in 1860, but this was a relatively short-lived body, lasting only until 1866. The influence

of the Knights of Labor, founded in the United States in 1869, was also relatively brief. The TUC is generally considered the earliest permanent national trade union federation. Similar federations were subsequently set up in other countries, including the AFL in the United States (1886), the Unión General de Trabajadores (1888) in Spain, the General Committee of Trade Unions (1890) in Germany, the Nationaal Arbeids-Secretariaat (1893) and the Nederlandsch Verbond van Vakvereenigingen (1906) in the Netherlands, and the Canadian Labour Union Congress (1883) (van der Linden, 1988).

[2]The precise number of international trade secretariats that had been formed prior to the outbreak of war in 1914 is disputed by a number of historical accounts. Busch (1983: 15) lists the number at 33, whereas a publication of the International Confederation of Free Trade Unions (ICFTU, 1957: 19) lists the number as 28, 24 of which were headquartered in Germany. Windmuller (1980: 23) states that by 1914 "there were almost 30 International Trade Secretariats."

[3]The current secretariats are: the Education International (EI), the International Federation of Chemical, Energy, Mine and General Workers' Unions (ICEM), the International Federation of Building and Wood Workers (IFBWW), the International Federation of Journalists (IFJ), the International Metalworkers' Federation (IMF), the International Textile, Garment and Leather Workers' Federation (ITGLWF), the International Transport Workers' Federation (ITF), the International Union of Food, Agricultural, Hotel, Restaurant, Catering, Tobacco and Allied Workers' Associations (IUF), the Public Services International (PSI), the Union Network International (UNI), and the Universal Alliance of Diamond Workers (UADW).

[4]Inspired by Pope Leo XIII's encyclical *Rerum Novarum*, Christian trade unions had been organized in several European countries since the 1890s to counter the growth of socialist, anticlerical unions. The CISC was founded in 1920 as an explicitly Christian entity. However, during the 1960s it underwent a dramatic secularization as virtually all mention of Christianity was dropped from its basic constitutional documents and the organization renamed itself the World Confederation of Labour. These changes were part of an effort to reflect a more interdenominational basis for international trade union organizing. Furthermore, whereas the CISC was initially established for the purpose of opposing socialist domination of earlier international labor organizations, in the 1970s the reborn WCL adopted a program that rejected capitalism and argued for the socialization of the means of production and workers' control of industry.

[5]These agencies are not dealt with in this chapter.

[6]One of the earliest indications of interest in international affairs was the adoption of resolutions at the AFL's 1887 National Convention condemning British policies in Ireland and the persecution of Jews in Russia.

[7]Although these three federations are not the only U.S. trade union organizations to have operated internationally (e.g., see Dubovsky, 1969, on the Industrial Workers of the World), they have been by far the most influential, and so it is their activities with which this chapter concerns itself.

[8]The ILGWU merged in 1995 with the Amalgamated Clothing and Textile Workers' Union to form the Union of Needletrades, Industrial and Textile Employees (UNITE).

[9]Indeed, unless explicitly stated otherwise, this chapter refers primarily to the

activities of the labor bureaucrats who make foreign policy, and in that sense it is largely an institutional account of U.S. labor's international activities.

[10]This self-interest was made abundantly clear by John L. Lewis of the United Mine Workers of America in a 1939 Labor Day address when he stated that "Central and South America are capable of absorbing all of our excess and surplus commodities" (quoted in Scott, 1978a: 201).

[11]For more on this, see Barry and Preusch (1990), Hardy (1936), Larson (1975), Morris (1967), Radosh (1969), Roberts (1995), Scott (1978a), and Simms (1992).

[12]Scott (1978a) suggests that, despite acting as the AFL's spokesperson in the Anti-Imperialist League, Gompers did support U.S. territorial annexation, provided that the subsequent incorporation of large numbers of low-paid workers into the U.S. economic system did not threaten the living standards of AFL members. Thus, in reference to the war with Spain and the annexation of the Philippines, Gompers declared that it was vital to save "American labour from the evil influence of close and open competition of millions of semi-barbaric labourers in the Philippine Islands" but that "the government may annex any old thing as long as the laws relating to labour are observed" (quoted in Scott, 1978a: 93–94).

[13]For example, in a 1965 speech to the Council for Latin America, AFL-CIO President George Meany stated: "We believe in the capitalist system. . . . We are dedicated to the preservation of this system, which rewards the workers, which is one in which management also has such a great stake. . . . It is, perhaps, not a perfect device, but it is the best that the world has ever produced, and it operates effectively in a free society. We are not satisfied, no, but we are not about to trade in our system for any other. We are content to make our methods work a little bit better in behalf of the people" (quoted in *AIFLD Report*, April 1966c: 4 and 6).

[14]Gompers believed that the term "Federation" implied a more loosely organized body than did the word "Secretariat," a semantic shift that more closely followed his voluntaristic philosophy that the AFL not be obliged to follow policies with which it did not agree, even though they had been adopted by the international labor body. He argued instead that each country's labor movement should be free to "decide upon its own policy, tactics, and tendencies" (Gompers, 1910: 131).

[15]The Federation rejoined the IFTU in 1937. Whereas publicly the AFL leadership portrayed its actions as a response to fascist intimidation of European unions, the growing challenge to the craft-based Federation's domestic political hegemony mounted by the industrially organized CIO was perhaps a more crucial shaper of the leadership's attitudes toward reaffiliation. In particular, IFTU rules permitted membership by only one trade union federation from each country. For the AFL, reaffiliating with the IFTU would shut out the nascent CIO and thereby preserve the Federation's right to speak as the sole voice of U.S. workers in this international arena.

[16]Roberts (1995) argues that the AFL's turn toward isolationism in the 1920s and 1930s was a reflection both of the wider concern in the United States that "foreign entanglements" had led the country into war and that they should thus be avoided in the future, and of the AFL's "economic nationalism," which argued that the problems of the Depression would be best served by focusing on national recovery policies. Hence, the AFL not only opposed imports as a danger to U.S. workers' jobs, but it also opposed exports, at least to the extent that these diverted support from what

the Federation considered the more crucial campaign for a 30-hour workweek that would generate work for its members (AFL leaders argued that an increase in exports would allow for surplus commodities in the United States to be absorbed abroad without reducing working hours domestically; a 30-hour week, rather than a workweek of 40 or more hours, they argued, would allow more workers to be hired and would thus solve the problem of unemployment).

[17]The French Confédération Générale du Travail (CGT) and the Italian Confederazione Generale Italiana del Lavoro (CGIL), both Communist-dominated, remained affiliated with the WFTU until the 1990s, when they both left the Federation. MacShane (1992) argues that the split of the WFTU had less to do with the emergence of the Cold War than with longer-standing differences within the trade union centers that made up the Federation, a position vigorously critiqued by Howard (1995: 374), who argues that, in fact, "MacShane's thesis that the Cold War was not the fundamental cause of the WFTU split simply falls flat on its face."

[18]Support for the Marshall Plan was a point of much contention in the CIO and led to a split between the Congress's left- and right-wing factions. However, after the expulsion of 11 more radical unions in 1949 and 1950, the CIO lined up solidly behind the Plan.

[19]For more on other labor links with the CIA, see Langley (1972) and Morris (1967).

[20]Although this split had been a long time in the making, the actual catalyst for it was the overtures made by the United Auto Workers concerning the possibility of joining the ICFTU as a trade union unaffiliated with its national center. The UAW had recently left the AFL-CIO over the latter's hawkish policy toward the war in Vietnam. The AFL-CIO leadership opposed the UAW's (ultimately unsuccessful) efforts and was outraged that the Confederation would even entertain such an idea.

[21]Although primarily concerned with Europe, the FTUI also served as an umbrella organization to distribute governmental funds to the other three institutes. The FTUI was the largest single recipient of funds from the National Endowment for Democracy, a Reagan-initiated body for strengthening pro-U.S. forces abroad, and between 1984 and 1988 received almost 50 percent of the dollar total of the Endowment's grants (Simms, 1992). In 1997 the AFL-CIO's four regional institutes were merged into a single office, the American Center for International Labor Solidarity.

[22]In 1992 some 63% of all U.S. investment in the global south was in Latin America and the Caribbean (U.S. Department of Commerce, 1994).

[23]Representatives from four Caribbean islands were also present at the PAFL's founding convention.

[24]As it turned out, the signing of the Armistice in Europe two days before the PAFL's founding conference began undermined this secret aim.

[25]Gompers himself described the PAFL as "based upon the spirit of the Monroe Doctrine" (quoted in Simms, 1992: 37).

[26]See Barbash (1948) for examples of the tone of CTAL's language; also, Radosh (1969: 360–368).

[27]For more on the rivalry between the AFL and the CIO in Latin America and in Canada, see Roberts (1995) and Scott (1978b).

[28]The ORIT is now the ICFTU's regional organization for Latin America and the

Caribbean. For a fairly uncritical account of early ORIT activities, see Hawkins (1965).

[29]Although the AFL-CIO denies that the AIFLD effectively served as a branch of the CIA, several ex-CIA operatives and congressional hearings have provided ample evidence to the contrary. According to one ex-operative: "The real purpose of [the AIFLD] was to train cadres to organize new trade unions or to take over existing ones, in such a way that the unions would be controlled, directly or indirectly, by the CIA" (Philip Agee, quoted in Barry & Preusch, 1990: 6). During the 1980s, for example, the Reagan-initiated National Endowment for Democracy allegedly channeled several million dollars to the AFL-CIO for covert operations in Nicaragua and El Salvador. For an outline of the AIFLD's goals, at least as publicly presented, see Beirne (1970).

[30]AFL-CIO President George Meany, in a speech before the National Press Club in 1967 (reported in *AIFLD Report*, August 1967a), said the AIFLD had included business representatives

> on the theory that they [businesses] should have the same stake, that they should have the same interest in the building of free societies in Latin America as we [the AFL-CIO] have. . . . They want to do business there, they certainly want to do business with countries that have viable economies and we feel that you can't have a successful economy unless you have the cooperation of all segments of the society and especially those who are the most important assets insofar as production is concerned.

Only in 1981, after nearly two decades and continued criticism of this tripartite arrangement, was membership of the Institute's Board limited to members of the U.S. and Latin American trade union movements. Nevertheless, several of the highest AIFLD officials have been members of the Council of Foreign Relations and the Trilateral Commission, two bodies which are major players in formulating U.S. foreign policy. The Institute has also worked closely with the Council for Latin America, Inc., a business group founded by David Rockefeller that seeks to further U.S. corporate interests in the region.

[31]Meany described the AIFLD's Social Projects Department (which is the focus of Chapter 7) as "in a sense a working agency of the United States government—working with its expenses paid for by AID [the United States Agency for International Development] . . . and used to channel American aid under the Alliance [for Progress] to union-sponsored projects in Latin America" (*AIFLD Report*, August 1967a: 2).

[32]This was later transferred to Georgetown University in Washington, DC.

[33]Barry and Preusch (1990) provide an excellent summary of AIFLD activities in Guatemala, Nicaragua, El Salvador, Honduras, and Costa Rica. For other examples, see the autobiography of Serafino Romualdi (1967), who played a very active role in the AIFLD and in implementing the AFL's and the AFL-CIO's Cold War foreign policy more generally. In the case of Brazil, AIFLD-trained communications workers played an active role in the military coup that overthrew João Goulart. As Eugene Methvin (1966: 28) of the *Reader's Digest* noted, "The Communists called a general strike, with emphasis upon communications workers. But to their dismay, the wires kept

humming, and the army was able to coordinate troop movements." The Brazilian telephone workers' union cooperative later received a small loan from AIFLD to help construct a new building (*AIFLD Report*, December 1967).

[34]The U.S. withdrawal from the ILO came largely as a result of pressure from the AFL-CIO, which accused the Organisation of pandering to Soviet foreign policy and failing to criticize strongly enough Soviet human rights' abuses. In fact, the Ford administration had made clear its intent to withdraw two years previously, as required by ILO bylaws. Thus, to all intents and purposes, AFL-CIO activities in the ILO had effectively ended sometime earlier. The United States rejoined the ILO in 1982.

[35]Both the WFTU and the WCL had their own version of trade secretariats organized on the basis of industrial sector, but with the end of the Cold War the WFTU lost many of its former members and much of its influence, particularly in Eastern Europe. Some former WFTU affiliates have joined ITSs, raising issues of the geographical expansion of the ICFTU and the ITSs into the countries of the former Soviet bloc (see Herod, 1998d, 1998e).

[36]The ICEF merged in 1995 with the Miners' International Federation (MIF) to form the International Federation of Chemical, Energy, Mine and General Workers' Unions (ICEM). For more on the ICEM and its goals, see ICEM (1996).

[37]The periodical *Labor Notes* frequently carries information concerning such international campaigns in its "Solidarity Network" section. *Labor Notes* is published monthly by the Labor Education and Research Project, 7435 Michigan Avenue, Detroit, MI 48210 (tel: 313-842-6262/e-mail: labornotes@labornotes.org).

[38]Olle and Schoeller (1977) have, in fact, suggested that attempts to promote labor solidarity on the basis of reducing wage differentials is destined to failure because, in the final instance, this policy will always degenerate into competition between workers in different parts of the globe to keep their (lower) wage advantage in exchange for the promise of continued employment. On the basis of such belief, for example, during the mid-1970s the Shop-Stewards' International Steering Committee at the Dunlop-Pirelli World Industry Council decided that pursuing wage parity was not simply an unrealistic goal but that it was, in fact, a divisive issue internationally. The Committee consequently argued against seeking international wage contracts but sought, instead, parity and improvements on a number of nonwage issues.

[39]The declaration (reproduced in IMF, 1991: 26) adopted by the WACs in Detroit on June 3, 1966, stated:

> Without neglecting the specially urgent problems that exist in specific countries and companies, we agree upon the need for coordinated worldwide concentration by the IMF affiliated organizations upon these problems that are of high priority:
>
>> —full recognition of the right to organize and the right to bargain collectively
>> —standardization of wages and social benefits at the highest level made possible by technological progress
>> —immediate introduction of paid recreation and rest periods
>> —ending of excessive overtime working and the introduction of adequate allowances for overtime
>> —payment of additional vacation allowance

—adequate retirement pensions

—guaranteed income for workers affected by production fluctuations and technological changes

—reduction of working hours without loss of earnings, more paid holidays, longer vacations, and earlier retirement.

[40]Though they are organizationally independent, the ITSs are themselves closely linked with the ICFTU.

[41]I do not want to give the impression that the spirit of internationalism has broken out in all corners of the AFL-CIO. Nevertheless, there is a growing movement within the Federation that is seeking to develop international links to pursue progressive policies of solidarity with workers abroad.

[42]Examples include Barnet and Muller (1974), Dixon et al., (1986), Dunning (1971, 1988), Radice (1975), Hymer (1971, 1979), Palloix (1973), Hennart (1982), Peet (1982), Taylor and Thrift (1982), Bornschier and Chase-Dunn (1985), Langdale (1985), Lipietz (1986), Fryer (1987), Jenkins (1987), Harvey (1988), Donaghu and Barff (1990), Schoenberger (1990), Thrift and Leyshon (1991), Daniels and Lever (1996), and Dicken (1998).

Chapter 7

[1]The quote attributed to the Earl of Shaftesbury comes from Porter (1994: 271). The slogan from the Dominican housing project appeared in *AIFLD Report* (January 1966b: 1). The quote attributed to Bill Levitt comes from an interview with Eric Larrabee (1948: 84) in the September 1948 issue of *Harper's Magazine*.

[2]The AIFLD's involvement began in the early 1960s and continued through to the late 1970s, when the global increase in interest rates made AIFLD's involvement no longer feasible (Doherty, 1997). However, local unions in the region had long been involved in providing housing for their members and have continued to do so after AIFLD's involvement ended.

[3]As the Institute's 1964 Country Plan for Bolivia stated (AIFLD, 1964b: iv, emphasis added):

> The . . . projects emphasized are specific impact programs for a number of non-Communist unions and cooperatives *in key geographic sections of the country*. These projects are also of a high priority nature. Since the benefits of Bolivian development and foreign aid have not directly reached the organized workers, despite their demands, we have chosen a number of strategic, low input, impact projects in the labor and cooperative areas as a means of correcting this situation. These projects will gain a great deal of free trade union sympathy even though their costs are low.

[4]There is a huge literature on how spatial engineering has been used to pursue social goals, by those both on the left and on the right. During the past three centuries many socialist, feminist, radical, religious, and utopian thinkers have advocated transforming spatial relations as a means of transforming the social relations of everyday life. One of the earliest such schemes was that put forward by the utopian socialist Robert Owen with his model mill community at New Lanark in Scotland in

which the community's physical layout was designed to improve the moral, economic, and social standing of its inhabitants. At about the same time, the French utopian socialist Charles Fourier had begun to sketch plans for cities designed according to humanist principals in which the physical environment would help shape the new social order (Frampton, 1969). Hayden (1976) argues that between 1820 and 1850 utopian settlements sprouted up in dozens of locations across the United States as abolitionists, prison reformers, labor unionists, feminists, and others engaged in programs to redesign society through a restructuring of the physical landscape of both city and countryside (see also Berry, 1992). Many feminists in particular saw the redesign of the physical layout of the home and of the city—especially the socializing of housework through the building of kitchenless houses, public daycare and kitchen facilities, and community dining halls—as central to liberating women from domestic drudgery. In the Soviet Union in the 1920s there were vigorous debates over urban planning and the form the new socialist landscape would have to take if it were to replace that of the old order and herald into being a rational and humane classless society (see Bliznakov, 1993; Hudson, 1994; Ruble, 1990; Ades et al., 1995). At less grand scales, Victorian social reformers in Scotland saw the physical constraints placed upon tenement dwellers not only as detrimental to their corporeal and mental health but also deleterious to their moral character (Rodger, 1986), although Benjamin Richardson's intention to solve the problem of "gutter children" in his 1876 plan for "Hygeia" (a "City of Health") by designing streets without gutters perhaps took the notion that social ills could be solved by spatial engineering a little too far (Rubinstein, 1974). However, not all such plans were for progressive purposes. Company towns such as Pullman, Illinois (established by George Pullman for the workers that made his railroad cars), Bourneville, England (established by George Cadbury of chocolate fame), and Potlatch, Idaho (established by the Potlatch Lumber Company for its workers), all used spatial planning to control what their inhabitants might do. Equally, the Nazis were well aware of the power of architecture and spatial planning in pursuing ideological goals, and it is perhaps no accident that Albert Speer, Reich Minister for Armaments, was an architect by training who planned to redesign Berlin in such a way as to "encourage discipline in the city's inhabitants and, moreover, to remind them of their 'heroic' past through [the] 'heroic' scale" of the new buildings and roadways he planned (Warner, 1983: 76; see also Helmer, 1985; Speer et al., 1985; and Thies, 1983). In Italy, Mussolini similarly planned to reconstruct the physical fabric of Rome and other cities as part of an effort to invoke associations between his fascist party and imperial Rome of the Caesars, particularly Augustus (Doordan, 1983; Ades et al., 1995). For more on other efforts to use such spatial engineering for purposes either of social liberation or social control, interested readers might consult the following: Mumford (1959 [1922]), Hayden (1976, 1981, 1984), Brauman (1976), Benevolo (1967), Meakin (1985 [1905]), and Kaufmann (1879) on a number of pre-twentieth-century utopian and model town and village schemes; Howard (1898) on the creation of "garden cities" designed to transcend the contradiction between town and country and bring about "an environment in which capitalism could be peacefully superseded" (Fishman, 1977: 65); Le Corbusier (1931: 289), who maintained that "revolution can be avoided" through good city planning; Holston (1989) on the construction of Brasilia as a "model city"; Lynch (1979/80) and Segre et al. (1997) on the remodeling of towns according to socialist principles in postrevolutionary Cuba; Buder (1967) on Pullman, Illinois; Cadbury (1912) on Bourneville; Rössler (1994) on the Nazis' Reich Office for Spatial Planning (*Reichsstelle für Raumordnung*) established to plan the Germanification of the East and create a spatial form that would facilitate the development of the *Volksgemeinschaft* ("People's Community") as part of the effort to create a sense of Aryan community based upon National Socialist ideology; and Sofsky (1993) on how concentration camps were laid out geographically so as to exert maximum social control and "efficiency" of operation.

[5]The cooperative's 1,450 member families were divided equally between the ACWA, the International Ladies' Garment Workers' Union (ILGWU), and other local unions.

[6]Such success spurred on other U.S. unions. In the 1950s, the ILGWU sponsored the "ILGWU village" housing project in the Lower East Side (Voorhis, 1964), while Local 3 of the International Brotherhood of Electrical Workers initiated a $20-million development of 2,100 apartments in the Pomonok section of Queens in New York City (Taylor, 1949; *New York Times*, May 14, 1949). Several more unions in the area quickly followed suit, including the Hat, Cap and Millinery Workers, the Brotherhood of Painters, Decorators and Paperhangers of America (New York District No. 9), the Amalgamated Meat Cutters and Butcher Workmen, the Painters Union, and the United Auto Workers, among others. In the post-World War II period, both the AFL and the CIO were vociferous supporters of suburbanization and urban renewal as a means of solving many of the housing problems faced by skilled (usually white) union members and to generate employment in the building trades (Parson, 1982, 1984). More recently, the AFL-CIO has been active in using its members' considerable pension funds to invest in various housing and commercial developments. In 1964 it established the Mortgage Investment Trust (MIT) to make funds available for the production of affordable housing units. Subsequently, the Federation set up its Housing Investment Trust (HIT) to provide a vehicle for channeling union and public employee pension funds into low-cost housing. Between 1964 and 1993 these two bodies produced more than 33,000 units of housing across the United States worth approximately $1.2 billion (AFL-CIO, no date a). (The MIT was merged into, and subsumed by, the HIT in 1984.) As part of its continued goal to encourage urban redevelopment and "put the pension funds of union members to productive use" (AFL-CIO, no date b: 4), in 1988 the AFL-CIO sought to move beyond its initial focus on housing by launching its Building Investment Trust (BIT), designed to invest pension funds in union-built commercial and industrial developments. During the first five years of operation, BIT produced more than 2.8-million square feet of commercial and industrial space. Using these two bodies as its urban redevelopment arm, the AFL-CIO put together a "National Partnership for Community Investment" in conjunction with the U.S. Department of Housing and Urban Development to invest $660 million (plus an additional $550 million leveraged from other public and private sources) in new housing and commercial projects in over 30 cities across the country, much of it in economically depressed areas, with the intent of "promot[ing] home ownership for working people and . . . help[ing to] renew these transitional areas of cities" (AFL-CIO, 1995: 3). The initial goal was to create 12,000 new affordable housing units, 20,000 new jobs in construction and related industries, and over 1-million square feet of commercial space in a five-year period (AFL-CIO, no date a).

[7]The AFL-CIO's role in the Alliance for Progress was publicly acknowledged by Secretary of State Dean Rusk in an address to a group of AIFLD students in November 1964 (quoted in *AIFLD Report*, December 1964a: 1–2):

> The Alliance for Progress has given the people of Latin America a new hope that the better life which they desire could and would come. It has offered an alternative to the status quo, on the one hand, and violent change, on the other. It has provided means to achieve economic progress and social justice through the peaceful transformation of societies. . . . Through it, in the Americas we have strengthened the hand of democratic leadership and are helping to build better societies. Because of it the forces of communism have been put on the defensive in Latin America. . . . The support which the American labor movement is giving to the AIFLD is only one example of its enormously helpful role in the worldwide struggle between freedom and tyranny. This struggle is the underlying crisis of our time. It will not end until freedom prevails throughout the world. As Secretary of State, I am particularly conscious of—and grateful for—the stalwart backing which the AFL-CIO has given to our foreign policies and the instruments necessary to make them effective—ranging from powerful military forces necessary to repel aggression through our foreign aid programs.

[8]The AIFLD's newsletter, *AIFLD Report*, contains a wealth of information on such projects stretching back to the 1960s. In this chapter I have selected smaller projects to illustrate the kinds of expenses made by AIFLD. Those smaller projects mentioned in the text are by no means the only ones in which AIFLD engaged. See also AIFLD (1965) for more information.

[9]After 1966 some of this work was conducted in cooperation with those international trade secretariats having affiliates in Latin America and the Caribbean and was part of the AFL-CIO's efforts to strengthen the secretariats as bastions against Communism (see Chapter 6).

[10]My choice of projects to examine has been shaped by an effort to provide examples from a broad spectrum of situations: Mexico (the first country in which an AIFLD housing project was located); Argentina and Uruguay (relatively developed, quite urbanized nations); Colombia (large-scale AIFLD involvement in consumer cooperatives as well as housing); Brazil (involvement in rural community center schemes); the Dominican Republic (a Caribbean example, together with the fact that housing was tied to the instabilities after the 1965 U.S. invasion of the island); and Ecuador (a country in which the main labor federation was allied with the Moscow-backed World Federation of Trade Unions). Projects were also planned in other countries in the region, including: Chile—1,000 units in Valparaíso (for maritime and railroad workers), in Santiago (for telephone workers), and in four coastal cities, though due to political changes ultimately no AIFLD-sponsored housing was built (AIFLD, 1964f); Peru—as of 1972, AIFLD claimed involvement in 3,229 units (mainly in Lima but also in other cities such as Piura, Arequipa, Ica, Cuzco, and Chiclayo), though some of these were undoubtedly entirely financed by the Alianza Sindical Cooperativa (ASINCOOP/Cooperative Union Alliance) with whom the Institute was working (AIFLD, 1972); Venezuela—1,000 units in El Valle near the Caracas airport financed by a $6-million loan from AFL-CIO pension and welfare funds, and involvement in planning 800 houses in Maracaibo, Maracay, Ciudad Bolivar, and Valencia; Costa Rica—a 128-unit housing project in San José; El Salvador, Nicaragua, and Panama, though none was ultimately built in these countries; Bolivia—8 housing pro-

jects to include 2,000 units for telecommunications and postal workers, railroad workers, airline pilots, oil workers, and others (AIFLD, 1964b), though due to the 1965 coup d'état these were not ever built; Jamaica—a small project that also, ultimately, did not come to fruition.

[11]Unless otherwise stated, all monetary amounts indicated are in U.S. dollars.

[12]The Confederación de Trabajadores de México (CTM), the largest Mexican national labor federation, was not involved in this project, largely because the leadership did not want to work with the AFL-CIO (Anon, 1998).

[13]In 1972 the memberships of these unions were: light and power workers, 31,000; FOECYT, 29,000; railway workers' union, 200,000; municipal workers' union, 70,000 (*AIFLD Report*, February 1972a).

[14]The projects were as follows (*AIFLD Report*, February 1972a).

Barrio Almagro: 56 two-bedroom and 56 three-bedroom units built in a 15-story building serviced by elevator on the corner of Rivadavia Street and Boedo Avenue. Built for members of the railway workers' union (Unión Ferroviaria), the units cost between $7,000 to $9,000.

Rivadavia: 116 two-bedroom and 56 three-bedroom units constructed in a 15-story building. Located on Rivadavia Street close to La Plata Avenue, the units were for members of the municipal workers' union and sold for between $7,600 and $10,500.

Barrio Burzaco: 246 two-bedroom and 84 three-bedroom units were built in 27 three-story structures for members of the light and power workers' union. Prices ranged from $7,000 to $8,600.

Nuñez: A residential area of Buenos Aires close to the Rivadavia railway station, this project involved construction of 150 two-bedroom and 30 three-bedroom apartments in one 15-story building, the ground floor of which contained a nursery, a children's playground, and a parking lot. The apartments ranged in price from $7,600 to $9,600 and were for FOECYT members.

Barrio Parque Olivos: located some 10 miles from downtown Buenos Aires, the project involved construction of 187 two-bedroom and 45 three-bedroom units in 3 three-story buildings and 4 ten-story buildings, all for members of the light and power workers' union. Units cost between $7,500 and $9,800.

Barrio San Justo: 507 two-bedroom and 54 three-bedroom units were built in 12 three-story buildings and one ten-story building some 12 miles from the center of Buenos Aires. The units sold for between $6,900 and $8,200 and were for members of the light and power workers' union.

Bahía Blanca: 50 two-bedroom apartments in an 8-story building were built in this city located 500 miles southwest of Buenos Aires and sold to members of FOECYT for $6,000 to $7,400.

Pergamino: an industrial city some 130 miles from the capital but still within Buenos Aires Province, the Pergamino project involved construction of 10 two-bedroom and 20 three-bedroom apartments in a single 10-story building. The apartments were sold to members of FOECYT for between $7,500 and $11,500.

[15]These cities were: Bogotá, Cali, Medellín, Barranquilla, Buenaventura, Cartagena, Palmira, Popayán, Bucaramanga, Pasto, Manizales, Santa Marta, Quibdó,

Pereira, and Girardot. The sites would contain a total of 4,014 housing units, divided evenly between CTC and UTC projects (AIFLD, 1964g: 11).

[16]The objectives of the centers, as stated in *AIFLD Report* (August 1965), were:

1. Stress self and mutual help as a basis for raising the standard of living, education and welfare of rural workers.
2. Raise the productivity of human labor.
3. Raise farm income.
4. Promote maximum utilization of regional resources.
5. Improve conditions of family life.
6. Organize the development of community life.
7. Organize the development of rural youth.
8. Raise the cultural and economic levels of the rural population.
9. Work with young people both socially and technically (e.g., promote 4-H Clubs).
10. Extend medical services ranging from the professional care of a doctor and dentist to that of a nurse, midwife, and practical nurse to teach the basic principles of sanitation, vaccination, etc.
11. Provide legal counseling for the improvement of contracts between groups of tenant farmers and their landlords, based on peasant rights under existing land reform laws.
12. Promote an anti-illiteracy campaign among adult workers.
13. Teach principles of sound agriculture relating to animal husbandry, crop rotation, soil analysis and the cultivation of particular native crops such as mandioca root, black beans, corn, sorghum, etc., as well as various fruits.
14. Dispense information on the setting up and operation of all types of cooperatives including those of credit, consumers, marketing and farm supply varieties.
15. Teach basic skills and crafts including carpentry and masonry.
16. Teach home economics to young girls and women emphasizing the relationships between foods and nutrition, improved family garden plots for raising vegetables, better sanitation and house keeping.

This latter objective suggests that the centers were certainly not designed to challenge patriarchal family structures!

[17]CONATRAL had its origins in the Frente Obrero Unido Pro Sindicatos Autónomos Libre (FOUPSA Libre), which received support from the AFL-CIO and AIFLD. FOUPSA Libre itself had formed in 1962 after its leaders had split off from the original FOUPSA, which, by this time, had become dominated by a nationalist, pro-Castro group (Murphy, 1987).

[18]At first, AIFLD had difficulty attracting people to the housing project, largely as a result of opposition by some anti-AFL-CIO officials within CONATRAL. Only after one of these officials was caught embezzling union dues and thus discredited did the project proceed (Pita, 1997).

[19]Initially, the plan called for a total of 900 units to be built in San Pedro de Macorís and other parts of the island (*AIFLD Report*, January 1966b). However, much

of the IDB funding was lost after complaints that AIFLD was insisting upon restricting occupancy of the houses to members of ORIT-affiliated unions (U.S. Senate Committee on Foreign Relations, 1969: 89).

[20]In addition to indicating the connections between spatially engineering the urban landscape and building "free, democratic and independent labor unions," this quote also provides insights into the (perhaps unconsciously adopted) "ideal" model of the families with which AIFLD would be working, namely, that it was assumed these would be patriarchally organized with a male head of household who was, presumably, the main breadwinner while his wife stayed at home to take care of domestic responsibilities—the social programs would "provide social justice that [would] restore human dignity for [the worker] and *his* family." (see also note 16 above).

[21]Gale Varela was subsequently a member of AIFLD's Board of Trustees. SITRATERCO was formed on August 28, 1955, after the protracted banana workers' strike. SITRATERCO has been one of the most important unions in Honduras since that time, the other being SUTRASFCO (Sindicato de Trabajadores de la Standard Fruit Company/Standard Fruit Company Workers Union).

[22]FESITRANH is the largest of two federations that has historically made up the Confederación de Trabajadores de Honduras (CTH—Confederation of Honduran Workers). After the 69-day 1954 banana strike, labor leaders from SITRATERCO and other North Coast unions met at a conference in the port city of Tela and in 1957 formed the FESITRANH (Federación Sindical de Trajabadores Norteños de Honduras/Union Federation of Northern Workers of Honduras). In 1958 the union received official recognition from the Ministry of Labor and changed its name to Federación Sindical de Trajabadores Nacionales de Honduras (Union Federation of National Workers of Honduras), though it kept the same acronym FESITRANH.

[23]A *manzana* is a traditional unit of land area in Latin America. It is the area of a square 100 *varas* on each side. Generally, the *vara* was approximately 33 inches in length, although this varied throughout Spanish America, being a little longer in some places such as Texas, and a little shorter in others such as Mexico. The Honduran *manzana* is approximately 1.7 acres in size.

[24]In 1972 U.S.$1 equaled approximately 2 Lempiras. When I visited the *Colonía* in April 1997 the exchange rate was approximately U.S.$1 = Lps 13.

[25]ANACH is the Asociación Nacional de Campesinos de Honduras (National Association of Honduran Peasants). It was founded in San Pedro Sula in October 1962 to serve as a counter to the Communist-dominated Federación Nacional de Campesinos de Honduras (FENACH/National Federation of Honduran Peasants) (for more details, see Pearson, 1987). Its main strength was drawn from northern Honduras and was heavily supported by FESITRANH. In southern Honduras ANACH was much weaker and faced competition for members from the Federación Auténtica Sindical de Honduras (FASH/Authentic Union Federation of Honduras), affiliated with the CLASC, the Christian international union federation.

[26]British Guiana changed its name to Guyana upon independence. I use the name Guyana throughout unless referring specifically to an organization that uses the country's former name.

[27]The 1961 elections had been run under the British-style "first past the post" electoral system. Although the PPP received 42.7% of the popular vote, it took 20

(57%) of the 35 seats in the legislature. The PNC, on the other hand, received 41% of the vote but only 11 (31%) of the seats. Under pressure from Washington, DC, and feeling that Forbes Burnham was preferable to Jagan, in October 1963 the British government switched the colony to a system of proportional representation that was used for the first time in the 1964 elections.

[28]This calculation is based on an extrapolation of figures published by AIFLD for 1968, which stated that 10,882 AIFLD-sponsored units had either been completed or were under construction in nine countries, with financing totaling some $36 million, while an additional 4,654 units valued at $25.4 million in six countries were in various stages of preparation (*AIFLD Report*, May 1968b; see also *AIFLD Report*, February 1971a). This calculates out to approximately $3,952 each ($61.4 million/15,536 units). Multiplying this figure by the total constructed (18,048) by the end of the project gives a total of approximately $71,326,000 (not including inflation). I have been unable to locate a more precise figure for the value of AIFLD-sponsored housing stock as it stood at the end of the program. Population figures are calculated estimating a family size of 7 persons (including parents, children, grandparents, in-laws, and other extended family members), which, based upon my conversations with union officials in Guyana, Honduras, and Washington, DC, seems about the average for the projects in the region.

[29]In a similar vein, Ruble (1990) has suggested that the commissioning by Soviet trade unions in the 1920s of educational and cultural clubs for members was a way of instilling the appropriate "revolutionary consciousness" in workers while simultaneously attempting, as a practical goal, to combat alcoholism.

[30]In an interesting aside, Banham (1960) shows how debates in early-nineteenth-century Germany over the design of factories and the use of aesthetics was tied up with notions of encouraging the "Spiritualisation of German Production."

Chapter 8

[1]Jennings is quoted in *AFL-CIO News* (July 6, 1992: 2); Trumka is quoted in *Labor Notes* (1991: 4); Davis is quoted in Gerdel (1999); Becker is quoted in ICEM (1998: 2).

[2]The monthly periodical *Labor Notes*, for example, often carries articles about ongoing international labor solidarity campaigns (see note 37, Chapter 6, for contact details). The Labour and Society International organization (3rd floor, ITF building, 49–60 Borough Road, London SE1 1DS, UK; Tel. +44 171 378 1515; *http://www.lsi.org.uk/*) also operates the Labourstart web page (*http://www.labourstart.org*), which provides up-to-date information on various international labor campaigns.

[3]The IUD emerged out of the political struggles between George Meany, head of the American Federation of Labor, and Walter Reuther, head of the Congress of Industrial Organizations, at the time of the 1955 merger between these two labor federations. As a condition of merger, Reuther insisted upon the establishment of a separate and financially autonomous organization (the IUD). This, he hoped, would not only prevent the AFL from getting its hands on the sizable treasury that he had built up but would also allow him to use the IUD to promote his goal of industrywide

bargaining coordinated through the national center, a goal that Meany opposed in the belief that the prerogative for bargaining should be left to individual AFL-CIO affiliates. While conceding the IUD's autonomy, Meany insisted that it be called a "Department" and be housed in the AFL-CIO building in Washington, DC to give the illusion of AFL-CIO control. Thus, although they often work fairly closely with one another, the IUD is financially autonomous from the AFL-CIO (Uehlein, 1992).

[4]Unions often seek to develop national agreements as a means of equalizing conditions between plants and so of reducing the potential for concessionary whipsawing. Consequently, during the past two decades many companies have vigorously attempted to break up such master contracts (Herod, 1991a). This policy, however, may have unintended consequences, for the destruction of master contracts also has the potential to free more-militant local unions from the constraints of concessionary national agreements and may allow them independently to pursue more-confrontational labor relations with particular employers. For more on the personalities of some of those involved in negotiating with RAC, see Juravich and Bronfenbrenner (1999).

[5]Legislation in the form of a Workplace Fairness bill was introduced to the U.S. Congress in 1991 and 1993 to prevent the use of replacement workers during a strike. On both occasions the bill fell victim to Republican-led filibusters in the Senate.

[6]The IUD and the ICEF (now ICEM) have very close links. In 1987 the IUD was designated as the ICEF's North American operational center (Uehlein, 1989). Uehlein's research provided several hundred pages of computer printout concerning these companies.

[7]For more on Rich's business dealings, see Copetas (1985), *Forbes* (April 30, 1990), *Business Week* (November 11, 1991), and *Institutional Investor* (1992).

[8]Harmon (1959) and Casserini (1993) provide histories of the International Metalworkers' Federation since its founding in 1893.

[9]The "average" worker was 52 years old and had worked for 24 years at the plant.

[10]Even within the "developed" world differences among countries may stifle cooperation, as when the United Auto Workers in the mid-1960s attempted to get the West German metalworkers' union IG Metall to address issues of wage differentials so that the flow of auto capital from the United States to lower-wage West Germany might be limited. The refusal of IG Metall to lose its cost advantage by participating in the World Auto Councils through which such wage differentials were to be addressed led the UAW to try to make contacts at the factory level at Ford and Opel behind IG Metall's back (Olle & Schoeller, 1977). The greater wage differentials between "developed" and "developing" countries only magnify such difficulties.

[11]I do not mean to imply here that Johns is adopting such an either/or position, for she suggests that there is in fact a continuum of types of international solidarity in which workers may engage. Furthermore, Johns's argument is that international solidarity can be conducted for different reasons and with different results, which is a somewhat different argument than one that would maintain that workers will engage *either* in internationalist campaigns *or* in nationalist ones in the face of globalization.

Notes to Chapter 9

[12]The collection of essays in Herod (1998b) seeks to address this issue.

[13]For more on just-in-time production in the North American auto industry and what it entails, see Juárez Núñez & Babson (1998), Babson (1995), and Wells (1996).

[14]Part of the reason for this is that the union's collective bargaining agreement with the auto producers outlawed "national" strikes during the period of the contract, but "local" strikes over "local" issues were not outlawed. Furthermore, the union's ability to convince various judges and arbitrators that such strikes were "local" rather than "national" meant that any workers who happened to be laid off elsewhere as a result of them were usually entitled to claim unemployment benefit in their various states, thereby saving the UAW from having to disburse strike pay (see Herod, 2000, for more details).

Chapter 9

[1]The first quote comes from IMF (1992b: 2); the second quote comes from IMF (no date a: 3).

[2]The WCL is the successor to the old Christian Confédération Internationale des Syndicats Chrétiens (CISC) (see Windmuller, 1980, for more details).

[3]The degree to which various plants were horizontally or vertically integrated into enterprises varied throughout the region. In Czechoslovakia state-owned enterprises were characterized by a high degree of horizontal integration, with an "average" enterprise in the manufacturing sector consisting of 8 or 9 plants producing similar products. In East Germany, too, giant vertically integrated combines (*Kombinate*) dominated the industrial and construction sectors of the economy. In Hungary, by way of contrast, although prior to 1980 there was a high degree of organizational centralization, with about 45% of state-owned enterprises having more than 1,000 employees, during the early 1980s a deconcentration campaign led to the dissolving of many of these large conglomerates (called "trusts"). Furthermore, most of these large organizations were only integrated administratively and remained geographically and technologically separate units. Nevertheless, despite such differences, throughout the Soviet bloc the average size of enterprises was much larger than in the market economies, and their breakup has been seen as crucial to the success of the privatization process (United Nations, 1994).

[4]Daintith (1993) and Estrin and Stone (1996) discuss some of the different forms of privatization that were adopted.

[5]Šarčević (1992) provides case studies of several countries' early plans for privatization.

[6]Although unions seeking to have their employers privatized may seem strange to those familiar with Western trade union practices (given that privatization usually involves job loss), in Central and Eastern Europe many labor leaders and workers see this as a necessary part of restructuring the economy and enterprises so as to stimulate economic development and break the political stranglehold of the old *nomenklatura* (Volkov, 1992). In the case of the Czech power workers, the Klaus government alleged that their union's efforts to have the state monopoly broken up was merely a ploy to place union officials in administrative positions within the privatized plants, something that the antiunion government vociferously opposed.

[7]Despite efforts to separate the unions from the political parties, many unions are, nevertheless, extremely politically active (e.g., Solidarność in Poland, Podkrepa in Bulgaria, and Fratia in Romania).

[8]In 1998 the ČMKOS changed its name, though not its acronym, to the Českomoravská Konfederace Odborových Svazů (Czech-Moravian Confederation of Trade Unions).

[9]The text of the Romanian law is reproduced in English in "Law on Trade Unions," Law No. 54/August 1, 1991, as published in *Monitorul Oficial* (Official Gazette of Romania), Part I, No. 164/August 7, 1991.

[10]In Hungary, for example, works councils are required by law in all establishments with more than 50 workers. The first free shop-floor elections for these councils were held in May 1993.

[11]In Hungary the Labour Code requires employers to cooperate with trade unions and provide them with information necessary for collective bargaining but does not require them to actually bargain collectively. In Poland, the Trade Union Act prevents an employer or the government from contesting a union's registration but, again, does not require an employer to bargain collectively with a union. In Bulgaria, by way of contrast, the Labour Code does require an employer to negotiate a collective bargaining agreement in good faith with the workers' representatives (IMF, 1994).

[12]In Romania, for example, under the 1991 Law on Trade Unions there are no legal regulations regarding the bargaining procedure and scope of the bargaining unit, which are determined on a case-by-case basis. Likewise, in Hungary there are no legally enforceable guidelines relating to the composition of the bargaining committee. In Slovenia, the composition of the bargaining unit is not governed by either the law or the collective bargaining agreement but is determined by the parties involved (IMF, 1994).

[13]I say apparent because there is some debate as to what kind of economy is emerging in the region. Stark (1996), for instance, has argued that a distinctive type of capitalism is emerging in Central and Eastern Europe that is quite different from that in Western Europe.

[14]The differences between these countries become particularly pronounced when foreign direct investment is calculated on a per-capita basis. Hungary's predominance is quite clear. Between 1991 and 1996, Hungary received a cumulative net inflow of FDI valued at $1,256 per capita, the highest amount for any transition economy. Per-capita amounts for transition economies have varied tremendously for this period, as shown by the following examples, Azerbaijan $120; Belarus $5; Bulgaria $65; Croatia $123; Czech Republic $617; Estonia $558; Georgia $6; Lithuania $65; Moldova $43; Poland $126; Romania $61; Russia $42; Slovak Republic $128; Slovenia $325; Tajikistan $10; and Ukraine $23 (International Monetary Fund, 1997: 107).

[15]Ironically, the same fear is also expressed by many in the (relatively) higher-wage countries of Central and Eastern Europe who are concerned that local capital will seek to migrate east in search of cheaper labor and less rigorous governmental restrictions on health and safety, the environment, labor law, etc. Already a number of Czech companies, for example, have been purchasing interests in recently privatized state concerns in the Slovak Republic, leading to con-

cern among both Czech and Slovak trade union officials that wages will remain low in the Slovak Republic, and that this, in turn, will be used to undercut wages in the Czech Republic (Uhlíř, 1995; Machyna, 1996). Likewise, recently a number of Slovak companies have been looking to buy enterprises in Ukraine to take advantage of lower wages there.

[16]In Croatia, the ICFTU has affiliated the Union of Autonomous Trade Unions of Croatia (UATUC) and has links with Hrvatska Unija Sindikate (HUS), the Confederation of Independent Trade Unions of Croatia (KNSH), and others; in Serbia it has affiliated the Union of Independent Branch Unions (Nezavisnost); in Slovenia ICFTU works with the official Free Trade Union Association of Slovenia [ZSSS] and Neodvisnost, the new national center established in March 1990.

[17]The desire to ensure the emergence of a market economy can be seen in a motion passed by delegates at the ICFTU's 15th World Congress, for example, that argued that the

> narrowly conceived liberalisation and stabilisation policies, advocated by the theoreticians of the free market and of the I[nternational] M[onetary] F[und], disregarding the needs and interests of the working people, resulting in large-scale unemployment and aggravating poverty, and all too easily exploited by the still-entrenched nomenklatura are conducive neither to the operation of democratic government *nor an efficient market economy*, and constitute a serious threat to the further democratic revolutions. (quoted in ICFTU, 1996: 275, emphasis added)

[18]As part of an effort to overcome its Cold War past, in 1997 the AFL-CIO merged the FTUI with its three other regional institutes to form the American Center for International Labor Solidarity.

[19]Established by the Reagan administration, the National Endowment for Democracy is made up of four core entities, these being the AFL-CIO, the Center for International Private Enterprise established by the U.S. Chamber of Commerce, the National Democratic Institute for International Affairs, and the International Republican Institute. The NED's purpose is to funnel U.S. government money to various groups abroad with the goal of encouraging U.S.-style entities to promote "democracy" and the establishment of "free enterprise."

[20]This has been particularly so over the use of the term "works council" (Olsen, 1995). The European Union has recently issued a directive that compels companies to form such works councils, a move supported by both the IMF and the European Metalworkers' Federation. Under this EU directive, works councils serve as bodies in which workers (who do not have to be members of the union, though in practice usually are) are granted some say over some plant and production issues. Some works councils have adopted the "French" model in which a joint committee of management and workers' representatives meets with the managing director chairing the meeting, whereas others have adopted the "German" model made up only of workers' representatives that meets periodically with management counterparts (see Thelen, 1991, for a discussion of Germany's works councils system). Many trade unionists in Central and Eastern Europe, however, appear suspicious of works councils because they associate them with the types of joint management–union committees that operated in plants during the Communist period.

[21]Although such new and reformed unions often work together at the plant level, at the regional and national level in their home countries their political rivalries frequently make it impossible to conduct joint seminars (Mureau, 1995a).

[22]At a meeting of the IMF Central Committee held in Marseilles in May 1994 the Federation laid out a charter that included demands to stimulate economic growth, to respect political democracy, human rights, and environmental protection, to reverse the "shock therapies" being implemented in Central and Eastern Europe, to encourage international assistance to stabilize employment and production in the region, to make available capital to finance necessary imports into the region, to regulate international financial markets so as to prevent instability, to support social safety nets and encourage tripartism with regard to negotiating changes in such provisions, and to implement education and training programs for workers (IMF, no date b).

[23]As of December 1998, ICFTU affiliates in Central and Eastern Europe were (Oulatar, 1999):

Bulgaria: Confederation of Independent Trade Unions in Bulgaria (CITUB) (Membership: 607,883) and Confederation of Labour "PODKREPA" (Membership: 154,894).

Croatia: Union of Autonomous Trade Unions of Croatia (UATUC) (Membership: 421,806).

Czech Republic: Czech–Moravian Confederation of Trade Unions (ČMKOS) (Membership: 1,298,312).

Estonia: Association of Estonian Trade Unions, (Membership: 108,000).

Hungary: Autonomous Trade Union Confederation (ATUC) (Membership: 250,000), Democratic League of Independent Trade Unions (FSZDL) (Membership: 98,000), and National Confederation of Hungarian Trade Unions (MSZOSZ) (Membership: 895,000).

Latvia: Free Trade Union Federation of Latvia (LBAS) (Membership: 252,400).

Lithuania: Lithuanian Trade Union Unification (LPSS) (Membership: 41,601) and Lithuanian Workers' Union (LDS) (Membership: 78,000).

Moldova: General Confederation of Trade Unions of Moldova (FGSRM) (Membership: 702,214).

Poland: NSZZ Solidarność (Membership: 1,300,000).

Romania: Blocul National Sindical (BNS) (Membership: 350,000) and National Free Trade Union Confederation of Romania—CNSLR–Fratia (Membership: 2,500,000).

Slovak Republic: Slovak Confederation of Trade Unions (KOZ SR) (Membership: 1,255,960).

Yugolsavia: Nezavisnost (Membership: 153,650).

[24]For example, in 1998 miners pressured the Czech government into accepting that the unions would be allowed a significant role in privatization decisions concerning three coal mines in northern Bohemia. The unions, concerned that "wild privatization" would lead to unfriendly foreign control, feared that sale of the mines to German investors might result in these investors closing them so that they could then export higher-priced energy to the Czech Republic (*Prague Post*, 1998).

Chapter 10

[1]The WTO, for instance, suggests that one of the advantages of the global free-trade system it is attempting to implement is that having a single set of rules by which everyone must play in the global economy, and having a single forum (the WTO) in which to negotiate such rules, "makes life much simpler. . . . The alternative would be continuous and complicated bilateral negotiations with dozens of countries simultaneously. And each country could end up with different conditions for trading with each of its trading partners, making life extremely complicated for its importers and exporters. . . . The fact that there is a single set of rules applying to all members greatly simplifies the entire trade regime" (see *www.wto.org/english/thewto_e/whatis_e/10ben_e/10b03_e.htm*).

[2]For a more detailed description of the network's goals, see *www.irnetwork.wayne.edu/ourresearch.html*.

[3]See *www.uniteunion.org/sweatshops/sweatshoparchive/kathielee/kathielee.html*.

References

Abeles Schwartz Associates. (1986). *The Outer Borough Garment Industry Study*. Report prepared for the Garment Industry Development Corporation, New York City Office of Business Development.

Abeles, Schwartz, Hackel, & Silverblatt, Inc. (1983). *The Chinatown Garment Industry Study*. Report prepared for ILGWU Local 23–25 and the New York Sportswear Association.

Ades, D., Benton, T., Elliott, D., & Whyte, I. B. (1995). *Art and Power: Europe under the Dictators 1930–45*. London: Thames & Hudson.

AFL (American Federation of Labor). (1921). *Report of proceedings of the forty-first convention of the American Federation of Labor*, Denver, CO, June 13–25.

AFL (American Federation of Labor). (1952). *Report of proceedings of the seventy-first convention of the American Federation of Labor*, New York, NY, September 15–23.

AFL-CIO (American Federation of Labor–Congress of Industrial Organizations). (1987a). *The AFL-CIO's foreign policy*. Perspectives on Labor and the World Publication No. 181. Washington, DC: AFL-CIO.

AFL-CIO (American Federation of Labor–Congress of Industrial Organizations). (1987b). *The AFL-CIO abroad*. Perspectives on Labor and the World Publication No. 182. Washington, DC: AFL-CIO.

AFL-CIO (American Federation of Labor–Congress of Industrial Organizations). (No date a). *A Framework for Rebuilding America*. Pamphlet produced by AFL-CIO Investment Trusts, Washington, DC. (No date of publication but probably circa 1992).

AFL-CIO (American Federation of Labor–Congress of Industrial Organizations). (No date b). *The AFL-CIO Building Investment Trust*. Prospectus produced by AFL-CIO Building Investment Trust, Washington, DC. (No date of publication but probably circa 1992).

AFL-CIO (American Federation of Labor–Congress of Industrial Organizations). (1995). Housing and the American Dream. Unions have a role; So does government. *AFL-CIO Reviews the Issues, Report No. 85, October*. Washington, DC: AFL-CIO.

AFL-CIO News. (July 22, 1991). Ravenswood steelworkers get international support. p. 12.

AFL-CIO News. (March 2, 1992). Ravenswood buyers quit scab product. pp. 1 and 6.

AFL-CIO News. (July 6, 1992). Ravenswood shows global potential. p. 2.

Aglietta, M. (1979). *A Theory of Capitalist Regulation: The US Experience*. London: New Left Books.

Agnew, J. (1993). Representing space: Space, scale and culture in social science. In J. Duncan & D. Ley (Eds.), *Place/Culture/Representation* (pp. 251–271). London: Routledge.

Agnew, J. (1997). The dramaturgy of horizons: Geographical scale in the 'reconstruction of Italy' by the new Italian political parties, 1992–1995. *Political Geography*, 16, 99–121.

AIFLD (American Institute for Free Labor Development). (1964a). Country plan for Argentina. In AIFLD (Ed.), *Country Plans for Latin America*. Washington, DC: AIFLD, Social Projects Department.

AIFLD (American Institute for Free Labor Development). (1964b). Country plan for Bolivia. In AIFLD (Ed.), *Country Plans for Latin America*. Washington, DC: AIFLD, Social Projects Department.

AIFLD (American Institute for Free Labor Development). (1964c). Country plan for Brazil. In AIFLD (Ed.), *Country Plans for Latin America*. Washington, DC: AIFLD, Social Projects Department.

AIFLD (American Institute for Free Labor Development). (1964d). Country plan for Caribbean Area. In AIFLD (Ed.), *Country Plans for Latin America*. Washington, DC: AIFLD, Social Projects Department.

AIFLD (American Institute for Free Labor Development). (1964e). Country plan for Central America. In AIFLD (Ed.), *Country Plans for Latin America*. Washington, DC: AIFLD, Social Projects Department.

AIFLD (American Institute for Free Labor Development). (1964f). Country plan for Chile. In AIFLD (Ed.), *Country Plans for Latin America*. Washington, DC: AIFLD, Social Projects Department.

AIFLD (American Institute for Free Labor Development). (1964g). Country plan for Columbia. In AIFLD (Ed.), *Country Plans for Latin America*. Washington, DC: AIFLD, Social Projects Department.

AIFLD (American Institute for Free Labor Development). (1964h). Country plan for Ecuador. In AIFLD (Ed.), *Country Plans for Latin America*. Washington, DC: AIFLD, Social Projects Department.

AIFLD (American Institute for Free Labor Development). (1964i). Country plan for Mexico. In AIFLD (Ed.), *Country Plans for Latin America*. Washington, DC: AIFLD, Social Projects Department.

AIFLD (American Institute for Free Labor Development). (1964j). Country plan for Uruguay. In AIFLD (Ed.), *Country Plans for Latin America*. Washington, DC: AIFLD, Social Projects Department.

AIFLD (American Institute for Free Labor Development). (1965). *Report on Activities: Social Projects Department* (dated September 1, 1965). Washington, DC: AIFLD, Social Projects Department.

AIFLD (American Institute for Free Labor Development). (1972). *1962–1972: A Decade of Worker to Worker Cooperation*. Washington, DC: AIFLD.

AIFLD (American Institute for Free Labor Development). (1987). *Twenty-Five Years of Solidarity with Latin American Workers*. Washington, DC: AIFLD.

AIFLD Report. (December 1964a). Dean Rush sees Communism as losing in hemisphere, democratic leadership strengthened throughout Americas; lauds Institute. pp. 1, 2, and 7.

AIFLD Report. (December 1964b). 6,000 attend opening of Kennedy housing project in Mexico. pp. 3, 4, 5, and 6.

AIFLD Report. (August 1965). Rehabilitation program launched in Brazil's rural northeast, agreement signed for first area worker service centers. pp. 1, 2, 3, 5, and 6.

AIFLD Report. (January 1966a). First Brazilian rural centers approach completion; São Paulo urban housing site ready; education total at new peak. pp. 1, 3, and 8.

AIFLD Report. (January 1966b). First Kennedy project homes assigned in Dominican Republic. pp. 2 and 4.

AIFLD Report. (April 1966a). AIFLD's future: Country directors stress self-help projects, graduates. p1, 2, and 6.

AIFLD Report. (April 1966b). AIFLD speaks . . . on radio—about the D.R. pp. 4 and 7.

AIFLD Report. (April 1966c). "We . . . are in fact the revolutionaries. . . ." p. 4.

AIFLD Report. (May 1966a). "Asincoop": The story of a growing idea. pp. 1 and 8.

AIFLD Report. (May 1966b). Housing: Bogota ground broken; in Brazil President speaks. pp. 2 and 6.

AIFLD Report. (May 1966c). Housing . . . The need for it in Latin America. p. 4.

AIFLD Report. (January 1967a). Carlson's remarks. pp. 1 and 6.

AIFLD Report. (January 1967b). AIFLD in the hemisphere . . . education and social projects. p. 4.

AIFLD Report. (February 1967a). Bogota workers draw lots for 280 new apartments. pp. 1 and 6.

AIFLD Report. (February 1967b). AIFLD in the hemisphere . . . education and social projects. p. 4.

AIFLD Report. (April–May 1967a). Dominican workers dedicate Kennedy Housing Project. pp. 1 and 5.

AIFLD Report. (April–May 1967b). Morgan tells AIFLD role in broadcast from conference at Punta del Este. pp. 2, 5 and 6.

AIFLD Report. (April–May 1967c). AIFLD in the hemisphere . . . education and social projects. p.4.

AIFLD Report. (June 1967a). Bank completes $2.2 million loan for workers' housing in Honduras. p. 3.

AIFLD Report. (June 1967b). School opens at Kennedy Housing Project. p. 3.

AIFLD Report. (June 1967c). AIFLD in the hemisphere . . . education and social projects. p. 4.

AIFLD Report. (June 1967d). Construction begins in Guyana on workers' housing project. p. 5.

AIFLD Report. (June 1967e). AFL-CIO grant and self-help create new rural school for Brazilians. p. 6.

AIFLD Report. (August 1967a). Meany addresses National Press Club on work of AIFLD. pp. 1 and 2.

AIFLD Report. (August 1967b). Model house started for FESITRANH. p. 1.

AIFLD Report. (August 1967c). Chamber of Commerce of Americas hears report from AIFLD spokesman: Recife, Brazil. p. 3.

AIFLD Report. (August 1967d). Road safety signs protect children in rural Guyana. p. 5.

AIFLD Report. (August 1967e). Social action projects for free trade unions in Latin America. p. 6.

AIFLD Report. (December 1967). Brazilian telephone workers cooperative receives AFL-CIO impact project loan to help complete building. p. 1.

AIFLD Report. (April 1968a). AFL-CIO grant to Honduran union. p. 3.

AIFLD Report. (April 1968b). New hall spurs membership drive in rural Brazil. p. 3.

AIFLD Report. (April 1968c). AIFLD in the hemisphere . . . education and social projects. p. 4.

AIFLD Report. (April 1968d). CO-OP housing seminar in Rio. p. 5.

AIFLD Report. (May 1968a). Guyana workers receive keys to new homes in May Day rally; XX class hears Dubinsky call education labor's modern arm. pp. 1 and 2.

AIFLD Report. (May 1968b). AIFLD progress report: 1962–67. pp. 2 and 6.

AIFLD Report. (May 1968c). AIFLD in the hemisphere . . . education and social projects. p. 4.

AIFLD Report. (May 1968d). New playground at D.R. project. p. 5.

AIFLD Report. (June 1968). Argentine Unions obtain $13 million loan for housing. pp. 1 and 2.

AIFLD Report. (July 1968a). Argentine workers receive keys to new homes; Covey Oliver hails 4–union project as Alliance feat. pp. 1 and 2.

AIFLD Report. (July 1968b). AIFLD in the hemisphere . . . education and social projects. p. 4.

AIFLD Report. (July 1968c). AIFLD students build bridges, schools, and social centers in Honduras and Guatemala. p. 5.

AIFLD Report. (January 1969). Labor education, social and housing projects of AIFLD bring broad program of assistance to workers in Guyana. pp. 1, 2, and 6.

AIFLD Report. (February 1971a). In eight years 159,053 leaders trained: Housing completed for 88,700 workers. p. 3.

AIFLD Report. (February 1971b). AIFLD in the hemisphere . . . education and social projects. p. 4.

AIFLD Report. (April 1971). Workers' housing project underway in Montevideo. pp. 1 and 3.

AIFLD Report. (June 1971). AIFLD in the hemisphere . . . education and social projects. p. 4.

AIFLD Report. (August 1971). Union-sponsored community clinic opened in Uruguay; 2,500 attend inauguration. p. 2.

AIFLD Report. (December 1971a). Union of Columbian workers gets an RRLF loan to expand its cooperative. p. 2.

AIFLD Report. (December 1971b). AIFLD's achievements in providing low-cost worker housing since June, 1962—a people-to-people program. pp. 5 and 6.

AIFLD Report. (February 1972a). Argentine housing project named "most successful achievement." pp. 1, 2 and 4.

AIFLD Report. (February 1972b). Self-help is underlying basis of AIFLD's programs for Latin America and Caribbean trade unionists. pp. 3 and 6.

AIFLD Report. (May 1972). AIFLD's role in savings and loan field. p. 6.

AIFLD Report. (October 1972). FESITRANH to open labor education center with assistance of RRLF loan. p. 3.

Alcock, A. E. (1971). *History of the International Labor Organization.* New York: Octagon.

Alderman, D. H. (1996). Creating a new geography of memory in the south: (Re)naming of streets in honor of Martin Luther King, Jr. *Southeastern Geographer,* 36.1, 51–69.

Alderman, D. H. (1998). *Creating a New Geography of Memory in the South: The Politics of (Re)Naming Streets After Martin Luther King, Jr.* Unpublished PhD dissertation, Department of Geography, University of Georgia, Athens, GA.

Alexander, R. J. (1965). *Organized Labor in Latin America.* New York: Free Press.

Ambrose, S. E. (2000). *Nothing Like it in the World: The Men Who Built the Transcontinental Railroad, 1863–1869.* New York: Simon & Schuster.

American Technical Assistance Corporation. (1970a). *An Appraisal of Program Effectiveness and Management of the American Institute for Free Labor Development. Part I: General Findings and Recommendations.* Report prepared for United States Agency for International Development pursuant to Contract No. AID/La-633, July 1970. Washington, DC: American Technical Assistance Corporation.

American Technical Assistance Corporation. (1970b). *An Appraisal of Program Effectiveness and Management of the American Institute for Free Labor Development. Part III: Field Survey Report on AIFLD Program in Guyana.* Report prepared for United States Agency for International Development pursuant to Contract No. AID/La-633, July 1970. Washington, DC: American Technical Assistance Corporation.

American Technical Assistance Corporation. (1970c). *An Appraisal of Program Effectiveness and Management of the American Institute for Free Labor Development. Part IV: Field Survey Report on AIFLD Program in Honduras.* Report prepared for United States Agency for International Development pursuant to Contract No. AID/La-633, July 1970. Washington, DC: American Technical Assistance Corporation.

American Trucking Associations. (No date). Memorandum by the American Trucking Associations and Tidewater Motor Truck Association. Located in "Dolphin Forwarding Inc.," National Labor Relations Board case nos. 2–CC-1364 et al., Folder I, National Labor Relations Board, Washington, DC.

American Trucking Associations, Inc. v. National Labor Relations Board, 734 F.2d 966 (4th Cir. 1984).

Amin, A. (Ed.). (1994). *Post-Fordism: A Reader.* Blackwell: Oxford.

Anderson, J. (1980). Towards a materialist conception of geography. *Geoforum,* 11, 172–178.

Andrews, G. (1991). *Shoulder to Shoulder: The American Federation of Labor, the United States, and the Mexican Revolution 1910–1924.* Berkeley: University of California Press.

Anon. (1973). *Factual Memorandum on Containerization Under NYSA-ILA and CONASA-ILA Labor Agreements.* Draft memorandum dated July 23, 1973, located in "Consolidated Express Inc.," National Labor Relations Board case nos. 22–CC-541 et al., Folder I, National Labor Relations Board, Washington, DC.

Anon. (1998). [Correspondence between author and former high-level official, American Institute for Free Labor Development.]

Aronowitz, S. (1990). Writing labor's history. *Social Text*, 8 & 9 (double issue), 171–195.

Babson, S. (1995). *Lean Work: Empowerment and Exploitation in the Global Auto Industry*. Detroit: Wayne State University Press.

Balicer v. International Longshoremen's Association and New York Shipping Association, 364 F.Supp. 205 (1973).

Balicer v. International Longshoremen's Association and New York Shipping Association, 86 LRRM 2559 (D. NJ 1974).

Banham, R. (1960). *Theory and Design in the First Machine Age*. New York: Praeger.

Barbash, J. (1948). International labor confederations: CIT and CTAL. *Monthly Labor Review*, May, 499–503.

Barnet, R. J., & Muller, R. E. (1974). *Global Reach: The Power of the Multinational Corporations*. New York: Simon & Schuster.

Barnett, J. (1982). *An Introduction to Urban Design*. New York: Harper & Row.

Barry, T., & Preusch, D. (1990). *AIFLD in Central America: Agents as Organizers*. Albuquerque, NM: Inter-Hemispheric Education Resource Center.

Battista, A. (1991). Political divisions in organized labor, 1968–1988. *Polity*, 24.2, 173–197.

Beard, J. (1995). The geography of private service sector trade unionism: A regional comparison. *Area*, 27.3, 218–227.

Becker, G. S. (1975). *Human Capital* (2nd ed.). New York: National Bureau of Economic Research.

Bédarida, F. (1974). Perspectives sur le mouvement ouvrier et l'impérialism en France au temps de la conquête coloniale. [Perspectives on the workers' movement and imperialism in France at the time of the colonial conquest]. *Mouvement Social*, 86, 25–42.

Beirne, J. A. (1970). *Labor's Priorities for the Americas in the Seventies*. An address by Joseph A. Beirne, President, Communication Workers of America, Vice President, AFL-CIO, and Secretary-Treasurer, AIFLD, delivered at Graduation Ceremonies for the IV Labor Economists Training Program of the American Institute for Free Labor Development, September 29, 1970, Georgetown University, Washington, DC.

Bendiner, B. (1987). *International Labour Affairs: The World Trade Unions and the Multinational Companies*. Oxford: Clarendon Press.

Benevolo, L. (1967). *The Origins of Modern Town Planning*. Cambridge, MA: MIT Press.

Berman, L. L. (1998). In your face, in your space: Spatial strategies in organizing clerical workers at Yale. In A. Herod (Ed.), *Organizing the Landscape: Geographical Perspectives on Labor Unionism* (pp. 203–224). Minneapolis and London: University of Minnesota Press.

Berry, B. J. L. (1992). *America's Utopian Experiments: Communal Havens from Long-Wave Crises*. Hanover, NH: University Press of New England.

Berry, B. J. L., Conkling, E. C., & Ray, D. M. (1976). *The Geography of Economic Systems*. Englewood Cliffs, NJ: Prentice-Hall.

Bliznakov, M. (1993). Soviet housing during the experimental years, 1918 to 1933. In

W. C. Brumfield & B. A. Ruble (Eds.), *Russian Housing in the Modern Age: Design and Social History* (pp. 85–148). New York: Woodrow Wilson International Center for Scholars.

Board of Inquiry. (1968). *Report to the President on the Labor Dispute Involving the International Longshoremen's Association and the Maritime Industry.* Dated October 1, 1968. Washington, DC: United States Government.

Bornschier, V., & Chase-Dunn, C. (1985). *Transnational Corporations and Underdevelopment.* New York: Praeger.

Bowers, J. (1983). Answers of John Bowers, ILA Executive Vice President, to questions posed by the Honorable Gene Snyder, U.S. House Subcommittee on Merchant Marine, Committee on Merchant Marine and Fisheries hearings on HR 2562, a bill to amend Section 45 of the 1916 Shipping Act, 98th Cong., 1st Sess., April 27, 1983.

Bradshaw, M., Stenning, A., & Sutherland, D. (1998). Economic restructuring and regional change in Russia. In J. Pickles & A. Smith (Eds.), *Theorising Transition: The Political Economy of Post-Communist Transformations* (pp. 147–171). New York and London: Routledge.

Brauman, A. (1976). *Le Familistère de Guise ou les Equivalents de la Richesse* [Guise's Familistère or the Equivalents of Wealth]. Brussels: Archives d'Architecture Moderne.

Braverman, H. (1974). *Labor and Monopoly Capital: The Degradation of Work in the Twentieth Century.* New York: Monthly Review Press.

Brenner, N. (1997). State territorial restructuring and the production of spatial scale: Urban and regional planning in the FRG 1960–1990. *Political Geography,* 16, 273–306.

Brenner, N. (1998). Between fixity and motion: Accumulation, territorial organization and the historical geography of spatial scales. *Environment and Planning D: Society and Space,* 16, 459–481.

Brewster, C. (1992). Starting again: Industrial relations in Czechoslovakia. *The International Journal of Human Resource Management,* 3.3, 555–574.

Bridge, G. (1998). Excavating nature: Environmental narratives and discursive regulation in the mining industry. In A. Herod, G. Ó Tuathail, & Susan Roberts (Eds.), *An Unruly World? Globalization, Governance and Geography* (pp. 219–243). London and New York: Routledge.

Bronfenbrenner, K. L. (1993). *Seeds of Resurgence: Successful Union Strategies for Winning Certification Elections and First Contracts in the 1980s and Beyond.* PhD Dissertation, Cornell University, Ithaca, NY.

Bryan, L., & Farrell, D. (1996). *Market Unbound: Unleashing Global Capitalism.* New York: Wiley.

Buckwalter, D. W. (1995). Spatial inequality, foreign investment, and economic transition in Bulgaria. *Professional Geographer,* 47.3, 288–298.

Buder, S. (1967). *Pullman: An Experiment in Industrial Order and Community Planning 1880–1930.* New York: Oxford University Press.

Burawoy, M. (1985). *The Politics of Production: Factory Regimes under Capitalism and Socialism.* London: Verso.

Burawoy, M. (1992). The end of Sovietology and the renaissance of modernization theory. *Contemporary Sociology,* 21.6, 744–785.

Busch, G. K. (1983). *The Political Role of International Trades Unions.* New York: St. Martin's Press.

Business Week. (July 23, 1984). The New York colossus. pp. 98–112.

Business Week. (November 11, 1991). Making Marc Rich squirm. pp. 120–122.

Business Week. (May 11, 1992). How the USW hit Marc Rich where it hurts. p. 42.

Cadbury, E. (1912). *Experiments in Industrial Organization.* London: Longmans, Green.

Carr, M. (1983). A contribution to the review and critique of behavioural industrial location theory. *Progress in Human Geography,* 7, 386–401.

Casserini, K. (1993). *International Metalworkers' Federation 1893–1993: The First Hundred Years.* Geneva: International Metalworkers' Federation.

Castells, M. (1977). *The Urban Question: A Marxist Approach.* Cambridge, MA: MIT Press.

Castells, M. (1978). *City, Class and Power.* London: Macmillan.

Castells, M. (1983). *The City and the Grassroots: A Cross-Cultural Theory of Urban Social Movements.* Berkeley and Los Angeles: University of California Press.

Castree, N. (2000). Geographic scale and grass-roots internationalism: The Liverpool dock dispute, 1995–1998. *Economic Geography,* 76.3, 272–292.

Chapman, J. (1992). [Interview by author with Joe Chapman, staff representative for District 23, United Steelworkers of America, Charleston, West Virginia, July 27].

Charleston [West Virginia] Gazette. (December 21, 1991). Ravenswood fined $600,000 on safety. pp. 1A and 9A.

Chilcote, P. (1988). The containerization story: Meeting the competition in trade. In M. J. Hershman (Ed.), *Urban Ports and Harbor Management: Responding to Change Along U.S. Waterfronts* (pp. 125–145). New York: Taylor and Francis.

Chopin, A. (1963). Statement by Alexander Chopin, NYSA Chairman, before the U.S. House Committee on Merchant Marine and Fisheries, hearings on HR 1897, HR 2004, HR 2331, bills to amend the 1936 Merchant Marine Act, 88th Cong., 1st Sess., 1963.

City Almanac. (1985). The redevelopment of 42nd street (special issue). *City Almanac,* 18.4, 1–28.

Clark, E. C. (1966). *Soviet Trade Unions and Labor Relations.* Cambridge, MA: Harvard University Press.

Clark, G. L. (1985). Restructuring the U.S. economy: The National Labor Relations Board, the Saturn Project, and economic justice. *Economic Geography,* 61, 289–306.

Clark, G. L. (1988). A question of integrity: The National Labor Relations Board, collective bargaining, and the relocation of work. *Political Geography Quarterly,* 7.3, 209–227.

Clark, G. L. (1989a). The context of federal regulation: Propaganda in U.S. union elections. *Transactions of the Institute of British Geographers, New Series,* 14.1, 59–73.

Clark, G. L. (1989b). *Unions and Communities Under Siege: American Communities and the Crisis of Organised Labour.* Cambridge, UK: Cambridge University Press.

Clarke, S., Fairbrother, P., & Borisov, V. (1995). *The Workers' Movement in Russia.* Aldershot, UK: Elgar.

Cobble, S. (1991). Organizing the postindustrial work force: Lessons from the history of waitress unionism. *Industrial and Labor Relations Review*, 44.3, 419–436.

Cockburn, C. (1983). *Brothers: Male Dominance and Technological Change*. London: Pluto Press.

Container News. (1969). Solving the plight of small shippers. May, pp. 24 and 40.

Conway Co., Inc. (1986). *Garment Center Real Estate Analysis*. Report prepared for New York City Department of City Planning, Office for Economic Development, and New York City Public Development Corporation by the Conway Company, Inc.

Cooke, P. (1985). Class practices as regional markers: A contribution to labour geography. In D. Gregory & J. Urry (Eds.), *Social Relations and Spatial Structures*, (pp. 213–241). New York: St. Martin's Press.

Coolidge, J. (1999). The art of attracting foreign direct investment in transition economies. *Transition Newsletter: The Newsletter About Reforming Economies*. October issue. Published by the World Bank. (Available at: *www.worldbank.org/html/prddr/trans/so99/pgs5–7.htm*).

Cope, M. (1998). 'Working steady': Gender, ethnicity, and change in households, communities, and labor markets in Lawrence, Massachusetts, 1930–1940. In A. Herod (Ed.), *Organizing the Landscape: Geographical Perspectives on Labor Unionism* (pp. 297–323). Minneapolis and London: University of Minnesota Press.

Copetas, A. C. (1985). *Metal Men: Marc Rich and the 10–Billion-Dollar Scam*. New York: Putnam's.

Council of North Atlantic Steamship Associations v. Federal Maritime Commission, 672 F.2d 171 (D.C. Cir., 1982), *cert. denied*, 459 U.S. 830 (1982).

Cowie, J. (1999). *Capital Moves: RCA's Seventy-Year Quest for Cheap Labor*. Ithaca, NY: Cornell University Press.

Cox, K. R. (1993). The local and the global in the new urban politics: A critical review. *Environment and Planning D: Society and Space*, 11, 433–448.

Cox, K. R. (1996). The difference that scale makes. *Political Geography*, 15, 667–670.

Cox, K. R. (1998a). Spaces of dependence, spaces of engagement and the politics of scale, or: Looking for local politics. *Political Geography*, 17.1, 1–23.

Cox, K. R. (1998b). Representation and power in the politics of scale. *Political Geography*, 17.1, 41–44.

Cox, K. R., & Mair, A. (1988). Locality and community in the politics of local economic development. *Annals of the Association of American Geographers*, 78, 307–325.

Cox, R. W. (1971). Labor and transnational relations. *International Organization*, 25.3, 554–584 (Special issue on "Transnational Relations and World Politics").

Cumiford, W. L. (1987). Guyana. In G. M. Greenfield & S. L. Maram (Eds.), *Latin American Labor Organizations* (pp. 432–447). New York: Greenwood Press.

Curtin, W. J. (1973). The multinational corporation and transnational collective bargaining. In D. Kujawa (Ed.), *American Labor and the Multinational Corporation* (pp. 192–222). New York: Praeger.

Cyert, R. M., & March, J. G. (1963). *A Behavioral Theory of the Firm*. Englewood Cliffs, NJ: Prentice-Hall.

Daily Telegraph (London). (December 27, 1995). Sacked dockers find support from across the water. p. 14.

Daintith, T. C. (1993). Legal forms and techniques of privatisation. In *Legal Aspects of Privatisation*, Proceedings, XXIst Colloquy on European Law, Budapest October 15–17, 1991 (pp.50–87). Strasbourg: Council of Europe Press.

Daniels, P. W., & Lever, W. F. (1996). *The Global Economy in Transition*. Harlow, Essex: Longman.

Davis, M. (1986). *Prisoners of the American Dream: Politics and Economy in the History of the US Working Class*. New York: Verso.

Davis, T. (1995). Cross border organizing comes home. *Labor Research Review*, 23 (Spring/Summer), 23–29.

De Luce, D. (1993). Premier charges extortion by union. *Prague Post*, February 17–23, p. 9.

Delaney, D., & Leitner, H. (1997). The political construction of scale. *Political Geography*, 16, 93–97.

Department of Labor. (1980). *Wage Chronology: North Atlantic Shipping Associations and the International Longshoremen's Association (ILA) 1934–80*. Bulletin No. 2063, Bureau of Labor Statistics, U.S. Department of Labor. Washington, DC: U.S. Government Printing Office.

Devreese, D. E. (1988). An inquiry into the causes and nature of organization: Some observations on the International Working Men's Association, 1864–1872/1876. In F. van Holthoon & M. van der Linden (Eds.), *Internationalism in the Labour Movement 1830–1940, Volume 1* (pp. 283–303). London: Brill.

Dicken, P. (1971). Some aspects of decision-making behavior of business organizations. *Economic Geography*, 47, 426–437.

Dicken, P. (1998). *Global Shift: Transforming the World Economy* (3rd ed.). New York: Guilford Press.

Dicken, P., & Lloyd, P. E. (1990). *Location in Space: Theoretical Perspectives in Economic Geography* (3rd edition). New York: HarperCollins.

Dickman, J. (1973). Testimony of James Dickman, President of the New York Shipping Association, in the matter of *"Balicer v. International Longshoremen's Association,"* Civ. A. No. 1155–73, U.S. District Court (D. NJ), opinion reported at 364 F.Supp. 205 (1973).

Dickman, J. (1980). Statement by James Dickman before the U.S. House Committee on Merchant Marine and Fisheries, Subcommittee on Merchant Marine hearings on HR 6613, a bill to amend the 1916 Shipping Act to prevent regulation of collective bargaining agreements by the Federal Maritime Commission, March 11, 1980, 96th Cong., 2nd Sess.

DiFazio, W. (1985). *Longshoremen: Community and Resistance on the Brooklyn Waterfront*. South Hadley, MA: Bergin and Garvey.

Dixon, C. J., Drakakis-Smith, D., & Watts, H. D. (Eds.). (1986). *Multinational Corporations and the Third World*. Boulder, CO: Westview Press.

Doherty, Jr., W. C. (1966). The American Institute for Free Labor Development. In *Proceedings of the First Interprofessional Forum for Priorities for Peace* held in New York City, December 6, 1966. Published by National Strategy Information Center, Inc., New York City.

Doherty, Jr., W. C. (1997). [Telephone interview by author with William Doherty Jr., Former Executive Director, AIFLD, February 5, 1997.]

Donaghu, M. T., & Barff, R. (1990). Nike just did it: International subcontracting and flexibility in athletic footwear production. *Regional Studies*, 24, 537–552.

Doordan, D. P. (1983). The political content in Italian architecture during the fascist era. *Art Journal*, Summer, 121–131.

Dreyfus, H. L., & Rabinow, P. (1983). *Michel Foucault, Beyond Structuralism and Hermeneutics*. Chicago: University of Chicago Press (2nd ed.).

Dubovsky, M. (1969). *We Shall Be All: A History of the Industrial Workers of the World*. Urbana and Chicago: University of Illinois Press.

Dunning, J. H. (1971). *The Multinational Enterprise*. London: Allen and Unwin.

Dunning, J. H. (1988). *Explaining International Production*. London: Unwin Hyman.

Earle, C. (1992). The last great chance for an American working class: Spatial lessons of the general strike and the Haymarket riot of early May 1886. In C. Earle (Ed.), *Geographical Inquiry and American Historical Problems* (pp. 378–399). Stanford, CA: Stanford University Press.

Ellem, B., & Shields, J. (1999) Rethinking "regional industrial relations": Space, place and the social relations of work. *Journal of Industrial Relations*, 41.4, 536–560.

Estrin, S., & Stone, R. (1996). A taxonomy of mass privatization. *Transition: The Newsletter about Reforming Economies*, 17.11–12 (November–December) (published by the World Bank, New York).

Fainstein, S. S. (1985). The redevelopment of 42nd street: Clashing viewpoints. *City Almanac*, 18.4, 2–8.

Falbr, R. (1996). [Interview by author with Richard Falbr, President, Czech-Moravian Confederation of Trade Unions (Českomoravská Konfederace Odborových Svazů [ČMKOS]), Prague, Czech Republic, August 20.

Federal Maritime Commission. (1973). *New York Shipping Association—International Longshoremen's Association Man-Hour/Tonnage Method of Assessment*, 16 FMC 381 (1973), *aff'd sub nom. New York Shipping Association v. Federal Maritime Commission*, 495 F.2d 1215 (2d Cir.), *cert. denied*, 419 U.S. 964 (1974).

Federal Maritime Commission. (1978). *Sea-Land Services, Inc. and Gulf Puerto Rico Lines, Inc.—Proposed Rules on Containers*. Initial Decision 21 F.M.C. 7 (1978), *aff'd in part and remanded in part sub nom. Council of North Atlantic Steamship Associations v. Federal Maritime Commission*, 672 F.2d 171 (D.C. Cir., 1982), *cert. denied*, 459 U.S. 830 (1982).

Federal Maritime Commission. (1987). *'50 Mile Container Rules' Implementation by Ocean Common Carriers Serving U.S. Atlantic and Gulf Coast Ports*. Federal Maritime Commission docket No. 81–11, served August 3, 1987.

FIOST (International Federation of Trade Unions of Transport Workers). (1997). *Activities Report for 1993–1997*. Brussels: International Federation of Trade Unions of Transport Workers.

Fishman, R. (1977). *Urban Utopias in the Twentieth Century: Ebenezer Howard, Frank Lloyd Wright, and Le Corbusier*. New York: Basic Books.

FMTI (Fédération Mondiale des Travailleurs de l'Industrie). (1997). *Activities*. Brussels: Fédération Mondiale des Travailleurs de l'Industrie.

Foner, P. S. (1989). *U.S. Labor and the Viet-Nam War*. New York: International Publishers.

Forbes. (April 30, 1990). Mutual convenience. pp. 36–37.

Forbes. (January 13, 1992). U.S. Mint makes Marc Rich richer. p. 85.

Frampton, K. (1969). Labour, work and architecture. In C. Jencks & G. Baird (Eds.), *Meaning in Architecture* (pp. 150–168). New York: Braziller.

Frank, D. (1994). *Purchasing Power: Consumer Organizing, Gender, and the Seattle Labor Movement, 1919–1929.* New York: Cambridge University Press.

Frankland, E. G., & Cox, R. H. (1995). The legitimation problems of new democracies: Postcommunist dilemmas in Czechoslovakia and Hungary. *Environment and Planning C: Government and Policy,* 13, 141–158.

Freeman, R. B. (1992). Getting there from here: Labor in the transition to a market economy. In B. Silverman, R. Vogt, & M. Yanowitch (Eds.), *Labor and Democracy in the Transition to a Market System: A U.S.-Post-Soviet Dialogue* (pp. 139–157). London: Sharpe.

Fröbel, F., Heinrichs, J., & Kreye, O. (1980). *The New International Division of Labour: Structural Unemployment in Industrialised Countries and Industrialisation in Developing Countries.* Cambridge, UK: Cambridge University Press.

Froehling, O. (1997). The cyberspace 'war of ink and internet' in Chiapas, Mexico. *The Geographical Review,* 87.2, 291–307.

Frundt, H. J. (1987). *Refreshing Pauses: Coca-Cola and Human Rights in Guatemala.* New York: Praeger.

Frundt, H. J. (1996). Trade and cross-border labor strategies in the Americas. *Economic and Industrial Democracy* 17: 387–417.

Fryer, D. W. (1987). The political geography of international lending by private banks. *Transactions of the Institute of British Geographers, New Series,* 12.4, 413–432.

FTUI (Free Trade Union Institute). (1994). *Annual Report of the Free Trade Union Institute.* Washington, DC: AFL-CIO.

Fukuyama, F. (1992). *The End of History and the Last Man.* New York: Free Press.

Galdámez, A. (1997a). [Letter from Armando Galdámez, General Secretary of the FESITRANH Executive Committee and Director of Housing Project, San Pedro Sula, Honduras, to Aida Buchhalter, AIFLD office, Tegucigalpa, Honduras, dated February 12.]

Galdámez, A. (1997b). [Interview by author with Armando Galdámez, General Secretary of the FESITRANH Executive Committee and Director of Housing Project, San Pedro Sula, Honduras, April 21.]

Gates, B., with Myhrvold, N., & Rinearson, P. (1995). *The Road Ahead.* New York: Viking.

Gaudier, M. (1991). Economic reform, social change and institutional perspectives in Central and Eastern Europe: An analysis of the literature. *Labour and Society,* 16.4, 439–466.

Gerdel, T. W. (1999). Goodyear unions form network for mutual aid. *Cleveland Plain Dealer,* March 12, p. 1C.

Gershman, C. (1975). *The Foreign Policy of American Labor.* The Washington Papers, Vol. III, No.29. Beverly Hills, CA: Sage.

Ghebali, V. Y. (1989). *The International Labour Organisation: A Case Study on the Evolution of U.N. Specialised Agencies.* Dordrecht, Netherlands: Nijhoff.

Gibb, R. A., & Michalak, W. Z. (1993). Foreign debt in the new East-Central Europe:

A threat to European integration? *Environment and Planning C: Government and Policy*, 11, 69–85.

Gibson-Graham, J. K. (1996). *The End of Capitalism (As We Knew It): A Feminist Critique of Political Economy*. Cambridge, MA: Blackwell.

Giddens, A. (1984). *The Constitution of Society: Outline of the Theory of Structuration*. Berkeley and Los Angeles: University of California Press.

Gingrich, N. (1995). *To Renew America*. New York: HarperCollins.

Gleason, T. (1955). Testimony by Thomas "Teddy" Gleason, General Organizer, International Longshoremen's Association, before the U.S. House Committee on Merchant Marine and Fisheries, hearings on HR 5734, a bill to amend Section 301(a) of the 1936 Merchant Marine Act, 84th Cong., 1st Sess., 1955.

Gleason, T. (1973). Testimony of Thomas "Teddy" Gleason, President, International Longshoremen's Association, in the matter of *Balicer v. International Longshoremen's Association*, Civ. A. No. 1155–73, U.S. District Court (D. NJ), opinion reported at 364 F.Supp. 205 (1973).

Gleason, T. (1980a). Testimony by Thomas Gleason, ILA President, before the U.S. House Committee on Merchant Marine and Fisheries, Subcommittee on Merchant Marine hearings on HR 6613, a bill to amend the 1916 Shipping Act to prevent regulation of collective bargaining agreements by the Federal Maritime Commission, 96th Cong., 2nd Sess.

Gleason, T. (1980b). Thomas Gleason, ILA President, interviewed by Debra Bernhardt, July 31, 1980, "New Yorkers at Work Series, New Series 1, tape #44," Robert F. Wagner Labor Archives, Tamiment Institute Library, New York University.

Godson, R. (1976). *American Labor and European Politics: The AFL as a Transnational Force*. New York: Crane Russak.

Godson, R. (1984). *Labor in Soviet Global Strategy*. New York: Crane Russak.

Goldberg, J. (1968). Containerization as a force for change on the waterfront. *Monthly Labor Review*, 91.1, 8–13.

Gompers, S. (1910). *Labor in Europe and America*. New York: Harper.

Gompers, S. (1925). *Seventy Years of Life and Labor: An Autobiography* (2 vols.). New York: Dutton.

Gordon, D. M. (1978). Capitalist development and the history of American cities. In W. K. Tabb & L. Sawers (Eds.), *Marxism and the Metropolis: New Perspectives in Urban Political Economy* (pp. 25–63). New York: Oxford University Press.

Green, J., & Tilly, C. (1987). Service unionism: Directions for organizing. *Labor Law Journal*, August, 486–495.

Greenfield, G. M. (1987). Brazil. In G. M. Greenfield & S. L. Maram (Eds.), *Latin American Labor Organizations* (pp. 63–128). New York: Greenwood Press.

Greenhut, M. (1956). *Plant Location in Theory and Practice*. Chapel Hill: University of North Carolina Press.

Greenhut, M. ,& Hwang, M.-J. (1979). Estimates of fixed costs and the sizes of market areas in the United States. *Environment and Planning A*, 11, 993–1009.

Gregory, D. (1978). *Ideology, Science and Human Geography*. London: Hutchinson.

Groom, P. (1965). Hiring practices for longshoremen: The diversity of arrangements shown by a Labor Department study of 10 East and Gulf Coast ports. *Monthly Labor Review*, 88, 1289–1296.

Haas, G., and the Plant Closures Project. (1985). *Plant Closures: Myths, Realities and Responses*. Boston: South End Press.

Hadjimichalis, C. (1984). The geographical transfer of value: Notes on the spatiality of capitalism. *Environment and Planning D: Society and Space*, 2, 329–345.

Hagemann, R. (1984). Affadavit of Robert Hagemann, Vice President of Traffic, Jayne's Motor Freight Inc., Joint Exhibit in the matter of *"National Labor Relations Board v. International Longshoremen's Association*, petition for writ of *certiorari* to the U.S. Court of Appeals for the Fourth Circuit" before the U.S. Supreme Court, case no. 84–861, October Term, 1984.

Hägerstrand, T. (1967). *Innovation Diffusion as Spatial Process*. Chicago: University of Chicago Press. Postscript and translation by Allan Pred.

Hanson, S., & Johnston, I. (1985). Gender differences in work-trip length: Explanations and implications. *Urban Geography*, 6, 193–219.

Hanson, S., & Pratt, G. (1995). *Gender, Work, and Space*. New York: Routledge.

Hardy, M. (1936). *The Influence of Organized Labor on the Foreign Policy of the United States*. Ph.D. Dissertation, University of Geneva. Liége, Belgium: H. Vaillant-Carmanne, Imprimerie de l'Académie (Thèse No. 30).

Harmon, J. L. (1959). *The International Metalworkers' Federation*. Washington, DC: U.S. Department of Labor, Office of International Labor Affairs.

Harrod, J. (1972). *Trade Union Foreign Policy: A Study of British and American Trade Union Activities in Jamaica*. Publications of the International Institute for Labour Studies. London: Macmillan.

Hart, J. F. (1982). The highest form of the geographer's art. *Annals of the Association of American Geographers*, 72.1, 1–29.

Hartshorne, R. (1939). *The Nature of Geography: A Critical Survey of Current Thought in the Light of the Past*. Lancaster, PA: Association of American Geographers.

Harvey, D. (1972). Revolutionary and counter revolutionary theory in geography and the problem of ghetto formation. *Antipode*, 4, 1–13.

Harvey, D. (1973). *Social Justice and the City*. London: Arnold.

Harvey, D. (1976). Labor, capital, and class struggle around the built environment in advanced capitalist societies. *Politics and Society*, 6.3, 265–295.

Harvey, D. (1978). The urban process under capitalism: A framework for analysis. *International Journal of Urban and Regional Research*, 2, 101–131.

Harvey, D. (1982). *The Limits to Capital*. Oxford: Blackwell.

Harvey, D. (1984). On the history and present condition of geography: An historical materialist manifesto. *Professional Geographer*, 36, 1–11.

Harvey, D. (1985a). *The Urbanization of Capital: Studies in the History and Theory of Capitalist Urbanization*. Baltimore, MD: Johns Hopkins University Press.

Harvey, D. (1985b). *Consciousness and the Urban Experience*. Baltimore, MD: Johns Hopkins University Press.

Harvey, D. (1985c). The Geopolitics of Capitalism. In D. Gregory & J. Urry (Eds.), *Social Relations and Spatial Structures* (pp. 128–163). New York: St. Martin's Press.

Harvey, D. (1988). The geographical and geopolitical consequences of the transition from Fordist to flexible accumulation. In G. Sternlieb & J. W. Hughes (Eds.),

America's New Market Geography (pp. 101–134). New Brunswick, NJ: Center for Urban Policy Research, Rutgers University.

Harvey, D. (1989). *The Condition of Postmodernity: An Enquiry into the Origins of Cultural Change.* Oxford: Blackwell.

Hatch, J. K., & Flores, A. L. (1977). *An Evaluation of the AIFLD/Histadrut Project Proposal to Assist Peasant Federations in Honduras, El Salvador, and Guatemala.* Report prepared by Rural Development Services for the Office of Multilateral Coordination and Regional Social Development Programs (LA/MRSD/L) of the Agency for International Development under Contract No. AID/La-C-1227, Project No. 598–0044, dated August 19, 1977.

Hawkins, C. (1965). The ORIT and the American trade unions—conflicting perspectives. In W. H. Form & A. A. Blum (Eds.), *Industrial Relations and Social Change in Latin America* (pp. 87–104). Gainesville: University of Florida Press.

Haworth, N., & Ramsay, H. (1984). Grasping the nettle: Problems in the theory of international labour solidarity. In P. Waterman (Ed.), *For a New Labour Internationalism: A Set of Reprints and Working Papers* (pp. 59–85). The Hague: International Labour Education, Research and Information Foundation.

Hayden, D. (1976). *Seven American Utopias: The Architecture of Communitarian Socialism, 1790–1975.* Cambridge, MA: MIT Press.

Hayden, D. (1981). *The Grand Domestic Revolution: A History of Feminist Designs for American Homes, Neighborhoods, and Cities.* Cambridge, MA: MIT Press.

Hayden, D. (1984). *Redesigning the American Dream: The Future of Housing, Work, and Family Life.* New York: Norton.

Haynes, J. M. (1977). Affadavit of John Haynes, Executive Vice President of the NYSA, in the matter of *"International Longshoremen's Association and New York Shipping Association v. Dolphin Forwarding Inc. and San Juan Freight Forwarding Inc."* before National Labor Relations Board Region 2. Located in "Dolphin Forwarding Inc.," National Labor Relations Board case nos. 2–CC-1364 et al., Folder I, National Labor Relations Board, Washington, DC.

Hayter, R., & Watts, H. D. (1983). The geography of enterprise: A reappraisal. *Progress in Human Geography*, 7, 157–181.

Heaps, D. (1955). Union participation in foreign aid programs. *Industrial and Labor Relations Review*, 9.1, 100–108.

Hecker, S., & Hallock, M. (Eds.). (1991). *Labor in a Global Economy: Perspectives from the U.S. and Canada.* Eugene: University of Oregon Labor Education and Research Center.

Heiman, M. K. (1988). *The Quiet Evolution: Power, Planning, and Profits in New York State.* New York: Praeger.

Helmer, S. D. (1985). *Hitler's Berlin: The Speer Plans for Reshaping the Central City.* Ann Arbor, MI: UMI Research Press.

Hennart, J.-F. (1982). *A Theory of the Multinational Enterprise.* Ann Arbor: University of Michigan Press.

Herod, A. (1991a). The production of scale in United States labour relations. *Area*, 23.1, 82–88.

Herod, A. (1991b). Homework and the fragmentation of space: Challenges for the labor movement. *Geoforum*, 22.2, 173–183.

Herod, A. (1991c). Local political practice in response to a manufacturing plant closure: How geography complicates class analysis. *Antipode*, 23.4, 385–402.

Herod, A. (1992). *Towards a Labor Geography: The Production of Space and the Politics of Scale in the East Coast Longshore Industry, 1953–1990*. Unpublished PhD dissertation, Department of Geography, Rutgers University.

Herod, A. (1994). Further reflections on organized labor and deindustrialization in the United States. *Antipode*, 26.1, 77–95.

Herod, A. (1995). The practice of international labor solidarity and the geography of the global economy. *Economic Geography*, 71.4, 341–363.

Herod, A. (1997a). Labor as an agent of globalization and as a global agent. In K. Cox (Ed.), *Spaces of Globalization: Reasserting the Power of the Local* (pp. 167–200). New York: Guilford Press.

Herod, A. (1997b). Labor's spatial praxis and the geography of contract bargaining in the US east coast longshore industry, 1953–1989. *Political Geography*, 16.2, 145–169.

Herod, A. (1997c). Reinterpreting organized labor's experience in the Southeast: 1947 to present. *Southeastern Geographer*, 37.2, 214–237.

Herod, A. (1997d). Notes on a spatialized labour politics: Scale and the political geography of dual unionism in the US longshore industry. In R. Lee & J. Wills (Eds.), *Geographies of Economies* (pp. 186–196). London: Arnold.

Herod, A. (1998a). The spatiality of labor unionism: A review essay. In A. Herod (Ed.), *Organizing the Landscape: Geographical Perspectives on Labor Unionism* (pp. 1–36). Minneapolis and London: University of Minnesota Press.

Herod, A. (Ed.). (1998b). *Organizing the Landscape: Geographical Perspectives on Labor Unionism*. Minneapolis and London: University of Minnesota Press.

Herod, A. (1998c). Discourse on the docks: Containerization and inter-union work disputes in US ports, 1955–85. *Transactions of the Institute of British Geographers, New Series*, 23, 177–191.

Herod, A. (1998d). Theorising unions in transition. In J. Pickles & A. Smith (Eds.), *Theorizing Transition: The Political Economy of Change in Central and Eastern Europe* (pp. 197–217). London: Routledge.

Herod, A. (1998e). Of blocs, flows and networks: The end of the Cold War, cyberspace, and the geo-economics of organized labor at the *fin de millénaire*. In A. Herod, G. Ó Tuathail, & S. M. Roberts (Eds.), *An Unruly World? Globalization, Governance and Geography* (pp. 162–195). London: Routledge.

Herod, A. (2000). Implications of Just-in-Time production for union strategy: Lessons from the 1998 General Motors–United Auto Workers dispute. *Annals of the Association of American Geographers*, 90.3, 521–547.

Herod, A. (2001). Labor internationalism and the contradictions of globalization: Or, why the local is sometimes still important in a global economy. *Antipode*, 33, 407–426.

Herod, A., Ó Tuathail, G., & Roberts, S. M. (Eds.). (1998). *An Unruly World? Globalization, Governance and Geography*. London: Routledge.

Herod, A., Peck, J., & Wills, J. (in press). Industrial relations and geography. In A. Wilkinson and P. Ackers (Eds.), *Reworking Industrial Relations: New Perspectives on Employment and Society*. Oxford: Oxford University Press.

Héthy, L. (1991). Towards social peace or explosion? Challenges for labour relations in Central and Eastern Europe. *Labour and Society*, 16.4, 345–58.

Hobsbawm, E. (1975). *The Age of Capital, 1848–1875*. New York: Mentor.

Hobsbawm, E., & Ranger, T. (Eds.). (1983). *The Invention of Tradition*. Cambridge, UK: Cambridge University Press.

Holmes, J., & Rusonik, A. (1991). The break-up of an international labour union: Uneven development in the North American auto industry and the schism in the UAW. *Environment and Planning A*, 23, 9–35.

Holston, J. (1989). *The Modernist City: An Anthropological Critique of Brasilia*. Chicago: University of Chicago Press.

Holway, J. (1972). AIFLD's programs in Argentina—1969–1972. *AIFLD Report*, October 1972, pp. 4 and 5.

Holway, J. (1997). [E-mail correspondence between author and James Holway, former Director, AIFLD Social Projects Program (1965–1969), dated May 15.]

Holway, J. (1998a). [Letter to author by James Holway, former Director, AIFLD Social Projects Program (1965–1969), dated April 8.]

Holway, J. (1998b). [E-mail correspondence between author and James Holway, former Director, AIFLD Social Projects Program (1965–1969), dated August 17.]

Homann, K., & Scarpa, L. (1983). Martin Wagner, the trades union movement and housing construction in Berlin in the first half of the nineteen twenties. *Architectural Design*, 53.11/12, 58–61.

Hood, N., & Young, S. (1979). *The Economics of Multinational Enterprises*. London: Longman.

Horne, W. W. (1991). Milgrim Thomajan strengthens ties to fugitive Rich. *American Lawyer*, July/August, 16.

Hotelling, H. (1929). Stability in competition. *Economic Journal*, 39, 41–57.

Howard, A. (1995). Global capital and labor internationalism in comparative historical perspective: A Marxist analysis. *Sociological Inquiry*, 65.3–4, 365–394.

Howard, E. (1898). *To-morrow: A Peaceful Path to Real Reform*. London: Swan Sonnenschein (republished in 1902 as *Garden Cities of To-morrow*).

Hudson, B. (1977). The new geography and the new imperialism: 1870–1918. *Antipode*, 9, 12–19.

Hudson, H. D., Jr. (1994). *Blueprints and Blood: The Stalinization of Soviet Architecture, 1917–1937*. Princeton, NJ: Princeton University Press.

Hudson, R., & Sadler, D. (1983). Region, class and the politics of steel closures in the European Community. *Environment and Planning D: Society and Space*, 1, 405–428.

Hudson, R., & Sadler, D. (1986). Contesting work closures in Western Europe's old industrial regions: Defending place or betraying class? In A. Scott & M. Storper (Eds.), *Production, Work, Territory: The Geographical Anatomy of Industrial Capitalism* (pp. 172–194). Boston: Allen and Unwin.

Hutchinson, J. (1970). *The Imperfect Union: A History of Corruption in American Trade Unions*. New York: Dutton.

Hyman, R. (1999). *An Emerging Agenda for Trade Unions?* Discussion paper DP/98/1999 published by the Labour and Society Programme, International Labour Organisation, Geneva, Switzerland.

Hymer, S. H. (1971). The multinational corporation and the law of uneven develop-
    ment. In R. B. Cohen, N. Felton, M. Nkosi, J. van Liere, & N. Dennis (Eds.), *The
    Multinational Corporation: A Radical Approach: Papers by Stephen Herbert
    Hymer* (pp. 54–74). New York: Cambridge University Press.
Hymer, S. H. (1976). *The International Operations of National Firms: A Study of Di-
    rect Involvement.* Cambridge, MA: MIT Press.
Hymer, S. H. (1979). *The Multinational Corporation: A Radical Approach.* Cam-
    bridge: Cambridge University Press.
ICEM (International Federation of Chemical, Energy, Mine and General Workers' Un-
    ions). (1996). *Power and Counterpoint: The Union Response to Global Capital.*
    London: Pluto Press.
ICEM (International Federation of Chemical, Energy, Mine and General Workers' Un-
    ions). (1998). Rubber unions go global. *ICEM Update,* No. 65, June 26, 1–2.
    (Available at *http://www.icem.org/update/upd1998/upd98–65.html*).
ICFTU (International Confederation of Free Trade Unions.). (1957). *Yearbook of the
    International Free Trade Union Movement 1957–1958.* London: Lincolns-
    Prager.
ICFTU (International Confederation of Free Trade Unions). (1996). *Report on Activ-
    ities of the Confederation and Financial Reports, 1991–1994.* Brussels: ICFTU.
ICFTU (International Confederation of Free Trade Unions). (1998). Internationally-re-
    cognised core labour standards in Hungary. *Report for the WTO General Coun-
    cil Review of the Trade Policies of Hungary,* dated July 7–8. Geneva: ICFTU.
IFTC (International Federation of Textile and Clothing Workers). (1997). *Activities
    Report 1993–97.* Brussels: IFTC. (Available at *http://www.cmt-wcl.org/en/
    fedtextile.htm* report).
ILA (International Longshoremen's Association). (1977). Brief by the International
    Longshoremen's Association for U.S. Court of Appeals (D.C. Cir.) in the matter
    of *International Longshoremen's Association and Council of North Atlantic
    Steamship Associations v. National Labor Relations Board,* Civ. A. 77–1735, and
    *International Longshoremen's Association and Council of North Atlantic Steam-
    ship Associations v. National Labor Relations Board,* Civ. A. 77–1758. Located in
    "Dolphin Forwarding Inc.," National Labor Relations Board case nos. 2–CC-1364
    et al., National Labor Relations Board, Washington, DC.
ILA (International Longshoremen's Association). (1980). Response brief of the Inter-
    national Longshoremen's Association, the New York Shipping Association and
    the Council of North Atlantic Shipping Associations in the matter of *"Interna-
    tional Longshoremen's Association et al.,* NLRB case nos. 5–CC-791 etc. and 5–
    CE-48 etc.; 2–CC-1364, 1365, and 2–CE-75; 12–CC-1002, 1004; 4–CC-1133 and
    4–CE-55." Located in "Dolphin Forwarding Inc.," National Labor Relations
    Board case nos. 2–CC-1364 et al., Folder I, National Labor Relations Board,
    Washington, DC.
ILA-IND (International Longshoremen's Association—Independent). (1956). Brief of
    Intervenor, International Longshoremen's Association, Independent before Na-
    tional Labor Relations Board Region 2 in the matter of *New York Shipping Asso-
    ciation et al.,* NLRB case no. 2–RC-8388. Located in box 691, National Labor Re-
    lations Board 1935–, Administrative Division Files and Dockets Section, Selected

Taft–Hartley Cases (1947–59), Records of the National Labor Relations Board, RG 25, National Archives, Washington, DC.

ILA/NYSA (International Longshoremen's Association/New York Shipping Association). (1959). *Containers—Dravo Size or Larger.* Memorandum of Settlement, dated December 3, 1959.

Institutional Investor. (1992). Smoking out Marc Rich. August, pp. 40–46.

International Association of Non-Vessel Operating Common Carriers. (1981). Brief of the International Association of Non-Vessel Operating Common Carriers, Twin Express Inc., and the Custom House Brokers' and Forwarders' Association of Miami Inc., dated July 3, 1981, before National Labor Relations Board Region 2 in the matter of *"International Longshoreman's Association and New York Shipping Association v. Dolphin Forwarding Inc. and San Juan Freight Forwarders Inc."* Located in "Dolphin Forwarding Inc.," National Labor Relations Board case nos. 2–CC-1364 et al., Folder I, National Labor Relations Board, Washington, DC.

International Longshoremen's Association, 221 NLRB 956 (1975), *enf'd sub nom. International Longshoremen's Association v. National Labor Relations Board,* 537 F.2d 706 (2d Cir., 1976), *cert. denied* 429 U.S. 1041 (1977).

International Longshoremen's Association Local 1408. 245 NLRB 1320 (1979).

International Longshoremen's Association v. National Labor Relations Board, 613 F.2d 890 (D.C. Cir., 1979).

IMF (International Metalworkers' Federation). (No date a). *Metalworkers in Eastern Europe.* Geneva: International Metalworkers' Federation.

IMF (International Metalworkers' Federation). (No date b). *A Metalworkers' Charter for Social and Economic Alternatives.* Geneva: International Metalworkers' Federation.

IMF (International Metalworkers' Federation). (1991). *The IMF and the Multinationals: The Role of the IMF World Company Councils* (Report, dated May 23–24). Lisbon: International Metalworkers' Federation Central Committee.

IMF (International Metalworkers' Federation). (1992a). *Investissements Etrangers et Droits Syndicaux en Europe Centrale et Orientale.* Geneva: International Metalworkers' Federation.

IMF (International Metalworkers' Federation). (1992b). *Industrial Restructuring and East-West Migration: The Impact on the Metal Industry.* Geneva: International Metalworkers' Federation.

IMF (International Metalworkers' Federation). (1993). *Metalworkers and Privatisation Worldwide.* Geneva: International Metalworkers' Federation.

IMF (International Metalworkers' Federation). (1994). *Collective Bargaining in Transition Economies.* Geneva: International Metalworkers' Federation.

IMF (International Metalworkers' Federation). (2000a). IMF to gain new members. *IMF News,* February 18. Geneva: International Metalworkers' Federation.

IMF (International Metalworkers' Federation). (2000b). Stability pact is focus of Balkan meeting. *IMF News,* May 11. Geneva: International Metalworkers' Federation.

International Monetary Fund. (1997). *World Economic Outlook.* Washington, DC: International Monetary Fund.

Isard, W. (1956). *Location and Space-Economy: A General Theory Relating to Industrial Location, Market Areas, Land Use, Trade, and Urban Structure.* Cambridge, MA: MIT Press.

Jacobs, R. M. (1973). Testimony of Roy Jacobs, Executive Vice President of Consolidated Express Inc., in the matter of *Balicer v. International Longshoremen's Association*, Civ. A. No. 1155–73, U.S. District Court (D. NJ), opinion reported at 364 F.Supp. 205 (1973).

Jagan, C. (1972 [1966]). *The West on Trial: The Fight for Guyana's Freedom.* New York: International Publishers.

Jagan, C. (1979). *The Caribbean Revolution.* Prague: Orbis Press Agency.

Jarley, P., & Maranto, C. L. (1990). Union corporate campaigns: An assessment. *Industrial and Labor Relations Review,* 43.5, 505–524.

Jencks, C. (1969). Semiology and architecture. In C. Jencks & G. Baird (Eds.), *Meaning in Architecture* (pp. 10–25). New York: Braziller.

Jenkins, R. (1987). *Transnational Corporations and Uneven Development: The Internationalization of Capital and the Third World.* London: Methuen.

Jensen, V. H. (1974). *Strife on the Waterfront: The Port of New York since 1945.* Ithaca, NY: Cornell University Press.

Johns, R. (1998). Bridging the gap between class and space: U.S. worker solidarity with Guatemala. *Economic Geography,* 74.3, 252–271.

Johns, R., & Vural, L. (2000). Class, geography, and the consumerist turn: UNITE and the Stop Sweatshops Campaign. *Environment and Planning A,* 32.7, 1193–1213.

Johnson, K. M., & Garnett, H. C. (1971). *The Economics of Containerisation.* London: Allen and Unwin.

Johnston, G. A. (1970). *The International Labour Organisation: Its Work for Social and Economic Progress.* London: Europa.

Jonas, A. E. G. (1994). The scale politics of spatiality. *Environment and Planning D: Society and Space,* 12, 257–264.

Jonas, A. E. G. (1998). Investigating the local–global paradox: Corporate strategy, union local autonomy, and community action in Chicago. In A. Herod (Ed.), *Organizing the Landscape: Geographical Perspectives on Labor Unionism* (pp. 325–350). Minneapolis and London: University of Minnesota Press.

Jones, B. (1982). Destruction or redistribution of engineering skills? The case of numerical control. In S. Wood (Ed.), *The Degradation of Work? Skill, Deskilling and the Labour Process* (pp. 179–200). London: Hutchinson.

Jones, K. T. (1998). Scale as epistemology. *Political Geography,* 17.1, 25–28.

Journal of Commerce. (October 19, 1956). ILA to push for all-port wage accord. pp. 1 and 24.

Journal of Commerce. (November 27, 1956). ILA appeals outport issue. pp. 1 and 32.

Journal of Commerce. (January 7, 1957). Teamsters act to renew ILA alliance. p. 1.

Journal of Commerce. (January 14, 1957). US mediators gird to avert pier impasse. pp. 1 and 13.

Journal of Commerce. (January 31, 1957). Southern, Gulf pier pact signed. pp. 1 and 13.

Journal of Commerce. (February 18, 1957). Agreement reached to end pier strike. pp. 1 and 40.

Journal of Commerce. (December 16, 1964). Hopes rising for docker pact accord. pp. 1 and 21.

Journal of Commerce. (December 18, 1964). Local docker talks begin on high note. pp. 1 and 26.

Journal of Commerce. (December 22, 1964). Wildcats pose threat to pier labor peace. pp. 1 and 23.

Journal of Commerce. (January 5, 1965). New York pier talks hit surprising snag. pp. 1 and 25.

Journal of Commerce. (March 3, 1965). South Atlantic pier talks lag. p. 24.

Journal of Commerce. (March 22, 1965). National dock contract urged. p. 24.

Journal of Commerce. (August 7, 1968). Mediators join dock labor talks. pp. 1 and 2.

Journal of Commerce. (October 1, 1968). ILA strike ordered on East, Gulf Coasts. pp. 1 and 2.

Journal of Commerce. (January 14, 1969). Dock labor talks set for other ports. pp. 1 and 19.

Journal of Commerce. (January 28, 1969). Dock union leaders will map strategy. pp. 1 and 10.

Journal of Commerce. (March 7, 1972). East, Gulf ports ready dock pacts. pp. 1 and 20.

Journal of Commerce. (June 16, 1972). Wildcat strike looms in Britain. pp. 25 and 26.

Journal of Commerce. (May 28, 1980). ILA contract signed for East, Gulf coasts. pp. 1 and 17.

Journal of Commerce. (January 17, 1984). NYSA talks produce little new. pp. 1A and 12A.

Journal of Commerce. (June 9, 1986). S. Atlantic ILA accepts concessions. pp. 1A and 14A.

Journal of Commerce. (October 9, 1986). Seeds of ILA problems sown in past successes. pp. 1A and 14A.

Journal of Commerce. (June 26, 1989). 50–mile ruling pleases truckers. p. 8B.

Journal of Commerce. (June 27, 1989). Plan to create dock jobs starts on NYC pier. p. 1B.

Journal of Commerce. (December 4, 1990). Dockers reach tentative agreement in S. Atlantic. p. 8B.

Juárez Núñez, H., & Babson, S. (Eds.). (1998). *Confronting Change: Auto Labor and Lean Production in North America/Enfrentando el Cambio: Obreros del Automóvil y Producción Esbelta en América del Norte.* Puebla, Mexico: Autonomous University of Puebla.

Judd, D. R. (1998). The case of the missing scales: A commentary on Cox. *Political Geography* 17.1, 29–34.

Juravich, T., & Bronfenbrenner, K. (1999). *Ravenswood: The Steelworkers' Victory and the Revival of American Labor.* Ithaca, NY: Cornell University Press.

Kanter, R. M. (1995). *World Class: Thriving Locally in the Global Economy.* New York: Simon & Schuster.

Katz, S. (1986). Towards a sociological definition of rent: Notes on David Harvey's *The Limits to Capital. Antipode,* 18, 64–78.

Katznelson, I. (1992). *Marxism and the City.* Oxford: Clarendon Press.

Kaufmann, M. (1879). *Utopias; or Schemes of Social Improvement, from Sir Thomas More to Karl Marx*. London: Paul.

Keeble, D. (1968). Industrial decentralization and the metropolis: The North-West London case. *Transactions of the Institute of British Geographers*, 44, 1–54.

Kelly, P. (1997). Globalization, power and the politics of scale in the Philippines. *Geoforum*, 28, 151–171.

Kern, S. (1983). *The Culture of Time and Space, 1880–1918*. Cambridge, MA: Harvard University Press.

Kimeldorf, H. (1988). *Reds or Rackets? The Making of Radical and Conservative Unions on the Waterfront*. Los Angeles: University of California Press.

Krumme, G. (1969). Towards a geography of enterprise. *Economic Geography*, 45, 30–40.

Krumme, G., & Hayter, R. (1975). Implications of corporate staregies and product cycle adjustments for regional employment changes. In L. Collins & D. F. Walker (Eds.), *Locational Dynamics of Manufacturing Activity* (pp. 325–356). New York: Wiley.

Labor Notes. (1991). NewsWatch. November, p. 4.

Labor Notes. (1992). NewsWatch. March, p. 4.

Labor Notes (1994). Labor in cyberspace. August, p. 2.

Labor Research Association. (1992). Elements of victory 1: An interview with the AFL-CIO's Joe Uehlein on the Steelworkers Ravenswood campaign. *LRA's [Labor Research Association] Economic Notes*, May–June, 3–6.

Lamson, R. (1954). The 1951 New York wildcat dock strike: Some consequences of union structure for management-labor relations. *Southwestern Social Science Quarterly*, 34.4, 28–38.

Langdale, J. V. (1985). Electronic funds transfer and the internationalisation of the banking and finance industry. *Geoforum*, 16.1, 1–13.

Langley, D. (1972). The colonization of the international union movement. In B. Hall (Ed.), *Autocracy and Insurgency in Organized Labor* (pp. 296–309). New Brunswick, NJ: Transaction Books.

Lapointe, J. (1999). [Letter to the author from Jean Lapointe, Regional Representative, Regional Office for Central and Eastern Europe, International Metalworkers' Federation, Budapest, Hungary, dated December 14.]

Larrabee, E. (1948). The six thousand houses that Levitt built. *Harper's Magazine*, September, 79–88.

Larrowe, C. P. (1955). *Shape-Up and Hiring Hall: A Comparison of Hiring Methods and Labor Relations on the New York and Seattle Waterfronts*. Berkeley: University of California Press (reprinted in 1976 by Greenwood Press, Westport CT).

Larson, S. (1975). *Labor and Foreign Policy: Gompers, the AFL, and the First World War, 1914–1918*. Cranbury, NJ: Associated University Presses.

Latham, W. R., III. (1978). Measures of locational orientation for 199 manufacturing industries. *Economic Geography*, 54, 53–65.

Lattek, K. (1988). The beginnings of socialist internationalism in the 1840s: The "Democratic Friends of all Nations" in London. In F. van Holthoon & M. van der Linden (Eds.), *Internationalism in the Labour Movement 1830–1940*, Vol. 1 (pp. 259–282). London: Brill.

Le Corbusier (Charles-Édouard Jeanneret). (1931). *Towards a New Architecture.* London: Rodker.

Lee, E. (1997). *The Labour Movement and the Internet: The New Internationalism.* London: Pluto Press.

Lee, R. W. (1980). Answers of Dolphin Forwarding Inc. to Interrogatories of National Labor Relations Board, by Richard W. Lee, dated June 10, 1980. Located in "Dolphin Forwarding Inc.," National Labor Relations Board case nos. 2–CC-1364 et al., Folder I, National Labor Relations Board, Washington, DC.

Lefebvre, H. (1970). *La Revolution Urbaine.* Paris: Gallimard.

Lefebvre, H. (1976 [1973]). *The Survival of Capitalism: Reproduction of the Relations of Production.* London: St. Martin's Press.

Lefebvre, H. (1991 [1974]). *The Production of Space.* Oxford, UK: Blackwell.

Leitner, H. (1997). Reconfiguring the spatiality of power: The construction of a supranational migration framework for the European Union. *Political Geography,* 16, 123–143.

Lenin, V. I. (1939 [1900]). *Imperialism: The Highest Stage of Capitalism.* New York: International Publishers.

Lever, W. F. (1975). Manufacturing decentralization and shifts in factor costs and external economies. In L. Collins & D. F. Walker (Eds.), *Locational Dynamics of Manufacturing Activity* (pp. 295–324). New York: Wiley.

Lipietz, A. (1986). New tendencies in the international division of labor: Regimes of accumulation and modes of regulation. In A. Scott & M. Storper (Eds.), *Production, Work, Territory: The Geographical Anatomy of Industrial Capitalism* (pp. 16–40). Boston: Allen and Unwin.

Lochhead, C. (1987). Rag trade: With alterations, New York has it buttoned up. *Washington (DC) Times,* March 16, 1987, pp. 44–45.

Lorwin, L. L. (1929). *Labor and Internationalism.* New York: Macmillan.

Lösch, A. (1954). *The Economics of Location.* New Haven: Yale University Press.

Louis Harris Associates, Inc. (1986). *Survey of Garment Center Apparel Firms.* Report prepared for Department of City Planning, Office for Economic Development, and New York City Public Development Corporation.

Low, M. (1997). Representation unbound: Globalization and democracy. In K. Cox (Ed.), *Spaces of Globalization: Reasserting the Power of the Local* (pp. 240–280). New York: Guilford Press.

Lynch, J. (1979/80). Cuban architecture since the revolution. *Art Journal,* 39.2, 100–106.

Lytel, R. (1969). Comments made by Robert Lytel, Executive Director of the New York State Motor Truck Association, in *Journal of Commerce* (January 17), ILA contract upsets NY truck firms. pp. 1, 25–26.

McCain, M. (1987). A new zoning shelter for harried manufacturers. *New York Times,* April 5, 1987, p. R-43.

McCarthy, J. D., & Mayer, N. Z. (1977). Resource mobilization and social movements: A partial theory. *American Journal of Sociology,* 82, 1212–1241.

McNee, R. B. (1960). Towards a more humanistic economic geography: The geography of enterprise. *Tijdschrift voor Economische en Sociale Geografie,* 51, 201–206.

McNeil, A. (1983). Testimony of Allie McNeil, Executive Vice President of D. D. Jones Transfer and Warehouse Company, before the U.S. House Subcommittee on Merchant Marine, Committee on Merchant Marine and Fisheries hearings on HR 2562, a bill to amend Section 45 of the 1916 Shipping Act, 98th Cong., 1st Sess., 1983.

Machyna, E. (1996). [Interview by author with Emil Machyna, President, Slovak Metalworkers' Federation (Odborový Zväz KOVO), Bratislava, Slovak Republic, August 28.]

MacShane, D. (1992). *International Labour and the Origins of the Cold War.* Oxford, UK: Clarendon Press.

MacShane, D. (1996). *Global Business: Global Rights.* London: Fabian Society (Pamphlet No. 575).

Madden, J. (1981). Why women work closer to home. *Urban Studies,* 18, 181–194.

Mahon, M. (1984). Affidavit of Mathew Mahon Jr., President of Mahon's Express, Joint Appendix in the matter of *National Labor Relations Board v. International Longshoremen's Association,* petition for writ of *certiorari* to the U.S. Court of Appeals for the Fourth Circuit before the U.S. Supreme Court, case no. 84–861, October Term, 1984.

Maier, C. S. (1987). *In Search of Stability: Explorations in Historical Political Economy.* New York: Cambridge University Press.

Mair, A., Florida, R., & Kenney, M. (1988). The new geography of automobile production: Japanese transplants in North America. *Economic Geography,* 64, 352–373.

Mankoff, W. (1998). [Telephone interview by author with Walter Mankoff, formerly of the Research Department, International Ladies' Garment Workers' Union, New York City, January 20.]

Mann, T. (1967 [1923]). *Tom Mann's Memoirs.* London: MacGibbon & Kee. (1967 ed., with a preface by Ken Coates).

Marston, S. A. (1990). Who are 'the people'?: Gender, citizenship, and the making of the American nation. *Environment and Planning D: Society and Space,* 8, 449–458.

Marston, S. A. (2000). The social construction of scale. *Progress in Human Geography,* 24.2, 219–242.

Martin, R. (1999). The new "geographical turn" in economics: Some critical reflections. *Cambridge Journal of Economics,* 23, 65–91.

Martin, R., & Sunley, P. (1996). Paul Krugman's "geographical economics" and its implications for regional development theory: A critical assessment. *Economic Geography,* 72, 259–292.

Martin, R., Sunley, P., & Wills, J. (1993). The geography of trade union decline: Spatial dispersal or regional resilience? *Transactions of the Institute of British Geographers, New Series,* 18.1, 36–62.

Martin, R., Sunley, P., & Wills, J. (1994). Labouring differences: Method, measurement and purpose in geographical research on trade unions. *Transactions of the Institute of British Geographers, New Series,* 19.1, 102–110.

Marx, K. (1967 [1867]). *Capital,* Vol. 1. New York: International Publishers (1987 printing).

Massey, D. (1973). Towards a critique of industrial location theory. *Antipode,* 5, 33–39.

Massey, D. (1978). Regionalism: Some current issues. *Capital and Class*, 6, 106–125.

Massey, D. (1984). Introduction: Geography matters. In D. Massey & J. Allen (Eds.), *Geography Matters! A Reader* (pp. 1–11). New York: Cambridge University Press.

Massey, D. (1994). The geography of trade unions: Some issues. *Transactions of the Institute of British Geographers, New Series*, 19.1, 95–98.

Massey, D. (1995 [1984]). *Spatial Divisions of Labour: Social Structures and the Geography of Production*. London: Macmillan.

Massey, D., & Meegan, R. (1982). *The Anatomy of Job Loss: The How, Why and Where of Employment Decline*. New York: Methuen.

Massey, D., & Meegan, R. (1985). *Politics and Method*. New York: Methuen.

Massey, D., & Painter, J. (1989). The changing geography of trade unions. In J. Mohan (Ed.), *The Political Geography of Contemporary Britain* (pp. 130–150). London: Macmillan.

Mazur, J. (1987). Comments made by Jay Mazur, President, ILGWU, at the meeting of the City of New York Board of Estimate. See "Transcript of the stenographic record on Calendar nos. 2 and 41 held at the meeting of the Board of Estimate on March 26, 1987." Copy located at Department of City Planning, City of New York, 22 Reade Street, New York, NY.

Meakin, B. (1985 [1905]). *Model Factories and Villages: Ideal Conditions of Labour and Housing*. New York: Garland Publishing (originally published in London by Unwin).

Mergner, G. (1988). Solidarität mit den 'Wilden'? Das Verhältnis der Deutschen Sozialdemokratie zu den Afrikanischen Widerstandskämpfen in den Ehemaligen Deutschen Kolonien um die Jahrhundertwende (Solidarity with the 'Wild ones'? The relationship of German social democracy towards African opposition fighters in the former German colonies at the turn of the century). In F. van Holthoon & M. van der Linden (Eds.), *Internationalism in the Labour Movement 1830–1940, Vol. 1* (pp. 68–86). London: Brill.

Merrifield, A. (1993). Place and space: A Lefebvrian reconciliation. *Transactions of the Institute of British Geographers, New Series*, 18, 516–531.

Mers, G. (1988). *Working the Waterfront: The Ups and Downs of a Rebel Longshoremen*. Austin: University of Texas Press.

Methvin, E. H. (1966). Labor's new weapon for democracy. *Reader's Digest*, October, 21–28.

Meurs, M., & Begg, R. (1998). Path dependence in Bulgarian agriculture. In J. Pickles & A. Smith (Eds.), *Theorising Transition: The Political Economy of Post-Communist Transformations* (pp. 243–261). New York and London: Routledge.

Michalak, W. Z. (1993). Foreign direct investment and joint ventures in East-Central Europe: A geographical perspective. *Environment and Planning A*, 25, 1573–1591.

Miller, B. (1997). Political action and the geography of defense investment: Geographical scale and the representation of the Massachusetts Miracle. *Political Geography*, 16, 171–185.

Mitchell, D. (1996). *The Lie of the Land: Migrant Workers and the California Landscape*. Minneapolis: University of Minnesota Press.

Mitchell, D. (1998). The scales of justice: Localist ideology, large-scale production, and agricultural labor's geography of resistance in 1930s California. In A. Herod (Ed.), *Organizing the Landscape: Geographical Perspectives on Labor Unionism* (pp. 159–194). Minneapolis: University of Minnesota Press.

Moberg, D. (1989). Workers challenged by the global marketplace. *In These Times*, April 26–May 2, p. 6.

Moberg, D. (1992). Lock-out knock-out. *In These Times*, June 24–July 7, pp. 12–13.

Moody, K. (1988). *An Injury to All: The Decline of American Unionism.* New York: Verso.

Moody, K. (1997). *Workers in a Lean World: Unions in the International Economy.* London: Verso.

Morris, G. (1967). *CIA and American Labor: The Subversion of the AFL-CIO's Foreign Policy.* New York: International Publishers.

Mumford, L. (1959 [1922]). *The Story of Utopias.* Gloucester, MA: Smith.

Murdoch, J., & Marsden, T. (1995). The spatialization of politics: Local and national actor-spaces in environmental conflict. *Transactions of the Institute of British Geographers, New Series*, 20, 368–380.

Mureau, A.-M. (1994). [Interview by author with Anne-Marie Mureau, Coordinator for Central and Eastern Europe, International Metalworkers' Federation, Geneva, August 30.]

Mureau, A.-M. (1995a). [Personal communication by telephone with Anne-Marie Mureau, Coordinator for Central and Eastern Europe, International Metalworkers' Federation, Geneva, February 20.]

Mureau, A.-M. (1995b). [Letter to the author by Anne-Marie Mureau, Coordinator for Central and Eastern Europe, International Metalworkers' Federation, Geneva, dated May 2.]

Murphy, A. B. (1992). Western investment in east-central Europe: Emerging patterns and implications for state stability. *Professional Geographer*, 44.3, 249–259.

Murphy, M. F. (1987). Dominican Republic. In G. M. Greenfield & S. L. Maram (Eds.), *Latin American Labor Organizations* (pp. 265–288). New York: Greenwood Press.

Musil, J. (1991). New social contracts: Responses of the state and the social partners to the challenges of restructuring and privatisation. *Labour and Society* 16.4, 381–399.

Nadler, J. (1987). Comments made by New York State Assemblyman Jerrold Nadler, New York City Planning Commission public meeting, January 14, 1987.

Nagy, L. (1984). *The Socialist Collective Agreement.* Budapest: Akadémiai Kiadó.

National Labor Relations Board v. International Longshoremen's Association, 447 U.S. 490 (1980).

National Labor Relations Board v. International Longshoremen's Association, 473 U.S. 61 (1985).

National Labor Relations Board v. Mackay Radio and Telegraph Co, 304 U.S. 333 (1938).

National Labor Relations Board v. Pipefitters, 429 U.S. 507 (1977).

New York. (1952). *Final Report to the Industrial Commissioner, State of New York, from Board of Inquiry on Longshore Industry Work Stoppage, October–November, 1951, Port of New York.* Albany, NY: State of New York.

New York City. (1985). *Summary of the Census: New York City's Garment Center.*

New York: New York City Office for Economic Development, Department of City Planning, and Public Development Corporation.

New York City. (1986). *New York City Garment Center Study: Program and Zoning Recommendations*. New York: New York City Office for Economic Development, Department of City Planning, and Public Development Corporation.

New York Shipping Association v. Federal Maritime Commission, 854 F.2d 1338 (D.C. Cir., 1988), *cert. denied sub nom. International Longshoremen's Association v. Federal Maritime Commission*, 57 U.S.L.W. 3486 (US January 23, 1989).

New York State Crime Commission. (1953). *Fourth Report of the New York State Crime Commission (Port of New York Waterfront) to the Governor, the Attorney General and the Legislature of the State of New York* (dated May 20). Albany: New York State Crime Commission.

New York Times. (May 14, 1949). A new housing partnership (editorial). p. 13.

New York Times. (January 23, 1998). U.S. labor leader seeks union support in Mexico. p. A3.

New York Times. (December 1, 1999). A chaotic intersection of tear gas and trade talks. p. 14A.

Novacek, A. C. (1973). Testimony of A. C. Novacek, Executive Vice President of Seatrain Lines Container Division, in the matter of *Balicer v. International Longshoremen's Association*, Civ. A. No. 1155–73, U.S. District Court (D. NJ), opinion reported at 364 F.Supp 205 (1973).

NYSA (New York Shipping Association). (1959). Contractual demands of August 12, 1959 (quoted in Anon [1973], *Factual Memorandum on Containerization Under NYSA-ILA and CONASA-ILA Labor Agreements*. Draft memorandum dated July 23, 1973, located in "Consolidated Express Inc.," National Labor Relations Board case nos. 22–CC-541 et al., Folder I, National Labor Relations Board, Washington, DC).

NYSA (New York Shipping Association). (1972). Minutes of meeting between the NYSA and the ILA concerning NYSA–ILA Rules on Containers, dated November 20, 1972. Located in "Consolidated Express Inc.," respondent's exhibit R-8, National Labor Relations Board case nos. 22–CC-541 et al., National Labor Relations Board, Washington, DC.

Ohmae, K. (1990). *The Borderless World: Power and Strategy in the Interlinked Economy*. New York: HarperBusiness.

Ohmae, K. (1995). *The End of the Nation State: The Rise of Regional Economies*. New York: Free Press.

Olle, W., & Schoeller, W. (1977). World market competition and restrictions upon international trade-union policies. *Capital and Class*, 2, 56–75.

Olsen, P.-E. (1995). [Interview by author with Poul-Erik Olsen, Director, Education and Working Environment, Eastern Europe, International Metalworkers' Federation, Geneva, September 24.]

OS KOVO. (No date). *Základní Informace* ("Basic Information"). Leaflet published by OS KOVO (Czech Metalworkers' Federation), Prague (in Czech).

Ost, D. (1989). The transformation of Solidarity and the future of Central Europe. *Telos*, 79, Spring, 69–94.

Ost, D. (1995). Labor, class, and democracy: Shaping political antagonisms in post-communist society. In B. Crawford (Ed.), *Markets, States, and Democracy: The*

Political Economy of Post-Communist Transformation (pp. 177–203). Boulder, CO: Westview Press.

Oulatar, A. (1999). [Email correspondence between author and Anna Oulatar, Head, Coordinating Unit for Central and Eastern Europe, International Confederation of Free Trade Unions, Brussels, dated January 27.]

Page, B. (1998). Rival unionism and the geography of the meatpacking industry. In A. Herod (Ed.), *Organizing the Landscape: Geographical Perspectives on Labor Unionism* (pp. 263–296). Minneapolis and London: University of Minnesota Press.

Painter, J. (1991). The geography of trade union responses to local government privatization. *Transactions of the Institute of British Geographers, New Series*, 16.2: 214–226.

Painter, J. (1994). Trade union geography: Alternative frameworks for analysis. *Transactions of the Institute of British Geographers, New Series*, 19.1, 99–101.

Palloix, C. (1973). *Les Firmes Multinationales et le Procès d'Internationalisation* [Multinational Firms and the Process of Internationalization]. Paris: Maspero.

Park, R. E. (1936). Human ecology. *American Journal of Sociology*, 42, July, 1–15.

Parson, D. (1982). The development of redevelopment: Public housing and urban renewal in Los Angeles. *International Journal of Urban and Regional Research*, 6.2, 393–413.

Parson, D. (1984). Organized labor and the housing question: Public housing, suburbanization, and urban renewal. *Environment and Planning D: Society and Space*, 2, 75–86.

Pavlínek, P. (1992). Regional transformation in Czechoslovakia: Towards a market economy. *Tijdschrift voor Economische en Sociale Geografie*, 83.5, 361–371.

Pearson, N. J. (1987). Honduras. In G. M. Greenfield & S. L. Maram (Eds.), *Latin American Labor Organizations* (pp. 463–494). New York: Greenwood Press.

Peet, R. (1981). Spatial dialectics and Marxist geography. *Progress in Human Geography*, 5, 105–110.

Peet, R. (1982). International capital, international culture. In M. Taylor & N. Thrift (Eds.), *The Geography of Multinationals: Studies in the Spatial Development and Economic Consequences of Multinational Corporations* (pp. 275–302). New York: St. Martin's Press.

Peet, R. (1983). Relations of production and the relocation of United States manufacturing industry since 1960. *Economic Geography*, 59, 112–143.

Peet, R. (1985). The social origins of environmental determinism. *Annals of the Association of American Geographers*, 75, 309–333.

Peet, R. (1987). (Ed.). *International Capitalism and Industrial Restructuring: A Critical Analysis*. Boston: Allen and Unwin.

Pehe, J. (1997). Czech rail workers' strike ends, ushers in a new era. *Analytical Briefs: Analyses of Breaking News in the Countries of Eastern Europe and the Former Soviet Union*, Vol. 1, No. 545 (February 10).

Peijnenburg, J. (1984). Workers in transnational corporations: Meeting the corporate challenge. In P. Waterman (Ed.) *For a New Labour Internationalism* (pp. 108–117). Hague: International Labour Education, Research and Information Foundation.

Petersen, K. C. (1987). *Company Town: Potlatch, Idaho, and the Potlatch Lumber Company*. Pullman: Washington State University Press.

Philadelphia Inquirer. (October 23, 1991). Why the world is closing in on the U.S. economy.

Philadelphia Inquirer. (December 5, 1991). Who aided fugitive's mint biz?

Phillips, G. A., & Whiteside, N. (1985). *Casual Labour: The Unemployment Question in the Port Transport Industry, 1880–1970*. Oxford University Press: New York.

Pickles, J. (1998). Restructuring state enterprises: Industrial geography and Eastern European transitions. In J. Pickles & A. Smith (Eds.), *Theorising Transition: The Political Economy of Post-Communist Transformations* (pp. 172–196). New York and London: Routledge.

Pickles, J., & Smith, A. (Eds.). (1998). *Theorising Transition: The Political Economy of Post-Communist Transformations*. New York and London: Routledge.

Piore, M. J. (1992). The limits of the market and the transformation of socialism. In B. Silverman, R. Vogt, & M. Yanowitch (Eds.), *Labor and Democracy in the Transition to a Market System* (pp. 171–182). Armonk, NY: Sharpe.

Pita, M. (1997). [Telephone interview by author with Mario Pita, former American Institute for Free Labor Development official in the Dominican Republic, January 16.]

Plunz, R. (1986). Reading Bronx housing, 1890–1940. In Bronx Museum of the Arts (Ed.), *Building a Borough: Architecture and Planning in The Bronx* (pp. 70–76). New York: Bronx Museum of the Arts.

Pollydore, J. H. (1997). [Interview by author with Joseph H. "Polly" Pollydore, General Secretary, Guyanese Trades Union Council, Georgetown, Guyana, April 1, 1997.]

Pope, J. E. (1905). *The Clothing Industry in New York*. New York: Franklin.

Porter, R. (1994). *London: A Social History*. London: Penguin.

Prague Post. (1998). State averts mining strike. *Prague Post*, April 8. (Available at *www.praguepost.cz/archive/busi40898c.html*).

Pravda, A., & Ruble, B. A. (Eds.). (1986). *Trade Unions in Communist States*. Boston: Allen and Unwin.

Premdas, R. R. (1993). Race, politics, and succession in Trinidad and Guyana. In A. Payne & P. Sutton (Eds.), *Modern Caribbean Politics* (pp. 98–124). Baltimore, MD: Johns Hopkins University Press.

Price, J. (1945). *The International Labour Movement*. London: Oxford University Press.

Puette, W. J. (1992). *Through Jaundiced Eyes: How the Media View Organized Labor*. Ithaca, NY: Cornell University Industrial and Labor Relations Press.

RAC (Ravenswood Aluminum Corporation). (1992). Complaint for determination of rightful Board of Directors in the matter of *Ravenswood Aluminum Corporation, A Delaware Corporation, Rinoman Investment B.V., I-Suk Oh, Willy R. Strothotte, Craig A. Davis and Jean L. Loyer v. R. Emmett Boyle and Charles E. Bradley*, Case No. 12528, Court of Chancery for New Castle County, Delaware. Complaint filed April 14, 1992.

Radice, H. (Ed.). (1975). *International Firms and Modern Imperialism: Selected Readings*. Harmondsworth, England: Penguin.

Radosh, R. (1969). *American Labor and United States Foreign Policy*. New York: Random House.

Rath, E. (1973). *Container Systems*. New York: Wiley.

Rees, J. (1974). Decision-making, the growth of the firm and the business environment. In F. E. I. Hamilton (Ed.), *Spatial Perspectives on Industrial Organization and Decision-Making* (pp. 189–211). London: Wiley.

Roberts, J. W. (1995). *Putting Foreign Policy to Work: The Role of Organized Labor in American Foreign Relations, 1932–1941*. New York and London: Garland.

Robleda Castro, A. (1995) *40 Años Después: La Verdad de la Huelga de 1954 y de la Formación del SITRATERCO* (40 Years After: The Truth about the Strike of 1954 and the Formation of SITRATERCO). Tegucigalpa, Honduras: Servicio Documental del Arte y la Literatura.

Rodger, R. (1986). The Victorian building industry and the housing of the Scottish working class. In M. Doughty (Ed.), *Building the Industrial City* (pp. 152–206). Leicester, England: Leicester University Press.

Romualdi, S. (1967). *Presidents and Peons: Recollections of a Labor Ambassador in Latin America*. New York: Funk and Wagnalls.

Rosenbaum, E. (1954). *The Expulsion of the International Longshoremen's Association from the American Federation of Labor*. Unpublished PhD Dissertation, University of Wisconsin.

Ross, P. (1970). Waterfront labor response to technological change: A tale of two unions. *Labor Law Journal*, 21.7, 397–419.

Rössler, M. (1994). 'Area research' and 'spatial planning' from the Weimar Republic to the German Federal Republic: Creating a society with a spatial order under National Socialism. In M. Renneberg & M. Walker (Eds.), *Science, Technology and National Socialism* (pp. 120–138). New York: Cambridge University Press.

Rubinstein, D. (1974). *Victorian Homes*. Newton Abbot, UK: David and Charles.

Ruble, B. A. (1990). Moscow's revolutionary architecture and its aftermath: A critical guide. In W. C. Brumfield (Ed.), *Reshaping Russian Architecture: Western Technology, Utopian Dreams*, (pp. 111–144). New York: Woodrow Wilson International Center for Scholars.

Salt, B., Cervero, R., & Herod, A. (2000). Workers' education and neo-liberal globalization: An adequate response to the TNCs? *Adult Education Quarterly*, 50.4, 9–31.

Samuelson, P. A. (1948). International trade and the equalization of factor prices. *Economic Journal*, 58, 163–184.

Sanchez, D. C. (1995). LRR Focus: Solidarity NOT charity. *Labor Research Review*, 23 (Spring/Summer), 30–33.

Sanjuro, N. (No date). Affidavit of Nestor Sanjuro, President of Twin Express. Located in "Consolidated Express Inc.," petitioner's exhibit P-12, National Labor Relations Board case nos. 22–CC-541 et al., National Labor Relations Board, Washington, DC.

Šarčević, P. (Ed.). (1992). *Privatization in Central and Eastern Europe*. London: Graham and Trotman.

Sassen, S. (1988a). New York City's informal economy. Paper prepared for the Social Science Research Council, Committee on New York City.

Sassen, S. (1988b). *The Mobility of Capital and Labor: A Study in International Investment and Labor Flows*. Cambridge: Cambridge University Press.

Satkin, R. (1987). Affidavit of Richard Satkin, City of New York Department of City Planning, Supreme Court of the State of New York, County of New York, dated December 17, 1987.

Satkin, R. (1988). Memo, Housing and Economic Planning Division, Department of City Planning, City of New York, dated August 12, 1988.

Satkin, R. (1989). [Personal Communication with Richard Satkin, Housing and Economic Planning Division, Department of City Planning, City of New York, April 14.]

Satkin, R. (1998). [Telephone interview by author with Richard Satkin, Housing and Economic Planning Division, Department of City Planning, City of New York, January 14.]

Savage, L. (1998). Geographies of organizing: Justice for Janitors in Los Angeles. In A. Herod (Ed.), *Organizing the Landscape: Geographical Perspectives on Labor Unionism* (pp. 225–252). Minneapolis and London: University of Minnesota Press.

Sayer, A. (1984). *Method in Social Science: A Realist Approach*. London: Hutchinson.

Schaffer, R., & Smith, N. (1986). The gentrification of Harlem? *Annals of the Association of American Geographers, 76*, 347–365.

Schoenberger, E. (1990). U.S. manufacturing investments in Western Europe: Markets, corporate strategy, and the competitive environment. *Annals of the Association of American Geographers 80*, 379–393.

Schwantes, C. A. (1979). *Radical Heritage: Labor, Socialism, and Reform in Washington and British Columbia, 1885–1917*. Seattle: University of Washington Press.

Schwartz, B. (1987). *George Washington: The Making of an American Symbol*. New York: Free Press.

Scott, A. J. (1988). *Metropolis: From the Division of Labor to Urban Form*. Berkeley: University of California Press.

Scott, J. (1978a). *Yankee Unions, Go Home! How the AFL Helped the U.S. Build an Empire in Latin America*. Vancouver: New Star.

Scott, J. (1978b). *Canadian Workers, American Unions: How the American Federation of Labour Took Over Canada's Unions*. Vancouver: New Star.

Scott, J. C. (1985). *Weapons of the Weak: Everyday Forms of Peasant Resistance*. New Haven, CT: Yale University Press.

Segal, M. J. (1953). The international trade secretariats. *Monthly Labor Review*, April, 372–380.

Segre, R., Coyula, M., & Scarpaci, J. L. (1997). *Havana: Two Faces of the Antillean Metropolis*. New York: Wiley.

Selna, J. (1969). Containerization and intermodal service in ocean shipping. *Stanford Law Review, 21*, 1077–1103.

Senft, P. (1995). [Interview by author with Peter Senft, Head, Economic Office, IG Metall (German Metalworkers' Union), Berlin, Germany, September 15.]

Sheppard, E. S., & Barnes, T. J. (1990). *The Capitalist Space Economy: Geographical Analysis after Ricadro, Marx and Sraffa*. Cambridge, MA: Unwin Hyman.

Shostak, A. B. (1991). *Robust Unionism: Innovations in the Labor Movement*. Ithaca, NY: Cornell University Industrial and Labor Relations Press.

Simms, B. (1992). *Workers of the World Undermined: American Labor's Role in U.S. Foreign Policy*. Boston: South End Press.

Smith, A. (1994). Uneven development and the restructuring of the armaments industry in Slovakia. *Transactions of the Institute of British Geographers, New Series*, 19.4, 404–424.

Smith, D. M. (1966). A theoretical framework for geographical studies of industrial location. *Economic Geography*, 42, 95–113.

Smith, D. M. (1970). On throwing out Weber with the bathwater: A note on industrial location and linkage. *Area*, 2, 15–18.

Smith, M. P. (1998). Looking for the global spaces in local politics. *Political Geography*, 17.1, 35–40.

Smith, N. (1979). Geography, science and post-positivist modes of explanation. *Progress in Human Geography*, 3, 356–383.

Smith, N. (1981). Degeneracy in theory and practice: Spatial interactionism and radical eclecticism. *Progress in Human Geography*, 5, 111–118.

Smith, N. (1988). The region is dead! Long live the region! *Political Geography Quarterly*, 7.2, 141–152.

Smith, N. (1989). Rents, riots and redskins. *Portable Lower East Side*, 6, 1–36.

Smith, N. (1990 [1984]). *Uneven Development: Nature, Capital and the Production of Space*. Oxford: Blackwell.

Smith, N. (1992). Geography, difference and the politics of scale. In J. Doherty, E. Graham, & M. Malek (Eds.), *Postmodernism and the Social Sciences* (pp. 57–79). New York: St. Martin's Press.

Smith, N. (1993). Homeless/Global: Scaling places. In J. Bird, B. Curtis, T. Putnam, G. Robertson, & L. Tucker (Eds.), *Mapping the Futures: Local Culture, Global Change* (pp. 87–119). London: Routledge.

Smith, N. (1995). Remaking scale: Competition and cooperation in prenational and postnational Europe. In H. Eskelinen & F. Snickars (Eds.), *Competitive European Peripheries* (pp. 59–74). Berlin: Springer-Verlag.

Smith, N. (1996). *The New Urban Frontier: Gentrification and the Revanchist City*. New York: Routledge.

Smith, N., & Dennis, W. (1987). The restructuring of geographical scale: Coalescence and fragmentation of the northern core region. *Economic Geography*, 63, 160–182.

Smith, N., & Godlewska, A. (1994). *Geography and Empire: Critical Studies in the History of Geography*. Cambridge, MA: Blackwell.

Snow, S. (1964). *The Pan-American Federation of Labor*. Durham, NC: Duke University Press.

Sofsky, W. (1993). *The Order of Terror: The Concentration Camp* (translated by W. Templer). Princeton, NJ: Princeton University Press.

Soja, E. (1980). The socio-spatial dialectic. *Annals of the Association of American Geographers*, 70.2, 207–225.

Soja, E. (1989). *Postmodern Geographies: The Reassertion of Space in Critical Social Theory*. New York: Verso.

Southall, H. (1988). Towards a geography of unionization: The spatial organization and distribution of early British trade unions. *Transactions of the Institute of British Geographers, New Series*, 13, 466–483.

Southall, H. (1989). British artisan unions in the New World. *Journal of Historical Geography*, 15.2, 163–182.

Spalding, H. A., Jr. (1988). US labour intervention in Latin America: The case of the American Institute for Free Labor Development. In R. Southall (Ed.), *Trade unions and the new industrialization of the Third World* (pp. 259–286). London: Zed Books.

Speer, A., Krier, L., & Olof, L. (1985). *Architecture, 1932–1942*. Brussels: Archives d'Architecture Mode.

Staddon, C. (1998). Democratisation and the politics of water in Bulgaria: Local protest and the 1994–5 Sofia water crisis. In J. Pickles & A. Smith (Eds.), *Theorising Transition: The Political Economy of Post-Communist Transformations* (pp. 347–372). New York and London: Routledge.

Staeheli, L. (1994). Empowering political struggle: Spaces and scales of resistance. *Political Geography*, 13, 387–391.

Stafford, H. (1991). Manufacturing plant closure selections within firms. *Annals of the Association of American Geographers*, 81, 51–65.

Stamp, L. D. (1937). *Chisholm's Handbook of Commercial Geography* (15th ed.). London: Longman's, Green. (Originally written by George Chisholm and published in 1889 as *A Handbook of Commercial Geography*).

Stark, D. (1996). Recombinant property in East European capitalism. *American Journal of Sociology*, 104.4, 993–1027.

Stidham, D. (1992). [Interview by author with Dan Stidham, President, Local 5668 United Steelworkers of America, Ravenswood, West Virginia, July 27.]

Storey, G. (1991). [Personal communication by author with Greg Storey, Secretary, New York Shipping Association, New York City, July 28.]

Storper, M., & Walker, R. (1983). The theory of labour and the theory of location. *International Journal of Urban and Regional Research*, 7.1, 1–43.

Storper, M., & Walker, R. (1984). The spatial division of labor: Labor and the location of industries. In W. Tabb & L. Sawers (Eds.), *Sunbelt/Snowbelt: Urban Development and Regional Restructuring* (pp. 19–47). Oxford: Oxford University Press.

Storper, M., & Walker, R. (1989). *The Capitalist Imperative: Territory, Technology, and Industrial Growth*. New York: Blackwell.

Sturmthal, A. (1950). The International Confederation of Free Trade Unions. *Industrial and Labor Relations Review*, 3.3, 375–382.

Swyngedouw, E. (1992). The mammon quest. 'Glocalisation,' interspatial competition and the monetary order: The construction of new scales. In M. Dunford & G. Kafkalas (Eds.), *Cities and Regions in the New Europe* (pp. 39–68). London: Belhaven Press.

Swyngedouw, E. (1996). Reconstructing citizenship, the re-scaling of the state and the new authoritarianism: Closing the Belgian Mines. *Urban Studies*, 33, 1499–1521.

Swyngedouw, E. (1997). Neither global nor local: 'glocalisation' and the politics of scale. In K. Cox (Ed.), *Spaces of Globalization: Reasserting the Power of the Local* (pp. 137–166). New York: Guilford Press.

Tabak, H. D. (1970). *Cargo Containers: Their Stowage, Handling and Movement*. Cambridge, MD: Cornell Maritime Press.

Tabb, W. K., & Sawers, L. (Eds.). (1978). *Marxism and the Metropolis: New Perspectives in Urban Political Economy*. New York: Oxford University Press.

Taylor, H. (1949). Letter from Harry Taylor, New York City Office of City Construction

Co-Ordinator to Robert Moses, Commissioner, dated May 16. Located in Mayor O'Dwyer Subject Files, Housing Projects 1946–1950, box 73, folder no. 739, Mayor's Archives, New York City.

Taylor, M., & Thrift, N. (Eds.). (1982). *The Geography of Multinationals: Studies in the Spatial Development and Economic Consequences of Multinational Corporations*. New York: St. Martin's Press.

Taylor, M., & Thrift, N. (Eds.). (1985). *Multinationals and the Restructuring of the World Economy*. London: Croom Helm.

Taylor, P. J. (1981). Geographical scales within the world-economy approach. *Review*, 5.1, 3–11.

Taylor, P. J. (1982). A materialist framework for political geography. *Transactions of the Institute of British Geographers, New Series*, 7.1, 15–34.

Taylor, P. J. (1987). The paradox of geographical scale in Marx's politics. *Antipode*, 19.3, 287–306.

Thelen, K. A. (1991). *Union of Parts: Labor Politics in Postwar Germany*. Ithaca, NY: Cornell University Press.

Theodorson, G. A. (Ed.). (1961). *Studies in Human Ecology*. Evanston, IL: Row, Peterson.

Thies, J. (1983). Nazi architecture—a blueprint for world domination: The last aims of Adolf Hitler. In D. Welch (Ed.), *Nazi Propaganda: The Power and the Limitations* (pp. 45–64). London: Croom Helm.

Thomson, D., & Larson, R. (1978). *Where Were You, Brother? An Account of Trade Union Imperialism*. London: War on Want.

Thompson, E. P. (1963). *The Making of the English Working Class*. New York: Vintage.

Thrift, N., & Leyshon, A. (1991). *Making Money*. London: Routledge.

Thrift, N., & Olds, K. (1996). Refiguring the economic in economic geography. *Progress in Human Geography* 20.3, 311–337.

Tichelman, F. (1988). Socialist 'internationalism' and the colonial world: Practical colonial policies of social democracy in Western Europe before 1940 with particular reference to the Dutch SDAP. In F. van Holthoon & M. van der Linden (Eds.), *Internationalism in the Labour Movement 1830–1940*, Vol. 1 (pp. 87–108). London: Brill.

Todd, G. (1997). [Interview by author with Gordon Todd, General Secretary, Guyanese Commercial and Clerical Workers' Union, Georgetown, Guyana, March 31.]

Toffler, A. (1980). *The Third Wave*. London: Collins.

Toll, S. I. (1969). *Zoned American*. New York: Grossman.

Townroe, P. M. (1975). Branch plants and regional development. *Town Planning Review*, 46, 47–62.

Uehlein, J. (1989). Using labor's trade secretariats. *Labor Research Review*, 8.1, 31–41.

Uehlein, J. (1992). [Interview by author with Joe Uehlein, Director of Special Projects, Industrial Union Department at the AFL-CIO, Washington DC, August 10.]

Uhlíř, J. (1994). [Interview by author with Jan Uhlíř, President, Odborovz Svaz KOVO (OS KOVO: Czech Metalworkers' Federation), Prague, Czech Republic, August 24.]

Uhlíř, J. (1995). [Interview by author with Jan Uhlíř, President, Odborový Svaz KOVO (OS KOVO: Czech Metalworkers' Federation), Prague, Czech Republic, September 7.]

Uhlíř, J. (1996). [Interview by author with Jan Uhlíř, President, Odborový Svaz KOVO (OS KOVO: Czech Metalworkers' Federation), Prague, Czech Republic, August 20.]

Ullman, G. (1968). Letter from Gerald Ullman, attorney for the Forwarders Intermodal Container Corporation to Leroy Fuller, Federal Maritime Commission, dated September 26. Located in "Consolidated Express Inc.," National Labor Relations Board case nos. 22–CC-541 et al., Folder I, National Labor Relations Board, Washington, DC.

Ullman, G. (1983). Testimony of Gerald Ullman, General Counsel, National Customs Brokers' and Forwarders' Association of America, Inc. before the U.S. House Subcommittee on Merchant Marine, Committee on Merchant Marine and Fisheries hearings on HR 2562, a bill to amend Section 45 of the 1916 Shipping Act, 98th Cong., 1st Sess., 1983.

United Nations. (1994). *Economic Survey of Europe in 1993–1994.* Report prepared by the Secretariat of the Economic Commission for Europe, United Nations, Geneva.

U.S. Department of Commerce. (1994). *Statistical Abstract of the United States.* Washington, DC: U.S. Department of Commerce, Bureau of the Census.

U.S. Senate Committee on Foreign Relations. (1968). *Survey of the Alliance for Progress—Labor Policies and Programs.* A Study Prepared at the Request of the Subcommittee on American Republics Affairs by the Staff of the Committee on Foreign Relations, United States Senate, Together with a Report of the Comptroller General, Ninetieth Congress, Second Session, July 15. Washington, DC: U.S. Government Printing Office.

U.S. Senate Committee on Foreign Relations. (1969). *American Institute for Free Labor Development.* Hearing before the Committee of Foreign Relations, United States Senate, Ninety-First Congress, First Session, with George Meany, President, AFL-CIO, August 1. Washington, DC: U.S. Government Printing Office.

USWA (United Steelworkers of America). (1991). RAC to Rich: The corporate web of the Ravenswood Aluminum Company. Memorandum dated May 1991. Pittsburgh, PA.: USWA.

USWA (United Steelworkers of America). (1992). *Summary: Proposed agreement between the United Steelworkers of America and the Ravenswood Aluminum Corporation.* Pittsburgh, PA.: USWA.

USWA (United Steelworkers of America). (No date). *Ravenswood Aluminum Corporation: A stakeholders' report.* Pittsburgh, PA.: USWA.

Van den Burg, G. (1975). *Containerisation and Other Unit Transportation.* London: Hutchinson Benham.

van der Linden, M. (1988). The rise and fall of the First International: An interpretation. In F. van Holthoon & M. van der Linden (Eds.), *Internationalism in the Labour Movement 1830–1940,* Vol. 1 (pp. 323–335). London: Brill.

van Holthoon, F., & van der Linden, M. (Eds.). (1988). *Internationalism in the Labour Movement 1830–1940,* Vols. 1 and 2. London: Brill.

Vernon, R. (1966). International investment and international trade. *Quarterly Journal of Economics*, 80, 190–207.

Vernon, R. (1971). *Sovereignty at Bay: The Multinational Spread of US Enterprises.* New York: Basic Books.

Vernon, R. (1979). The product cycle hypothesis in the new international environment. *Oxford Bulletin of Economics and Statistics*, 41.4, 255–267.

Volkov, I. (1992). The transition to a mixed economy and the prospects for the labor and trade-union movement. In B. Silverman, R. Vogt, & M. Yanowith (Eds.), *Labor and Democracy in the Transition to a Market System* (pp. 53–67). Armonk, NY: Sharpe.

Voorhis, J. (1964). Homes and neighborhoods. In J. Liblit (Ed.), *Housing—the Cooperative Way* (pp. 72–84). New York: Twayne.

Vural, L. F. (1994). *Unionism as a Way of Life: The Community Orientation of the International Ladies' Garment Workers' Union and the Amalgamated Clothing Workers of America.* Unpublished PhD dissertation, Department of Geography, Rutgers University, New Brunswick, NJ.

Waldinger, R. D. (1986). *Through the Eye of the Needle: Immigrants and Enterprise in New York's Garment Trades.* New York and London: New York University Press.

Waldinger, R. D. (1991). Immigrants rescue the rag trade. *NY: The City Journal*, 1.2, 46–51.

Walker, D. F. (1975). A behavioralist approach to industrial location. In L. Collins & D. F. Walker (Eds.), *Locational Dynamics of Manufacturing Activity* (pp. 135–158). New York: Wiley.

Walker, R. (1978). Two sources of uneven development under advanced capitalism: Spatial differentiation and capital mobility. *Review of Radical Political Economics*, 10, 28–37.

Walker, R. (1981). A theory of suburbanization: Capitalism and the construction of urban space in the United States. In M. Dear & A. J. Scott (Eds.), *Urbanization and Urban Planning in Capitalist Society* (pp. 383–429). London: Methuen.

Walker, R. (1998). Foreword. In A. Herod (Ed.), *Organizing the Landscape: Geographical Perspectives on Labor Unionism* (pp. xi–xvii). Minneapolis and London: University of Minnesota Press.

Walker, R., & Storper, M. (1981). Capital and industrial location. *Progress in Human Geography*, 5, 473–509.

Wallerstein, I. (1979). *The Capitalist World-Economy.* Cambridge, UK: Cambridge University Press.

Warner, B. (1983). Berlin—the Nordic homeland and the corruption of urban spectacle. *Architectural Design*, 53.11/12, 73–80.

Waterfront Commission of New York Harbor. (Various years). *Annual Reports.* New York: Waterfront Commission of New York Harbor.

Waterfront Commission of New York Harbor. (1970). *Special Report of the Waterfront Commission of New York Harbor to the Governors and Legislatures of the States of New York and New Jersey.* New York: Waterfront Commission of New York Harbor.

Waterman, P. (1993). Internationalism is dead! Long live global solidarity? In J. Brecher, J. Brown Childs, & J. Cutler (Eds), *Global Visions: Beyond the New World Order* (pp. 257–261). Boston: South End Press.

Waterman, P. (1998). The new social unionism: A new union model for a new world order. In R. Munck & P. Waterman (Eds), *Labour Worldwide in the Era of Globalization* (pp. 247–264). London: Macmillan.

Waterman, P. (1999). *Globalization, Social Movements and the New Internationalisms*. London: Mansell.

WCL (World Confederation of Labour). (1997a). SLOVAKIA—Trade union training. *Tele-Flash*, 33 (June 1). Brussels: World Confederation of Labour. (Available at *http://www.cmt-wcl.org/en/pubs/archive/teleflash33.html*).

WCL (World Confederation of Labour). (1997b). "Act 120" threatens free trade union movement. *Tele-Flash*, 33 (June 1). Brussels: World Confederation of Labour. (Available at *http://www.cmt-wcl.org/en/pubs/archive/teleflash33.html*).

WCL (World Confederation of Labour). (1997c). Czeck (sic) Parliament holds to trade union pluralism. *Tele-Flash*, 40 (September 15). Brussels: World Confederation of Labour. (Available at *http://www.cmt-wcl.org/en/pubs/archive/teleflash40.html*).

WCL (World Confederation of Labour). (1998a). Internal initiatives of the WCL. *World Federation of Labour "Labor Magazine,"* January. Brussels: World Confederation of Labour. (Available at *http://www.cmt-wcl.org/en/pubs/labor98–1.html#nr1*).

WCL (World Confederation of Labour). (1998b). New-style European section. *Tele-Flash*, 52 (April 1). Brussels: World Confederation of Labour. (Available at *http://www.cmt-wcl.org/en/pubs/archive/teleflash52.html*).

WCL (World Confederation of Labour). (1998c). WFIW will reflect on a world-wide social dialogue. *Tele-Flash*, 52 (April 1). Brussels: World Confederation of Labour. (Available at *http://www.cmt-wcl.org/en/pubs/archive/teleflash52.html*).

WCL (World Confederation of Labour). (1998d). WCL trade action in Croatia. *Tele-Flash*, 55 (May 15). Brussels: World Confederation of Labour. (Available at *http://www.cmt-wcl.org/en/pubs/archive/teleflash55.htm*).

WCL (World Confederation of Labour). (1998e). IFTC active in Croatia. *Tele-Flash*, 53 (April 15). Brussels: World Confederation of Labour. (Available at *http://www.cmt-wcl.org/en/pubs/archive/teleflash53.html*).

WCL (World Confederation of Labour). (1999). WFIW on the move in Central and Eastern Europe. *Tele-Flash*, 71 (January 15). Brussels: World Confederation of Labour. (Available at *http://www.cmt-wcl.org/en/pubs/teleflash71.html*)

Weber, A. (1929 [1909]). *Theory of the Location of Industries*. Chicago: University of Chicago Press.

Wedin, A. (1984). Harmful effects of so-called 'trade union solidarity' on the Latin American labour movement. In P. Waterman (Ed.), *For a New Labour Internationalism: A Set of Reprints and Working Papers* (pp. 13–35). The Hague, Netherlands: International Labour Education, Research and Information Foundation.

Weiler, P. (1988) *British Labour and the Cold War*. Stanford, CA: Stanford University Press.

Weinberg, P. J. (1978). *European Labor and the Multinationals*. New York: Praeger.

Wells, D. (1996) New dimensions for labor in a post-fordist world. In W. C. Green & E. J. Yanarella (Eds.) *North American Auto Unions in Crisis: Lean Production as Contested Terrain* (pp. 191–207). Albany, NY: SUNY Press.

West, L. A. (1991). U.S. foreign labor policy and the case of militant political unionism in the Philippines. *Labor Studies Journal*, 16.4, 48–75.

Wial, H. (1994). New bargaining structures for new forms of business organization. In S. Friedman, R. W. Hurd, R. A. Oswald, & R. L. Seeber (Eds.) *Restoring the Promise of American Labor Law* (pp. 303–313). Ithaca, NY: ILR Press.

Willis, E. (1988). *Technology and the Labour Process: Australasian Case Studies*. Sydney: Allen and Unwin.

Wills, J. (1996). Geographies of trade unionism: Translating traditions across space and time. *Antipode*, 28.4, 352–378.

Wills, J. (1998a). Space, place, and tradition in working-class organization. In A. Herod (Ed.), *Organizing the Landscape: Geographical Perspectives on Labor Unionism* (pp. 129–158). Minneapolis and London: University of Minnesota Press.

Wills, J. (1998b). Taking on the Cosmo Corps? Experiments in transnational labor organization. *Economic Geography* 74.2, 111–130.

Windmuller, J. P. (1954). *American Labor and the International Labor Movement, 1940 to 1953*. Ithaca, NY: Institute of International Industrial and Labor Relations, Cornell University.

Windmuller, J. P. (1970). Internationalism in eclipse: the ICFTU after two decades. *Industrial and Labor Relations Review*, 23.4, 510–527.

Windmuller, J. P. (1980). *The International Trade Union Movement*. Deventer, Netherlands: Kluwer.

Wójcik, T. (1998). Statement by Tomasz Wójcik, representative of Solidarność union, Poland, to 86th International Labour Conference of the World Confederation of Labour, June 2–18 (available at *http://www.cmt-wcl.org/en/act/86conf.html*).

Wolff, R. D., & Resnick, S. A. (1987). *Economics: Marxian Versus Neoclassical*. Baltimore, MD: Johns Hopkins University Press.

World Bank. (1996). *From Plan to Market: World Development Report 1996*. New York: Oxford University Press.

Zharikov, A. (1995). [Interview by author with Alexander Zharikov, General Secretary, World Federation of Trade Unions, Prague, Czech Republic, September 11.]

Zieger, R. H. (1986). *American Workers, American Unions, 1920–1985*. Baltimore, MD: Johns Hopkins University Press.

Zukin, S. (1982). *Loft Living: Culture and Capital in Urban Change*. Baltimore, MD: Johns Hopkins University Press.

Index

341

About the Author

Andrew Herod, Ph.D. is Associate Professor of Geography at the University of Georgia, Athens, GA, USA. His research focuses upon trying to understand how the geography of capitalism is made and how workers' lives are constituted geographically, particularly within the context of processes of economic globalization. His research has won a number of awards for creativity, including the University of Georgia Research Foundation's Creative Research Medal, the University of Georgia Franklin College of Arts and Sciences M.G. Michael Award, and the J. Warren Nystrom Award presented by the Association of American Geographers. He is editor of *Organizing the Landscape: Geographical Perspectives on Labor Unionism* (University of Minnesota Press, 1998) and co-editor (with Gearóid Ó Tuathail and Susan Roberts) of *An Unruly World?: Globalization, Governance and Geography* (Routledge, 1998), together with numerous articles on organized labor, globalization, and qualitative research methods. He teaches at the undergraduate and graduate level in the fields of general human geography, economic and cultural geography, and in the areas of social theory and the sociology of knowledge.